MASS SPECTROMETRY OF NATURAL
SUBSTANCES IN FOODS

RSC Food Analysis Monographs

Series Editor: P.S. Belton, *The Institute of Food Research, Norwich, UK.*

The aim of this series is to provide guidance and advice to the practising food analyst. It is intended to be a series of day-to-day guides for the laboratory worker, rather than library books for occasional reference. The series will form a comprehensive set of monographs providing the current state of the art on food analysis.

Dietary Fibre Analysis

by David A.T. Southgate, *Formerly of the AFRC Institute of Food Research, Norwich, UK.*

Quality in the Food Analysis Laboratory

by Roger Wood, *Joint Food Safety and Standards Group, MAFF, Norwich, UK*, Anders Nilsson, *National Food Administration, Uppsala, Sweden*, and Harriet Wallin, *VTT Biotechnology and Food Research, Espoo, Finland*

Chromatography and Capillary Electrophoresis in Food Analysis

by Hilmer Sørensen, Susanne Sørensen and Charlotte Bjergegaard, *Royal Veterinary and Agricultural University, Frederiksberg, Denmark,* and Søren Michaelsen, *Novo Nordisk A/S, Denmark*

Mass Spectrometry of Natural Substances in Foods

by Fred A. Mellon, *The Institute of Food Research, Norwich, UK*, Ron Self, *University of East Anglia, Norwich, UK*, and James R. Startin, *Central Science Laboratory, York, UK*

How to obtain future titles on publication

A standing order plan is available for this series. A standing order will bring delivery of each new volume immediately upon publication. For further information, please write to:

Turpin Distribution Services Ltd.
Blackhorse Road
Letchworth
Herts. SG6 1HN

Telephone: Letchworth (01462) 672555

RSC
FOOD
ANALYSIS
MONOGRAPHS

Mass Spectrometry of Natural Substances in Foods

Fred A. Mellon
Institute of Food Research, Norwich, UK

Ron Self
University of East Anglia, Norwich, UK

James R. Startin
Central Science Laboratory, Sand Hutton, York, UK

ROYAL SOCIETY OF CHEMISTRY

ISBN 0-85404-571-6

A catalogue record for this book is available from the British Library.

Published by The Royal Society of Chemistry,
Thomas Graham House, Science Park, Milton Road, Cambridge CB4 0WF, UK

For further information see our web site at www.rsc.org

Typeset by Paston Prepress Ltd, Beccles, Suffolk, NR34 9QG
Printed & bound in Great Britain by TJ International Ltd, Padstow, Cornwall

Preface

Mass spectrometry has been a pre-eminent analytical technique in food science from an early stage in its development. Flavour chemists, for example, were quick to realise the potential of combined gas chromatography/mass spectrometry (GC/MS) for determining the very complex mixtures of volatile compounds in food aromas. The subsequent introduction of powerful data acquisition and processing systems, including automated library search techniques, ensured that the information content of the large quantities of data generated by GC/MS instruments was exploited fully. These early successes were the foundation of an increasingly diverse range of applications, utilising many different mass spectrometric techniques, in other areas of food science.

Mass spectrometry's unparalleled combination of sensitivity and selectivity (especially when allied to separation techniques such as GC and HPLC) ensured its early adoption as a key analytical technique for monitoring chemical aspects of food safety. Initially, this was exemplified by the determination of anthropogenic compounds in foods (pesticide and veterinary drug residues, for example). These types of application are not discussed in this book, which is devoted to natural substances in foods. However, applications to the analysis of naturally occurring defensive compounds in food plants and naturally occurring substances that are either detrimental or beneficial to health are introduced and described.

Mass spectrometry has evolved rapidly in the last two decades. The introduction of cheap open-access instrumentation has ensured that mass spectrometry is no longer confined to research laboratories but is now used routinely in many analytical laboratories for compositional analysis. The last few years have also witnessed the revolution in analysis of molecules of biological interest brought about by the introduction of soft ionisation techniques such as matrix-assisted laser desorption (MALDI) and electrospray ionisation (ESI). This has particular implications for food quality and safety. The ability to analyse biopolymers, more or less routinely, is beginning to have a major impact on the study of food structure components (oligosaccharides, for example) and food proteins (including allergens). Furthermore, the rapidly developing field of proteomics, where advanced mass spectrometric techniques are pivotal, has tremendous potential in the study of food pathogens and food components produced from genetically modified organisms. The ability to measure high molecular weights accurately is also benefiting research into naturally occurring food antibiotics and food allergens. The development of ESI and the related atmospheric pressure chemical ionisation (APCI) source has also brought routine LC/MS techniques within the reach of many food and nutrition laboratories. This is revolutionising the study of micronutrient metabolism, of non-volatile taste and flavour components and of biologically active, naturally occurring non-nutrients, to identify just a few examples.

Inorganic mass spectrometry, particularly inductively coupled plasma mass spectrometry (ICP-MS), is another discipline that is making a major contribution to food science. In addition to its role in food toxicology (not discussed in this book), ICP-MS is used in a growing number of laboratories to study the metabolism of nutrient minerals, including iron, zinc, calcium, copper and selenium. This is helping to fill major gaps in our knowledge of the amounts of these minerals needed to maintain optimal health (not simply to prevent deficiencies). Furthermore, inorganic mass spectrometry is also helping to elucidate the influence of the food matrix on the absorption and metabolism of mineral nutrients.

Metabolic studies of organic food components have benefited from improvements in the performance and flexibility of high precision isotope ratio mass spectrometers (used for measuring ^{13}C, ^{18}O and ^{15}N isotopes). These developments are reflected in increased activity in the field of metabolic studies in humans, particularly those involving isotopically labelled macronutrients.

One of our main aims has been to convey the depth and breadth of mass spectrometric applications to natural substances in foods. After opening chapters that introduce the principles and practice of mass spectrometry, we cover applications in flavour analysis, and the determination of non-nutrient biologically active natural substances in foods. We go on to discuss the analysis and metabolism of amino acids, peptides, proteins, lipids, sugars, carbohydrates and vitamins, with separate chapters on mineral and macronutrient metabolism and techniques of pyrolysis mass spectrometry. Numerous references are given to specific analyses, to encourage the reader who wishes to pursue these applications in greater depth. We hope that this volume will be a useful resource to food scientists, food analysts and researchers into the composition of foods and into nutrition and food safety.

Fred A Mellon
Ron Self
James R Startin
Norwich and York, 1999

Contents

Glossary of Abbreviations

AE	Appearance Energy
AMS	Accelerator Mass Spectrometry
APCI	Atmospheric Pressure Chemical Ionisation
CAD	Collisionally Activated Decomposition ($\equiv$ CID)
CE/MS	Combined Capillary Electrophoresis/Mass Spectrometry
CI	Chemical Ionisation
CID	Collision Induced Dissociation
DCI	Desorption Chemical Ionisation
DEI	Desorption Electron Ionisation
DLI	Direct Liquid Introduction
EI	Electron Ionisation
ESI	Electrospray
FAB	Fast Atom Bombardment (also FABMS)
FD	Field Desorption (also FDMS)
FFR(1, 2. . .)	First (second. . ., *etc.*) Field Free Region
FI	Field Ionisation (also FIMS)
FTICR	Fourier Transform Ion Cyclotron Resonance
FTMS	Fourier Transform Mass Spectrometry
GC/C/IRMS	Gas Chromatography/Combustion/Isotope Ratio Mass Spectrometry
GC/MS	Combined Gas Chromatography/Mass Spectrometry
GIRMS	Gas Isotope Ratio Mass Spectrometry
ICP-MS	Inductively Coupled Plasma Mass Spectrometry
IE	Ionisation Energy
ITMS	Ion Trap Mass Spectrometer
LC/MS	Combined High Performance Liquid Chromatography/Mass Spectrometry
LD	Laser Desorption
LSIMS	Liquid Secondary Ion Mass Spectrometry
MALDI	Matrix Assisted Laser Desorption Ionisation
MECC	Micellar Electrokinetic Capillary Chromatography
MPI	Multiphoton Ionisation
MRM	Multiple Reaction Monitoring
MS	Mass Spectrometry (or Mass Spectrometer)
MS/MS	Tandem Mass Spectrometry

NICI	Negative Ion Chemical Ionisation
PDMS	Plasma Desorption Mass Spectrometry
Py/MS	Pyrolysis Mass Spectrometry
R	Resolution
REMPI	Resonance Enhanced Multiphoton Ionisation
SFC/MS	Combined Supercritical Fluid Chromatography/Mass Spectrometry
SFI/MS	Combined Supercritical Fluid Injection/Mass Spectrometry
SID	Surface Induced Dissociation
SIM	Selected Ion Monitoring
SIMS	Secondary Ion Mass Spectrometry
SRM	Selected Reaction Monitoring
TIC	Total Ion Current
TIMS	Thermal Ionisation Mass Spectrometry
TLC/MS	Combined Thin Layer Chromatography/Mass Spectrometry
ToF	Time of Flight
TSP	Thermospray
VOC	Volatile Organic Compounds

CHAPTER 1

Introduction to Principles and Practice of Mass Spectrometry

1 History of Mass Spectrometry

According to the International Union of Pure and Applied Chemistry definition, 'Mass spectrometry is the branch of science dealing with all aspects of mass spectroscopes and the results obtained with these instruments.'[1] It evolved from research in particle physics at the turn of the century, with Goldstein discovering positively charged 'rays' in 1886 and Wien (1898) studying their electric and magnetic properties. In the early 1900s J.J. Thomson built his 'parabola mass spectrograph' to measure the charge to mass ratio (z/m) for several ionic species. In the expression z/m, z is the charge number, *i.e.* the total charge on an ion divided by the elementary charge (e), and m is the nucleon number, *i.e.* the sum of the total number of protons and neutrons in an atom, molecule or ion. In modern mass spectrometry, the parameter measured is m/z, rather than z/m: the unit of m/z was recently designated the thomson (Th).

Aston continued the work at Cambridge and built instruments that helped him to establish the presence of isotopes. He was subsequently able to measure the atomic mass of most elements with sufficient accuracy to be able to calculate the 'packing fraction' of their atomic nuclei. The packing fraction is the difference between the accurate atomic mass of the isotope and the nearest whole number divided by the mass number, also known as the mass defect. Aston also obtained accurate measurements of the ratios of the stable isotopes of many of the known elements.

At the end of this exciting period of development, Aston was convinced that much of the potential of mass spectrometry had been exploited. It was not until the 1940s that the technique was put to work in elucidating organic structures in the petroleum industry. Ionisation was effected by electron 'impact' [now called electron ionisation (EI)] for those molecules that could withstand vaporisation into the heated and evacuated ion source without decomposition. This limited the practical mass range to less than 1000 daltons (Da)* but yielded useful

---

* The dalton (Da) is the unit of mass (also known as the mass unit, u) and is 1/12 of the mass of ^{12}C (defined as 12.000000).

fragmentations for structure elucidation (see Chapter 2). By choosing to work with 70 electron volt (eV) electrons many ions were formed with internal energies far in excess of the ionisation energy (IE)†. These ions decompose rapidly to produce lower mass (fragment) ions and neutral radicals or molecules.

During the 1950s, commercial instruments were being built and new applications discovered. One of the earliest of these was the identification of low molecular weight volatile food flavour compounds. Ten years later, the powerful combination of electron ionisation mass spectrometry (EIMS) with gas chromatography (GC/MS) led to an explosion of applications where mass spectrometry was used in qualitative and quantitative, chemical and biochemical studies. GC/MS instruments produced enormous amounts of data, which were best handled by computers, and data acquisition and processing methods were devised. In 1966, Munson and Field described chemical ionisation (CI). This technique increased the yield of ions representative of the molecular weight of volatile molecules through interactions with reagent gas ions (*e.g.* CH_5^+ ions from methane) with little excess energy.[2] Other 'soft' ionisation techniques such as field desorption (FD) and particle desorption methods based upon ion generation by Cf-252 fast fission products [plasma desorption, (PDMS)] were introduced during the 1970s for involatile compounds. At the same time (and in response to these developments) the instrumental mass range was increased to cope with the larger sample molecule ions now entering the gas phase. This process accelerated in the 1980s with the introduction of Fast Atom Bombardment (FAB) ionisation. FAB was the first ionisation technique to enable biologists and biochemists *routinely* to obtain molecular weight information on complex, labile biomolecules, including polypeptides and small proteins.

Ionisation from the liquid state, followed by evaporation/desolvation of charged droplets, includes techniques such as ion spray, thermospray (TSP) and electrospray ionisation (ESI). These methods differed mainly in the manner in which ionisation was initiated. Multiply charged molecular ions could be formed under ESI, facilitating the measurement of high molecular masses, even on conventional instruments (*i.e.* those with a mass range up to 2000 or 4000 Th). More efficient pumping systems were required to cope with the increased gas volumes generated by vaporising liquids.

Separation techniques such as liquid chromatography and capillary electrophoresis coupled to mass spectrometry (LC/MS and CE/MS respectively) have extended the advantages first associated with the analysis of volatile compounds by GC/MS to compounds of low volatility and high molecular weight. Tandem mass spectrometry (MS/MS) collision-induced dissociation (CID), focal-plane array detectors, ion traps and hybrid instruments are providing a high sensitivity structure elucidation facility for involatile compounds similar to that provided by EIMS of volatiles. Recently, Laser Desorption (LD), and especially Matrix Assisted Laser Desorption Ionisation (MALDI), combined

---

† Ionisation energy – minimum energy of excitation of an atom or molecule required to remove an electron in order to produce a positively charged ion.

with Time-of-Flight (ToF) mass analysis has extended the practical mass range to over 300 000 Da, producing mainly singly-charged molecular ions. Fourier transform mass spectrometry (FTMS), also known as Fourier transform ion cyclotron resonance (FTICR) mass spectrometry is only slowly entering the commercial area.

Only a brief introduction to the principles and practice of mass spectrometry is given herein. The reader is directed towards more general textbooks, *e.g.* Throck Watson[3] or Rose and Johnstone,[4] for a more detailed discussion of the principles and practice of mass spectrometry. Beynon and Brenton[5] lucidly introduce the physical aspects of mass spectrometry and ionisation in the gas phase. Finally, the latest volume by McLafferty and Turecek[6] is recommended for a very thorough introduction to the interpretation of organic mass spectra.

Mass Spectrometry – a definition

> Mass spectrometry is the study of systems causing the formation of gaseous ions, with or without fragmentation, which are then characterised by their mass to charge ratios (m/z) and relative abundances.

Mass spectrometry is unlike most other forms of spectroscopy or spectrometry that are concerned with non-destructive interactions between molecules and electromagnetic radiation. This is because mass spectrometry is the study of the effect of ionising energy on molecules. It depends upon chemical reactions in the gas phase in which sample molecules are consumed during the formation of ionic and neutral species. Although sample is consumed destructively by the mass spectrometer the technique is very sensitive and only trace amounts of material are used in the analysis. A mass spectrometer converts sample molecules into ions in the gas phase, separates them according to their mass to charge ratio (m/z) and sequentially records the individual ion current intensities at each mass – the mass spectrum. If these ion current intensities are drawn in histogram form taking the most intense ion current as 100%, the values of m/z versus percentage relative intensity (%RI) is called a line diagram, *e.g.* Figure 1.1.

2   Ionisation of Molecules

Several types of mass spectrometry can generate ions representative of the mass of the sample molecule. These are described below and, because ionisation is often intimately linked to sample introduction techniques, this topic is also discussed.

Production of Molecular Ions

If a quantity of energy equivalent to the IE of the molecule is supplied under EI conditions, a molecular ion is formed that is a radical ion denoted as $M^{+\cdot}$. There are several ways of forming molecular ions.

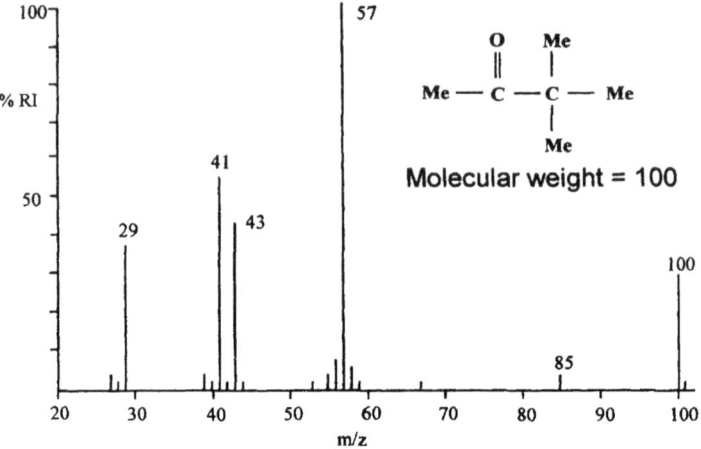

Figure 1.1 *Electron ionisation mass spectrum of methyl t-butyl ketone*

Electron Ionisation (EI)

70 eV electrons passing through the ionisation chamber from the filament to the trap, under the control of the magnetic field, interact with volatilised sample molecules, which enter their path (Figure 1.2).

The temperature of volatilisation can be varied from ambient to $> 400\,°C$ in a vacuum of about 10^{-3} Pa. Interaction of the sample molecule with an energetic electron removes one (and sometimes two or more) electrons from the valence

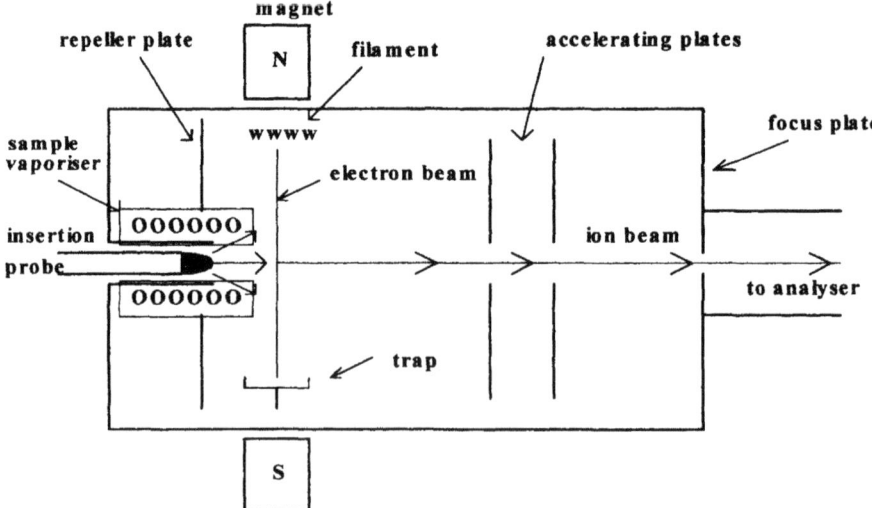

Figure 1.2 *Schematic diagram of an Electron Ionisation source*

orbitals. During this process, an excess of energy can be transferred to the newly formed positively charged 'radical ion'.

$$M \xrightarrow{e^-} M^{+\cdot} + 2e^- \tag{1.1}$$

Since the IEs of most organic molecules lie in the range 8–12 eV, the excess energy can cause bond dissociation (fragmentation) within the ion. The distribution of energy among the ions formed, and the ensuing pattern of fragmentation produced from different chemical structures, depends on several factors. Two of these, chemical bond lability and fragment ion stability, form a basis for the qualitative interpretation of mass spectra and are discussed in Chapter 2. The efficiency of the ionisation process in relation to the electron beam energy used is shown in Figure 1.3.

Most organic molecules are ionised by 8–12 eV and an additional 1–6 eV will dissociate any cleavable bonds. The choice of 70 eV electrons for conventional EI mass spectrometry (a) ensures efficient transfer of enough excess energy to induce structurally informative fragmentation and (b) is in the plateau region of the figure, where it is easier to generate reproducible mass spectra. Although 70 eV is the ionisation energy of choice in EI, it is important to note that not all this energy is transferred to molecules during ionisation.

In principle, a spectrum free from fragmentation can be obtained by lowering the energy of the bombarding electrons to values close to the IE. However, the ionisation efficiency is also lowered by a factor of 200–300, so that large quantities of sample would be required for successful analysis. Some molecules with little excess energy can dissipate this by stabilisation among their degrees of vibrational freedom. These ions will be detected as molecular ions in the mass spectrum. Other ions will apportion the energy to cleave bonds to produce well-stabilised fragment ions. Many organic compounds decompose at the elevated temperatures required for vaporisation into the conventional EI source and others do not produce stable molecular ions under normal EI conditions. In these cases, alternative ionisation methods are available.

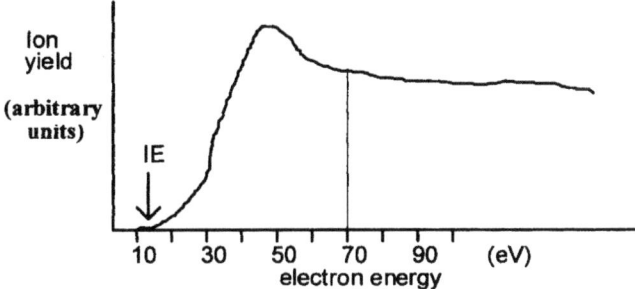

Figure 1.3 *Typical variation in ion yield from an organic compound with increasing electron beam energy*

'In-beam' (Desorption) Electron Ionisation

For the improved ionisation of relatively involatile but stable molecules, a special probe is used to place the sample very close to the electron beam. This technique, commonly known as desorption EI (DEI), has been successful with a limited range of previously intractable biomolecules, for example cyclic peptides and some glycosides.

Chemical Derivatisation

The preparation of chemical derivatives will increase the volatility and (sometimes) reduce the IE of polar compounds, allowing EI to generate stable molecular ions. Organic acids, fatty acids, *etc.* can be analysed by EIMS of their methyl esters. Many chemical classes can be rendered volatile through the preparation of a variety of silylated products that enable mass and structural information to be obtained. Special chemical derivatives can be made that yield characteristic fragmentation properties, thereby aiding analysis. An example is the preparation of *t*-butyldimethylsilyl derivatives of steroids. These yield an abundant $[M - 57]^+$ ion (generated by loss of C_4H_9) that is especially useful in quantitative measurements.

The Particle Beam Interface

The particle beam interface (Figure 1.4) is a sample introduction rather than an ionisation technique in which the incoming liquid sample, *e.g.* HPLC eluent, is nebulised with helium gas to form an aerosol of solvent droplets.

The stream of liquid droplets is allowed to desolvate at ambient temperature in a chamber at the reduced pressure provided by vacuum pump 1. The

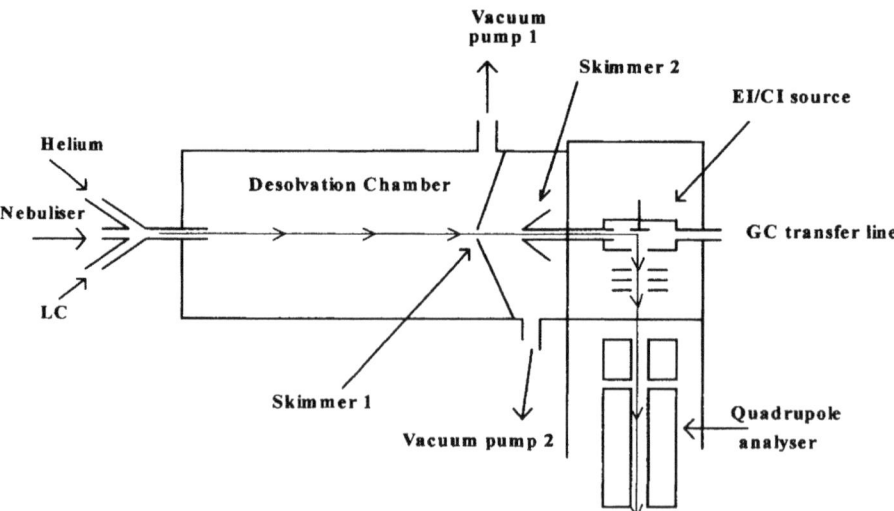

Figure 1.4   *Schematic diagram of a particle beam LC/MS interface*

remaining molecules then expand into the second evacuated region (provided by vacuum pump 2) through skimmer 1, where a supersonic jet (molecular beam) is formed. Skimmer 2 allows the molecular beam, containing heavy sample particles, to pass through into the ionisation chamber for subsequent electron, or chemical ionisation. The lighter helium and solvent particles are skimmed off and removed by vacuum pump 2. Although particle beam systems appear to be an ideal method for ionising involatile compounds, the technique lacks sensitivity and is unsuitable for polar, thermally labile molecules.

Production of Protonated Molecules and Adduct Ions

Methods that generate ions representative of the molecular weight of less volatile and thermally labile molecules are generally based on two main principles. Firstly, reaction and thermal equilibration with a reagent gas. Secondly, direct desorption from a liquid or solid matrix during or prior to ionisation. Under these conditions, ion/molecule reactions are generally responsible for the ionisation process.

Chemical Ionisation (CI)

In CI, chemical reactions occur between thermally equilibrated reagent ions and sample molecules. The conditions required are: a large excess (10^4:1) of reagent gas (R) to sample molecules (M), necessitating the use of higher energy electrons (500 eV) at the resultant source pressures in the range of 10–150 Pa. Primary ions are formed in the dense gas:

$$R + e^- \longrightarrow R^{+\cdot} \tag{1.2}$$

These primary ions react with more reagent gas to form stable, reactive secondary cations (even-electron species) by collisional hydrogen transfer:

$$R^{+\cdot} + R \longrightarrow RH^+ + (R - H)^{\cdot} \tag{1.3}$$

If methane is the reagent gas, CH_5^+ (a Lewis acid) is produced. This ion reacts strongly with organic molecules (at around 20 Pa pressure) by proton or hydride transfer reactions, generating stable, protonated molecules. Other reagent ions formed during methane CI include $C_2H_5^+$, $C_2H_3^+$ and $C_3H_5^+$, although CH_5^+ and $C_2H_5^+$ are the predominant species.

Protonation

Protonation occurs when the proton affinity of the sample molecules is higher than that of the reagent gas:

$$RH^+ + M \longrightarrow R + MH^+ \tag{1.4}$$

Hydride Ion Abstraction (Dissociative Proton Transfer)

This is common for samples with lower proton affinities than the reagent gas, *e.g.* alkanes:

$$RH^+ + C_{10}H_{22} \longrightarrow C_{10}H_{21}^+ + R + H_2 \tag{1.5}$$

Charge Exchange

Monoatomic reactant gas ions have no vibrational degrees of freedom and therefore the ionisation energy, *e.g.* 15.755 eV for argon, is all transferred to the colliding molecule; a useful property for energetic and kinetic studies. Charge exchange can also be used selectively to ionise particular compound classes.

$$R^{+\cdot} + M \longrightarrow R + M^{+\cdot} \tag{1.6}$$

Other popular CI reagent gases include isobutane, which yields mainly $C_4H_9^+$, and ammonia, which yields NH_4^+ ions. Depending on the acidity of the reactant secondary ion and the basicity of the sample molecule, adduct ions (electrophilic addition) can be formed

$$M + C_2H_5^+ \longrightarrow [M + C_2H_5]^+ \tag{1.7}$$

$$M + NH_4^+ \longrightarrow [M + NH_4]^+ \tag{1.8}$$

Negative Ion Chemical Ionisation

Various types of reaction can take place according to the nature of the reagent gas used, *e.g.* a mixture of hydrocarbon and water (95:5) can produce negatively-charged sample ions by the following reactions:

$$H_2O + e^- \longrightarrow OH^- + H \tag{1.9}$$

$$AH + OH^- \longrightarrow A^- + H_2O \tag{1.10}$$

$$M + HO^- \longrightarrow MOH^- \tag{1.11}$$

Other possible reagent gases include hydrocarbon/organic halide/oxygen (which produces Cl^- and O^- attachment ions) and fluorocarbons yielding negative ions by hydride abstraction and fluorine attachment. Negative ions can also be formed from suitable sample molecules by electron capture processes. In this case, the reagent gas acts as a moderator, generating thermal electrons that can then attach to molecules with high electron affinities forming negative radical ions.

$$M + e^- \ (thermal) \longrightarrow M^{-\cdot} \tag{1.12}$$

Negative ion CI, especially electron capture ionisation, can be two to three orders of magnitude more sensitive than positive ion CI for electronegative molecules. It is therefore especially useful in the quantitative determination of trace substances that have, or that through the production of suitable derivatives are induced to have, electron-capturing properties.

Alternate CI-EI (ACE)

Electron and chemical ionisation produce complementary information on volatilisable samples. This is particularly useful when analysing compounds in complex mixtures by high-resolution capillary column gas chromatography. Although it is possible to run the same mixture twice, once by CI and once by EI, it can be difficult to replicate the chromatograms exactly. Furthermore, the time for each analysis is at least doubled. However, if the conditions in the source can be changed rapidly from EI to CI and back again during the elution time of the GC peak then the two spectra are known to be related and consecutive. ACE sources have a cycle time of around 1 second to record two spectra. If the GC peak is at least 4 seconds wide at the base it is possible to obtain two sets of data as the component of interest elutes (Figure 1.5).

Chemical ionisation can be performed with carefully selected reagent gases to reveal, through specific chemical reactions, something of the nature of the unknown sample. However, the sample must withstand transfer to the gaseous phase and is therefore only suitable for thermally stable, volatile molecules.

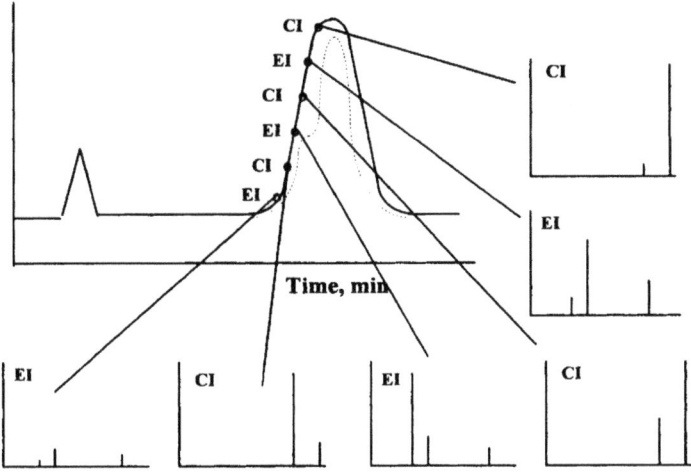

Figure 1.5   *A schematic representation of alternate Chemical Ionisation/Electron Ionisation*

Rapid Heating (Desorption CI)

Probes fitted with special rapid-heating elements produce gas-phase ions by rapid desorption or thermal evaporation while the sample is in intimate contact with the CI reagent gas. Using this technique it is possible to desorb labile samples into the gas phase whilst minimising thermal decomposition.

Atmospheric Pressure Ionisation

In a classical experiment, a ^{63}Ni β-source was used to generate primary electrons at atmospheric pressure.[7] These electrons were allowed to collide with, for example, benzene vapour. The ionised benzene reacts with very low pressures of sample molecules and the ensuing ions are leaked into the mass spectrometer's vacuum system for analysis. Very high sensitivities have been recorded for pure samples.

Atmospheric Pressure Chemical Ionisation (APCI)

In APCI a flow of liquid, typically an HPLC eluent, is induced to form a spray in a pneumatic (usually nitrogen powered) nebuliser. The emerging plume of liquid droplets is generated in atmosphere and is directed towards a corona discharge electrode that is maintained typically at 1–3.5 kV. This is positioned close to a small-diameter orifice leading to the high vacuum region containing the mass analyser. A nozzle and skimmer arrangement allows solvent ions to be pumped away, so that only desolvated ions are admitted into the analyser. A flow of warm nitrogen gas is passed across the front of the sampling orifice to ensure declustering of solvent/analyte ion complexes and to ensure that all the liquid has evaporated (see Figures 1.6 and 1.7). Ions formed from the solvent in the corona discharge plasma react with analyte molecules to generate protonated or deprotonated molecules ($[M + H]^+$ or $[M − H]^−$), in a similar manner to conventional CI.

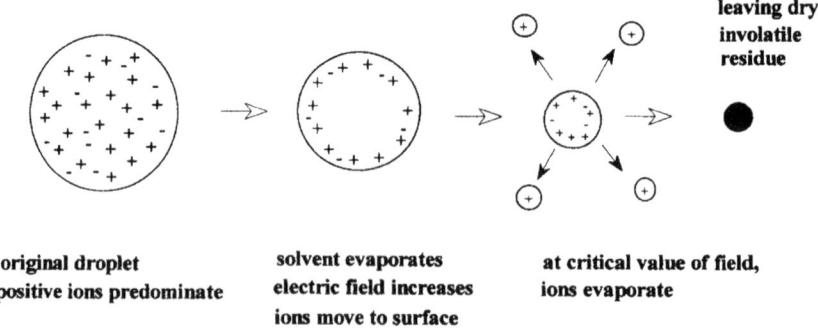

| original droplet positive ions predominate | solvent evaporates electric field increases ions move to surface | at critical value of field, ions evaporate | leaving dry, involatile residue |

Figure 1.6   *Schematic representation of the electrospray ionisation process*

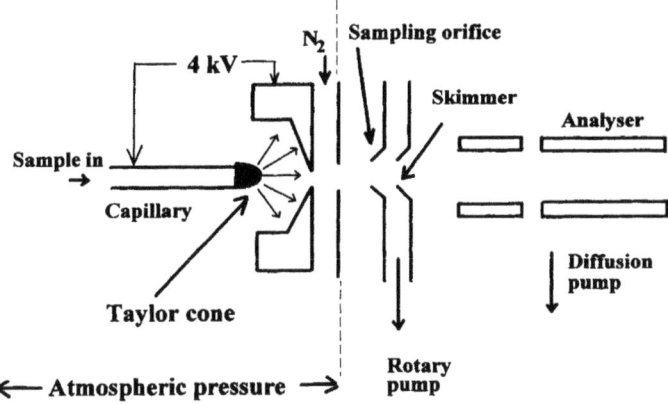

Figure 1.7 *Schematic diagram of an electrospray ionisation source*

Electrospray Ionisation (ESI)

The liquid flow from the HPLC pump enters a small diameter stainless steel capillary, which is maintained at a voltage of 3000–4000 V, A plume of charged liquid droplets is formed (the Taylor cone). Ions are produced by an evaporation mechanism and ionised particles are transported into the mass spectrometer vacuum via a pumped nozzle-skimmer arrangement (Figure 1.7). A curtain of nitrogen gas, which also aids evaporation of the charged droplets, prevents cluster ion formation. The ions are sampled through an orifice into a mass analyser. ESI and (to a lesser extent) its companion technique APCI have revolutionised mass spectrometry by allowing mass spectra to be obtained routinely on polar, involatile molecules. Even high molecular weight biomolecules can be analysed by ESI. Proteins, for example, yield a series of multiply charged ions well within the mass range of conventional mass spectrometers. These signals can be transformed mathematically to yield the molecular weight of the sample (Figure 1.8).

Thermospray Ionisation (TSP)

Thermospray was essentially a forerunner of APCI in which the eluent from the HPLC column was heated in a capillary tube until it formed a spray of liquid droplets that could be directed towards a discharge electrode for ionisation. Alternatively, addition of volatile buffers to the solvent enabled polar samples to be ionised directly from the solution phase, without application of an external ionising technique. The technique was never particularly routine (the phrase 'spray and pray' was commonplace in the mass spectrometry community) and has effectively been superseded by the more robust and sensitive ESI and APCI methods.

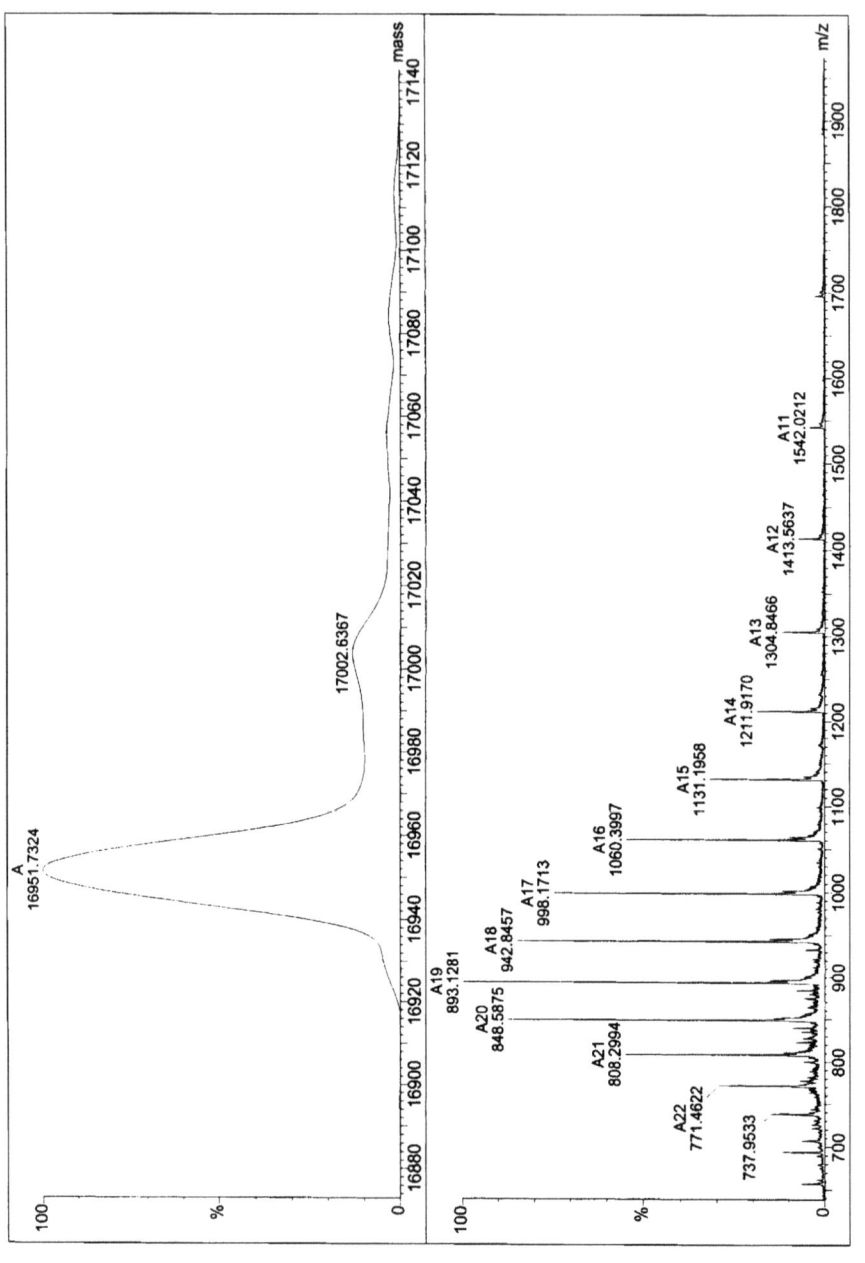

Figure 1.8 *Transformation of the mass spectrum of a multiply charged protein (horse heart myoglobin) to the molecular mass of the sample*

Laser Ionisation (Laser Desorption)

Short but intense pulses of photons cause very rapid surface heating (1000 °C in nanoseconds) with associated thermal ion formation. This effect is useful in surface studies since the beam tends not to penetrate below the surface of the sample.

Matrix Assisted Laser Desorption Ionisation (MALDI)

When used in conjunction with time-of-flight mass analysis (Section 3), MALDI is one of the favoured methods for the production of protonated molecules and adduct ions from large biomolecules (Figure 1.9).

The sample is mixed intimately with a liquid matrix solution of say, sinapinic acid, chosen for its ability to absorb and dissipate energy at the laser wavelength. It is thought that the matrix then passes transverse vibrational energy to the sample molecules in quanta sufficient to desorb them with virtually no excess energy for fragmentation. The technique is particularly useful for mass measuring large biomolecules, especially proteins, accurately and generates (mainly) $[M + H]^+$, $[M - H]^-$ or adduct ions.

Field Ionisation (FI) and Field Desorption (FD)

Molecules close to, or absorbed on surfaces of high curvature such as tips, blades, *etc.*, subjected to intense electric fields (10^7–10^8 V cm^{-1}) are ionised by

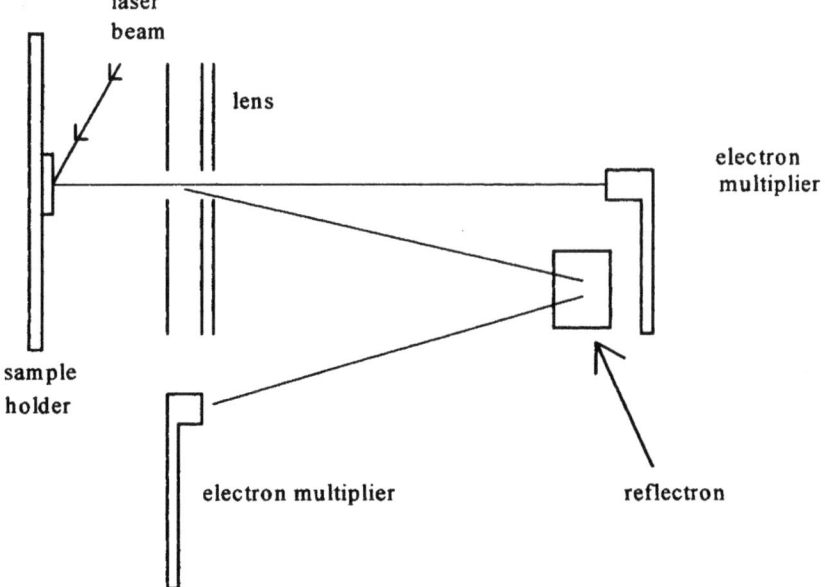

Figure 1.9 *Schematic representation of the MALDI-TOF mass spectrometer (reflectron type)*

quantum tunnelling of valence electrons from the molecule to the metal surface. The FI technique can be used for gas phase experiments, but lacks sensitivity, whereas the FD method for condensed phase samples has been highly successful in skilled hands. FD is not commonly used in modern mass spectrometry laboratories as MALDI and ESI are generally preferred for analysing polar and/or large molecules. However, the method is undergoing a new lease of life for selected applications because the introduction of wide-angle array detectors has allowed recording of the transient ion currents generated by many sample types under FD conditions.

Fission Fragment Ionisation or Plasma Desorption Mass Spectrometry (PDMS)

Pulses of californium-252 fission fragments pass through a nickel foil on which sample molecules have been deposited. The thermal shock vaporises mobile impurity ions (H^+, Na^+ and H^-). These ions interact with involatile sample molecules near the thermal shock, converting them into ions. ToF mass spectrometry, with its wide mass range and compatibility with pulse ionisation techniques, was an ideal partner for the PDMS ion source. PDMS was the most successful technique for large biomolecules (up to ~50 kDa) until the advent of MALDI and ESI. Samples adsorbed on nitrocellulose were found to yield singly- and multiply-charged ions with very little fragmentation. An additional advantage of the nitrocellulose surface was the ease of applying a water washing procedure, which removed inorganic impurities. PDMS has now been superseded by MALDI techniques.

Secondary Ion Mass Spectrometry (SIMS)

A beam of ions is made to collide with the sample deposited on a metal surface. Secondary ions ejected from the surface (sample) are accelerated and analysed. Protonated molecules and sodium adduct ions have been seen from a variety of organic compounds.

Liquid SIMS (LSIMS) or Fast Atom Bombardment (FAB)

Analyte samples are dissolved in a viscous, relatively involatile, liquid matrix and are bombarded with a beam of atoms or ions with kilovolt translational energies. The nature of the projectile is not particularly important and therefore the use of neutral atoms or the equivalent ion beams (*e.g.* Xe atoms or Cs ions) (Figure 1.10) produce similar spectra. The most common matrix used in FAB is glycerol, a useful solvent for many biomolecules.

In the high-pressure region above the surface of the matrix, ion–molecule reactions similar to those that occur during chemical ionisation create a variety of protonated molecules and adduct ions from the molecules emerging from the condensed phase. The method is suitable for polar molecules up to ~20 kDa. The higher-energy Cs ion beam used in LSIMS yields more abundant molecular

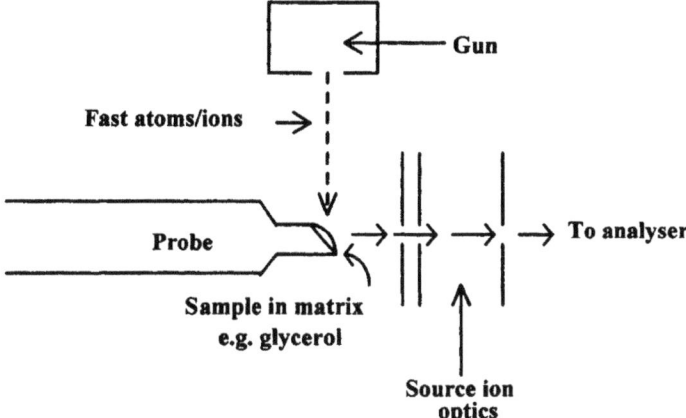

Figure 1.10 *Schematic representation of a FAB/LSIMS ion source*

and fragment ions than conventional FAB ion sources. However, in other respects the spectra are very similar.

Photon and Multiphoton Ionisation (MPI)

Photoionisation using classical light sources has been used for many years but suffers from the disadvantage of producing very low intensity ionisation, much lower than EI for example. MPI with pulsed tunable dye lasers demonstrates combined sensitivity and selectivity superior to other methods. MPI can be considered in two stages:-

(1) *n* photons combine (coherently) to excite the sample molecule to a real, intermediate electronic state (a resonance-enhanced process) and
(2) the irradiation of this intermediate with *m* photons (sequentially) leading to its ionisation.

When the laser is not tuned to a real intermediate state (non-resonant state), nearly negligible MPI occurs, demonstrating the high selectivity. When the sample molecules are concentrated into the laser beam, very high ionisation efficiencies (up to 100%) can be experienced. Finally, by increasing the power density of the laser it is possible to change from the 'soft' molecular ion spectrum to one resembling the classical EI spectrum, dominated by fragmentation.

Resonance Ionisation

A primary source of atomisation is subjected to irradiation from a tunable dye laser, thereby yielding selective multiphoton absorption and resonance-enhanced ionisation of only certain atoms in the matrix. For example, tuning a laser to 271.9 nm will produce ions of Fe out of a mixed organic and inorganic matrix. Such specificity is invaluable in elemental analyses of environmental samples, or in the determination of specific isotopes, for example the long-lived radioactive metabolic tracer isotope [41]Ca.

Inorganic Mass Spectrometry Ionisation Techniques

Although several different inorganic ionisation methods are available, only two have been widely used in food sciences, mainly in studies of mineral metabolism or in multi-element analysis. The relevant techniques are described below.

Thermal Ionisation

Thermal Ionisation Mass Spectrometry (TIMS) is based upon the generation of atomic or molecular ions at the surface of an electrically heated filament.[8] It is generally accepted to be the practical mass spectrometric method that yields the most precise and accurate measurements of stable isotope ratios of inorganic elements. Its use in the food and nutritional sciences is comparatively recent, compared with well-established applications in such fields as geology, geo-chronology and nuclear science. It is most commonly applied to studies of the absorption and metabolism of inorganic nutrients. However, it is also used as a reference technique for quantifying nutritional and toxic trace elements in foods by isotope dilution mass spectrometry (IDMS).

A double-filament thermal ionisation source is shown in Figure 1.11. The double-filament ion source enables separation of the evaporation and ionisation processes and is useful for determining elements that evaporate at low

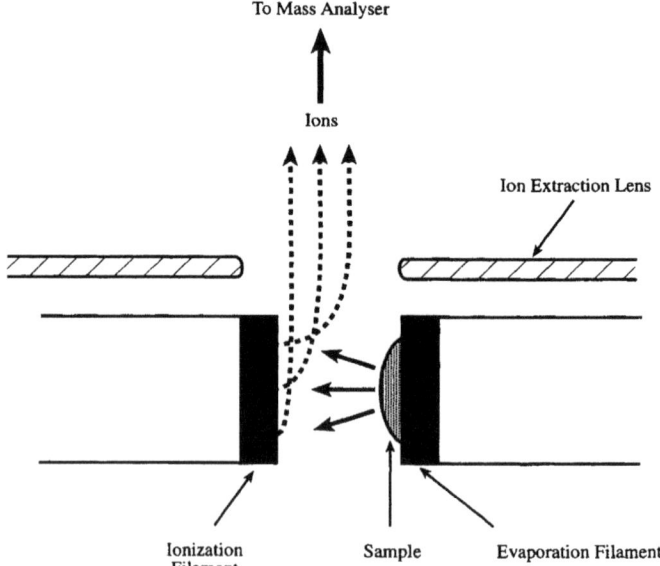

Figure 1.11   *Schematic diagram of a double-filament thermal ionisation source*

temperatures, but require high ionisation temperatures (Ca, for example). Single filament assemblies, where evaporation and ionisation take place from the same filament, are used for elements that have similar evaporation and ionisation temperatures. Positive or negative ions can be formed and mass analysed, depending on the nature of the analyte. Ions desorbed from the filament are extracted and focused by electrostatic lenses into either a magnetic sector or a quadrupole mass analyser.

High positive ion yields are obtained for elements with low ionisation energy and high negative ion yields for elements that possess a high electron affinity. Most inorganic elements evaporate and ionise inside a temperature range of 800–2000 °C. Consequently, filaments must be fabricated from high melting point filament materials such as rhenium, tantalum or tungsten. Metals with first ionisation energies greater than 7 eV frequently require the addition of chemicals to the filament to enhance ion yields. For example, silica gel and phosphoric or boric acid, or (for Fe) aluminium salts. Negative ion yields from elements such as selenium can be increased by the addition of $La(NO_3)_3$ or $Ba(OH)_2$.

Inductively Coupled Plasma

The ionisation region of an inductively coupled plasma mass spectrometer (ICP-MS) comprises a high temperature (typically *ca.* 8000 K) self-sustaining electrical discharge, applied *via* a high frequency induction coil, in a flow of argon gas at atmospheric pressure.[9-13] A schematic diagram of an ICP-MS ion source is shown in Figure 1.12.

Samples are most commonly introduced into the plasma in nebulised solutions flowing at rates of $0.5–1$ ml min^{-1} and are ionised efficiently by the high-temperature plasma. Alternative sample introduction techniques, including electrothermal vaporisation, laser desorption, hydride generation,

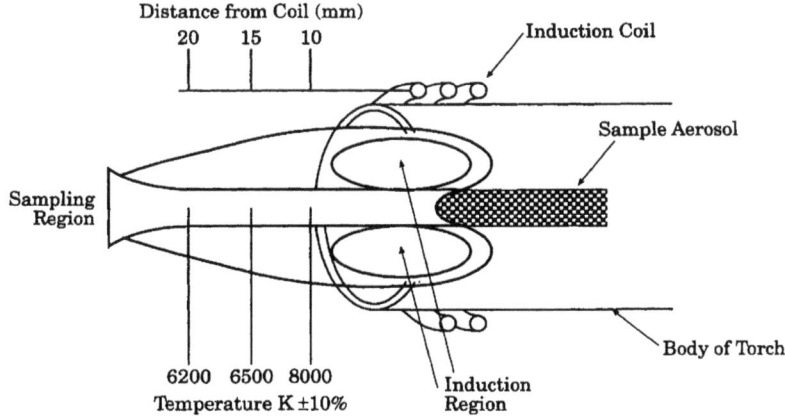

Figure 1.12 *Schematic diagram of an ICP/MS ion source*
(Redrawn from a diagram supplied by VG Elemental, with permission)

electrospray and several types of combined chromatography/mass spectrometry, are also available. A useful summary of alternative methods of sample introduction for plasma mass spectrometry has been published.[14]

Ionisation efficiency approaches 100% for most elements of interest to food and nutrition scientists. The plasma forms a supersonic jet in the differentially pumped region behind the sampling cone. The jet is sampled by a skimmer cone, yielding additional pressure reduction, and ions are then allowed to enter the mass analyser. This is usually a quadrupole device, however high resolution magnetic sector mass spectrometers may be used instead.[15-17] These are useful to resolve isobaric interferences, for example by separating $^{40}ArO^+$ and $^{56}Fe^+$ that nominally appear in the same mass channel (m/z 56).

A recent innovation has been the use of collision cells to remove argon and argide monatomic and polyatomic interference ions from the mass spectrum.[18,19] This technology has been used with both quadrupole and magnetic sector instruments and is particularly useful for increasing the accuracy and precision of measurement of certain elements and isotopes. For example, the important nutritional element selenium suffers from an abundant interference peak from $^{40}Ar_2^+$ in the ^{80}Se mass channel. A properly tuned collision cell can eliminate this and related peaks by charge exchange, allowing all isotopes of Se to be measured accurately. It is also possible to reduce the effects of other, non-argide polyatomics, although collision cell conditions for removing these interferents are different from those for removing argides, *i.e.* it is generally not possible to remove argide and non-argide species simultaneously.

ICP-MS permits multi-element detection and measurement of isotope ratios. The technique can be used to determine nutritional and toxic inorganic elements in foods, or to conduct metabolic studies using enriched stable isotopes. A schematic diagram of a complete ICP-MS instrument is shown in Figure 1.13.

3   Ion Separation

The most popular methods of separating the sample ion beam into beams of different masses are described briefly in this section. For a more comprehensive description of ion separation techniques, consult the recommended texts.[3,4,20]

Time-of-Flight (ToF)

The drift velocity of an ion is dependent on its mass and therefore, as the name suggests, the 'time-of-flight' in a field-free chamber is a measure of ion mass. ToF is ideally suited to very rapid sampling rates in, for example, fast GC/MS, and the analysis of large molecular weight peptides and proteins by pulsed ionisation methods such as PDMS or MALDI. Laser ionisation is well suited to ToF analysis since the start time of a laser pulse can be determined accurately and synchronised with arrival time of the pulse at the detector. Early ToF instruments were limited by their intrinsically low resolution. However, the development of reflectron and delayed extraction techniques has increased the performance (particularly the resolution) of ToF instruments.

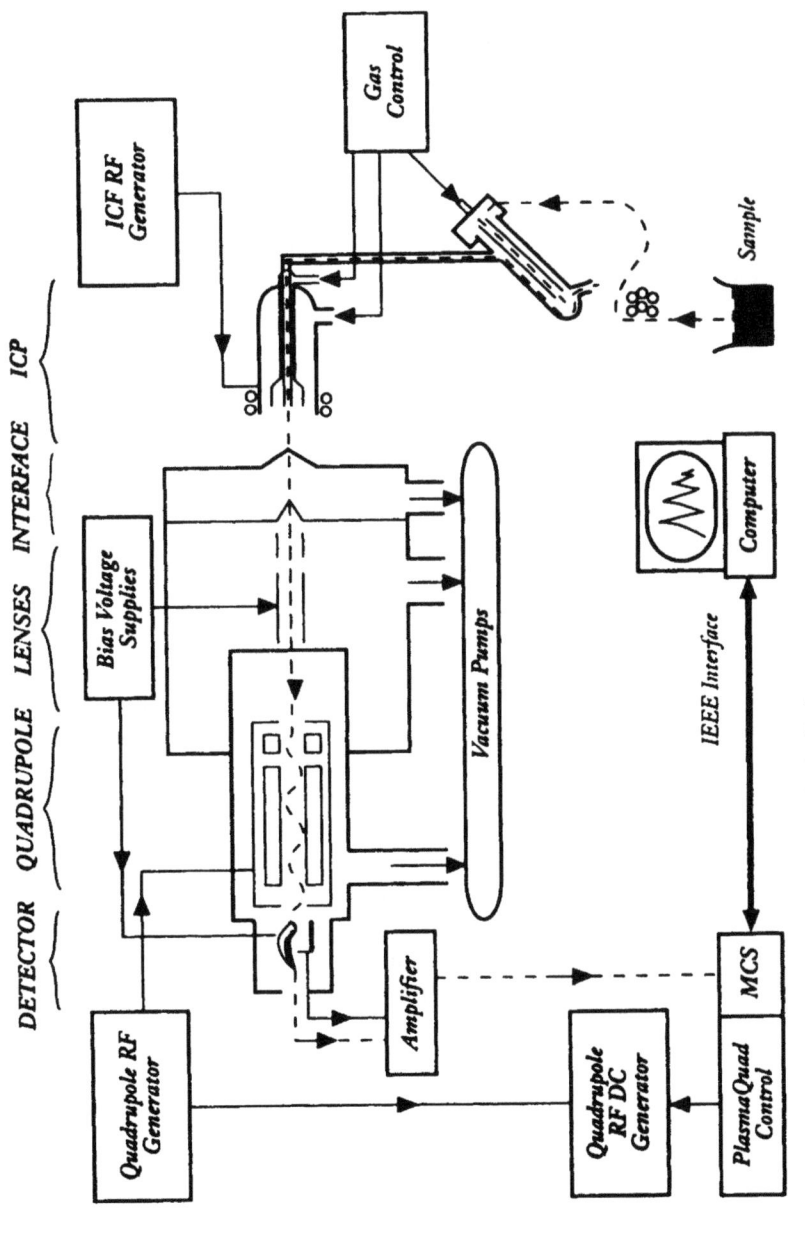

Figure 1.13   *Schematic diagram of an VG PlasmaQuad ICP/MS instrument* (Reproduced from material supplied by VG Elemental)

Quadrupoles

Four parallel metal rods of circular cross-section are electronically connected in pairs and a combination of DC and RF voltages are applied. By varying these voltages (scanning), at a fixed frequency, the mass spectrum is produced from low energy ions – a distinct advantage for coupling GC and HPLC, but the efficiency of transmission of ions falls off rapidly with increasing mass.

Ion Traps

Ion traps can be described as three-dimensional forms of a quadrupole, *i.e.* quadrupoles that have undergone a solid of rotation.[21] The advantage of the ion trap over magnetic analysers is that the focusing fields can be changed rapidly which is valuable in SIM experiments. They are more compact than quadrupole mass analysers, are capable of very high sensitivity and can be used to perform sequential MS/MS (MS^n) experiments.

Mass and Energy Analysis with Sector Fields

High-energy (*e.g.* 8 kV) ions focused onto the entrance slit of a magnetic sector field analyser are dispersed into discrete mass components according to the equation:

$$m/z = r^2 B^2 / 2V$$

where r = radius traversed by a singly charged ion (z) of mass (m) for constant values of electric potential (V) and magnetic field (B). In other words, by scanning either B or V ions of different m/z ratio will pass through the collimating exit slit to the detector. Modern sector instruments can transmit ions up to 20 000 Th. These high masses require post-separation accelerators to accelerate heavier ions to a sufficient impact energy that secondary electron emission can occur at the detector. Alternatively, array detection systems that yield high sensitivity because of the multi-channel advantage can be used (see below).

Newer ion separation methods based on earlier cycloidal mass spectrometers are being developed now that ion storage and induction detection methods are better understood.

Fourier Transform Mass Spectrometry (FTMS)

The advantages of using FTMS are becoming apparent and commercial instruments are now available. Ions of mass m and charge z when subjected to a strong magnetic field B move in circular orbits whose radius is proportional to the velocity of the ion. The ion will have a characteristic frequency (w) for its orbital motion, given by the cyclotron equation:

$$w = zB/m$$

Ions are excited to their cyclotron frequency by applying a broad band 'chirp' of electromagnetic radiation. The mass analyser (ion cyclotron cell) also acts as a detector by measuring the alternating current induced in the cell receiver plates by the orbiting ions. Fourier transformation of the resulting complex signal yields the masses of all ions in the spectrum. Very high resolving powers (in excess of 10^6) are obtainable. Other advantages of the method include compatibility with pulse techniques (*e.g.* laser desorption), multi-channel advantage (shared with TOF) and the ability to carry out MS^n experiments very simply. Disadvantages include the very high vacuum ($< 10^{-5}$ Pa) required for optimum performance, only moderate dynamic range and the high cost of the instrumentation. All these problems, apart from the high cost, are being overcome by novel designs of ion source and 'tailored' scan modes. After undergoing a relatively fallow period, FTMS has been reinvigorated through combination with the newer ionisation techniques of MALDI and (particularly) ESI. In the latter case, the very high resolution characteristics of FTMS in the lower mass regions (*i.e.* below 2000 Th) can be used to full advantage. This is especially useful in MS and MS/MS studies of proteins, where the charge state of ions can be determined very quickly from the resolved isotopic patterns of the multiply charged ions.

High Resolution Mass Spectrometers

A magnetic sector is a momentum selector requiring a monoenergetic source of ions to provide high resolution. Conversely, an electric sector is a velocity selector, which is of limited resolving power unless it receives a constant-momentum beam of ions. Therefore a sequential double sector combination of E and B fields, in any order, will enable both velocity and direction focusing (double focusing) to be achieved (Figure 1.14).

Such combinations are capable of resolving powers of in excess of 150 000. A reference compound with ions at regular intervals across the mass range is introduced with the sample and the mass of an unknown ion in the spectrum is calculated by interpolation between the two nearest reference ions, which bracket it. Isobaric ions, *i.e.* those with the same nominal mass but with different elemental composition, and therefore different nuclidic masses, can often be resolved and their atomic composition determined. If a reference compound exists which does not interfere with the ions of the sample, then an accurate mass value can be obtained at low resolution, *i.e.* with a single-focusing or quadrupole instrument. This principle has been applied to the accurate mass measurement on quadrupole mass spectrometers of ions generated by APCI.

4   Ion Detection

Faraday Cup

The ion beam is allowed to collide with the interior walls of an open-ended

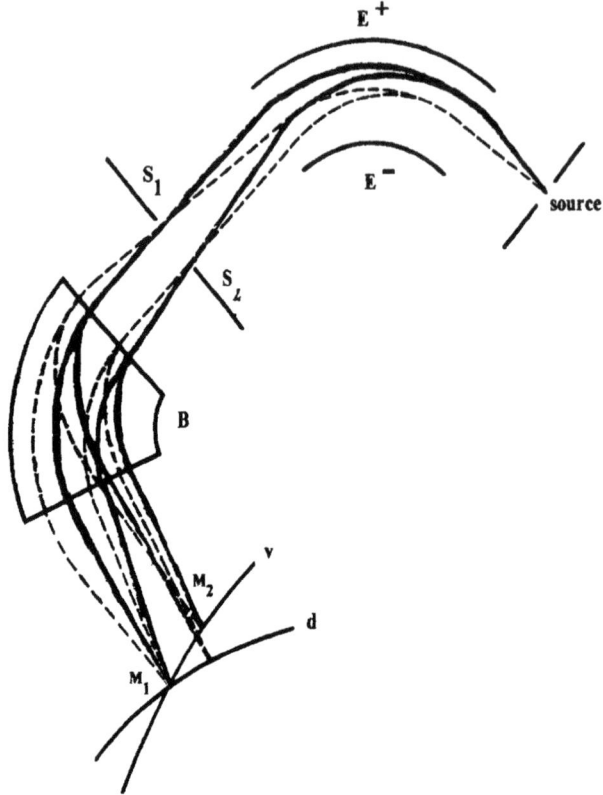

Figure 1.14   *Schematic diagram of a double-focusing magnetic sector mass spectrometer*

metal cup and all secondary electron emission is suppressed. An accurate measure of the charge deposited by the primary ion current is obtained by voltage amplification. The time constant limits the use to slow scanning or beam splitting applications such as accurate stable isotope ratio measurements (*e.g.* in Thermal Ionisation Mass Spectrometry or Gas Isotope Ratio Mass Spectrometry).

Electron and Photon Multipliers

Ions incident upon the conversion dynode produce secondary electrons that impinge on an array of dynodes. As little as 1–10 primary ions can generate reliable signals in modern instruments, where amplifications of 10^6 are routine. Photon multipliers comprise conversion dynodes, a phosphorescent screen and a photomultiplier. Ions incident on the conversion dynode generate a shower of secondary electrons. These strike the phosphorescent screen, producing bursts of photons that are detected by the photmultiplier. Photon multipliers are more robust than electron multipliers and are preferred as universal detectors by some manufacturers.

Array Detectors

Array detectors can attain very high sensitivities by collecting all the ions of a given mass range continuously. This contrasts with conventional scanning methods using point detectors (*e.g.* electron multipliers) that collect only a small fraction of the ions in each mass channel as the spectrum is scanned. Photodiode arrays and similar devices were limited by their low resolving powers and narrow mass ranges. However, recent improvements have extended the mass range by sequentially stepping 'scanning arrays' under computer control. Alternatively, specially designed ion-optical systems have been developed for use with wide-angle array detectors. Both developments allow reasonable scan rates for the total spectrum. Array detection can increase the apparent sensitivity by two orders of magnitude.

5   Metastable Ions, Collision Induced Dissociation and Tandem Mass Spectrometry

During electron ionisation at 70 eV, ions possessing a wide range of internal energies are generated. Those with large excess of internal energy will fragment rapidly and most do so in less than 10^{-7} s, *i.e.* before the ion leaves the source. Other ions with lower internal energies will survive the flight through the analyser, which takes $> 10^{-5}$ s, and will be collected at the detector as normal ions. However, ions with intermediate internal energies can decompose in transit. These ions are said to be 'metastable', having reactions with rate constants in the range 10^5–10^7 s^{-1}. There are three regions, denoted 1FFR, 2FFR and 3FFR in Figure 1.15 where the metastable decomposition process might occur in the geometry of a double-focusing instrument.

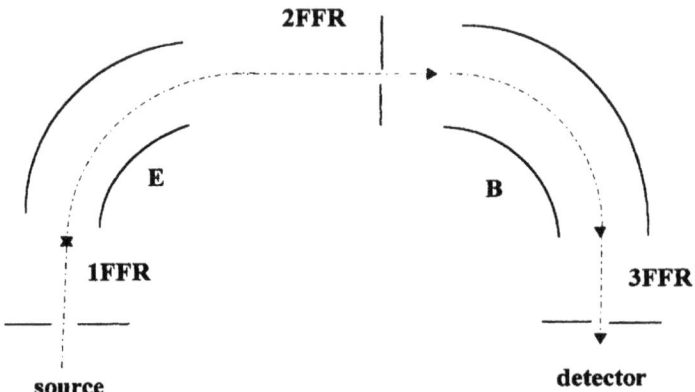

Figure 1.15   *The three field-free regions in a double focusing mass spectrometer*

First Field-free Region (1FFR)

Only by defocusing the electric sector (E) to normal ions can metastables, formed in the region between the source and the electric sector, be focused onto the detector. This was the first and simplest method for studying decomposition processes occurring immediately outside the source. Other methods were subsequently introduced, collectively referred to as 'linked scanning' techniques, that enable various combinations of values of E, B and V to be scanned in order to collect specific types of metastable ions.

Second Field-free Region (2FFR)

Ions produced from decomposition occurring between the electric (E) and magnetic (B) sectors of a double-focusing instrument will be recorded. They appear as diffuse peaks in the normal spectrum at non-integral masses, due to the momentum to charge selectivity of the magnetic analyser, and their actual position can be computed from the formula

$$m^* = m_2^2/m_1$$

where $m^* =$ the mass of the metastable ion, $m_1 =$ the mass of the precursor ion, and $m_2 =$ the mass of the product ion. For the loss of CH_3 from cholesterol $M = 386$,

$$m^* = \frac{371^2}{386} = 356.6$$

Another method of detecting these second FFR ions is to employ a momentum-selective detector in which a retarding voltage can be applied to differentiate between normal ion energies and those of metastable ions.

Third Field-free Region (3FFR)

Metastable ions formed after the magnetic analyser and before the detector of a double-focusing instrument will not be separated from the normal ions, but some broadening of the 'skirt' around the peak base may be visible.

Collisional Activation

So far we have considered ions which decompose spontaneously, *i.e.* those with critical rate constants. By deliberately inserting a zone of high-pressure gas molecules in the path of the ion beam, controlled gas phase reactions can be performed and collisionally induced decompositions (CID) monitored using any of the above techniques.

Instrumentation for Metastable Analysis and Tandem Mass Spectrometry

Metastable analysis began early in the history of mass spectrometry but because the number of metastable ion decompositions was small in relation to the 'normal' ions (< 5%), this remained an academic exercise. Improvements in instrument sensitivity subsequently showed that metastable ions were potentially very useful as an aid to interpreting mass spectra or improving analytical selectivity. Several types of instrument that generate data on metastable or CID processes are now available.

Reversed Geometry (BE) Double-focusing Instrument

There was no good reason why the electric sector was placed in front of the magnetic sector in the first double-focusing instruments and later versions were designed with 'reversed geometry' to perform certain specific experiments. Reversed fields allow kinetic energy analysis to be performed on ions of known mass, *viz.* mass analysed ion kinetic energy spectroscopy (MIKES).

Linked-scanning of EB Double-focusing Instrument

This technique allows metastable ions to pass through the sectors according to various formulae. The interested reader is referred to a recent book for a useful summary of the physics, mathematics and scanning methods available for various combinations of magnetic and electric sectors.[22]

Triple Sector Instruments (EBE)

If a collision cell is introduced into the third FFR of a conventional double-focusing instrument and an additional electric sector is introduced after this, ions of known atomic composition can be subjected to kinetic energy analysis.

Four-sector Instruments (EBEB)

This design allows the ions emerging from the first double-focusing (EB) region to be analysed in the second double-focusing region. In other words, a highly resolved ion beam selected from MS1, having a particular atomic composition, is decomposed in a collision chamber and the product ions analysed in MS2, again at high resolution. This, and the triple quadrupole instrument described below, is a true 'tandem' mass spectrometer or MS/MS instrument. Because the ion beam incident on the collision cell has high translational energy, large amounts of vibrational and electronic excitation can be transferred to analyte ions.

Triple Quadrupole Instruments

Quadrupoles are mass selective in comparison with magnetic sector instruments, which are momentum selective. In the triple quadrupole, the first sector acts as a mass filter for the middle (RF only) collision chamber section and the third mass filter is used to analyse the decomposition products. Energies transferred to analyte ions are lower than in multi-sector magnetic instruments.

Fourier Transform and Ion Trap Mass Spectrometers

These are capable of generating MS/MS spectra by temporal, rather than spatial, separation of precursor and product ions. Unlike the instruments described above, they can be used to perform sequential MS experiments, *e.g.* MS/MS/MS or MS^n. This is achieved by isolating precursor ions of interest in the trap or cyclotron cell, decomposing these by collisional activation, isolating a product ion of interest, decomposing this, and so forth.

The principles of tandem mass spectrometry on ion trap mass spectrometers are very similar to those applied to achieve tandem mass spectrometry on FTMS instruments, although the equipment is very different. All but the precursor ions of interest are swept from the ion trap by judicious choice of trapping voltages. The ions are activated by resonant excitation and the ionic products are mass analysed. As with FTMS, only product ion scans can be performed. However, further MS/MS experiments can be conducted in a similar manner to FTMS, by isolating product ions of interest and subjecting these to further stages of collisional activation, *i.e.* MS^n experiments may be carried out.

Quadrupole/Time of Flight Tandem Instruments

One of the main drawbacks of triple quadrupole mass spectrometers is that the product ion spectra of biopolymers, particularly peptides, are usually of low intensity. Thus, it is difficult to obtain usable product ion spectra containing sequence information from the femtomole amounts of sample isolated in the course of studies of, for example, protein expression by living organisms. This problem has been overcome by constructing hybrid quadrupole/time of flight mass spectrometers.[23,24] Precursor ions are mass selected by a quadrupole mass spectrometer and activated in a collision cell, as normal. However, the product ions formed are analysed by a Time of Flight mass spectrometer. ToF essentially detects all the ions in a mass spectrum simultaneously, instead of scanning the spectrum (and yielding only a weak ion current per mass channel because of the short dwell time). This yields a large increase in the sensitivity with which product ion spectra are detected, up to 100-fold in favourable cases. Although quadrupole/time-of-flight instruments have a variety of potential applications, their particular strength is in the generation of 'sequence tags', short, characteristic sequences of peptide fragments formed by enzymolysis of protein spots isolated from 2-D gels. By searching against a library of known peptide sequences the parent protein can be identified unequivocally. This methodology has helped to advance the new and rapidly expanding field of proteomics, which has many potential applications in food science. Although quadrupole/time-of-flight instruments were developed in the 1980s[23] it was only the advent of ESI that allowed their full potential to be exploited.[24]

Specialised Instruments

In addition to the inorganic mass spectrometric methods already described above, the following techniques have been used in food and nutrition studies.

Gas Isotope Ratio Mass Spectrometry (GIRMS)

Isotope ratios are frequently measured, in the course of quantitative or metabolic studies, by using 'conventional' organic mass spectrometers or inorganic mass spectrometers. However, the term Gas Isotope Ratio Mass Spectrometry (GIRMS) is, by custom and convention, applied to specialised instruments used for determining the isotope ratios of gases such as CO_2, N_2, SO_2 and H_2 to a very high degree of accuracy. Combustion or reduction of the sample of interest generates these gases. The main uses of GIRMS in food science are (i) to authenticate the sources of foods and food ingredients and (ii) to study the metabolism of stable isotope labelled organic micro- and macronutrients in circumstances where conventional organic mass spectrometry does not have sufficient precision (see Chapter 10). A schematic diagram of a GIRMS instrument is shown in Figure 1.16.

Ionisation is achieved using an electron ionisation source that is comparatively gas tight. This increases the residence time of the gas in the ion source, allowing longer measurement times and thus increasing precision. This design feature differs from the EI sources used in organic mass spectrometry where sample residence time is minimised to reduce thermal artefacts, ion molecule

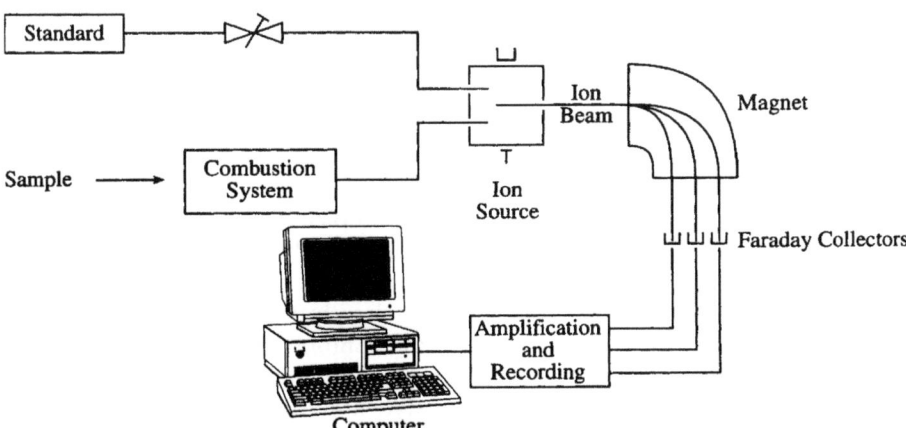

Figure 1.16 *Schematic diagram of a GIRMS instrument*

reactions, and potentially deleterious effects on electrically insulating components. These problems do not occur when measuring isotope ratios of clean, unreactive gases, when sensitivity and stable ion source conditions are the most important considerations. The mass analyser is specially constructed to ensure wide dispersion of the ion beam, so that Faraday cup collectors can be used to record each individual isotope. Combustion or reduction of samples to the appropriate gases is conducted under carefully controlled conditions and is often fully automated, with the entire measurement cycle under computer control. Samples are always measured against a standard of known isotopic composition to cancel time and instrument-dependent mass discrimination effects. Measurements are reported as 'δ-values', defined by the equation:

$$\delta = \left(\frac{r}{r_s} - 1\right) \times 1000$$

where r is the isotope ratio of the sample and r_s is the isotope ratio of the standard. The relative difference between the standard and sample isotope ratio is therefore given in parts per thousand (per mil, ‰). For example, the δ values for carbon are expressed relative to a reference standard, Pee Dee belemnite (PDB). Isotope data may also be expressed in atom % or atom % excess (APE), calculated directly from δ values.

The δ ^{13}C values for natural (unenriched) organic compounds of plant or animal origin usually lie in the approximate range -10 to -40, *i.e.* biological carbon sources are depleted in ^{13}C relative to the reference standard. The accuracy and precision of δ ^{13}C GIRMS measurements are typically $< \pm 0.4$.[25] Gas chromatographs may be coupled directly to GIRMS *via* a combustion chamber (GC/C/GIRMS), yielding compound specific isotope ratio measurements.[26] This type of instrument is now widely available from mass spectrometer manufacturers and provides a very powerful tool for conducting metabolic studies with enriched stable isotopes or for determining the authenticity of foods and food ingredients.

Pyrolysis Mass Spectrometry

Pyrolysis mass spectrometry (Py-MS) is based on the direct or indirect coupling of a 'flash pyrolyser' to a mass spectrometer. Direct Py-MS entails thermal decomposition of the analyte near to or, ideally, inside the mass spectrometer ion source. In indirect mode, the pyrolyser is coupled to the mass spectrometer *via* a chromatograph, usually a gas chromatograph (Py/GC/MS).

Although the word pyrolysis is derived from Greek roots that together mean 'to decompose by burning' this is an incorrect description of analytical pyrolysis techniques like Py-MS. Burning involves the combustion of molecules, in the presence of oxygen, to yield water, CO_2 and heteroatomic oxides, without generating any structural information. Analytical pyrolysis, on the other hand, is the thermal decomposition of an analyte in the presence of an inert gas to yield sizeable fragments. These fragments are often highly diagnostic of structural features in the original molecule(s) and can be regarded as a 'fingerprint' of the analyte. The sample must be heated very rapidly (usually within a few milliseconds) to a precisely controlled equilibrium temperature, typically in the range 600–800 °C. After this equilibrium temperature has been held for several seconds, the probe is cooled rapidly. It is very important that the pyrolysis fragments escape rapidly from the pyrolyser region to avoid overheating, rearrangement of the pyrolysis products and recombination into larger fragments. There are several different devices available for conducting pyrolysis, including platinum-coil heated filaments, Curie-point (inductive heating of ferromagnetic material by an applied radiofrequency field), furnace and laser pyrolysis. Direct pyrolysis techniques often require data analysis by sophisticated statistical and pattern recognition techniques to optimise information content. Mass spectrometric analysis of pyrolysis fragments is usually conducted using conventional mass spectrometers (magnetic sector or quadrupole) equipped with EI and/or CI ion sources. The ion sources are sometimes modified to generate pyrolysates 'in-source'. Sample sizes are typically in the range 1–200 μg,[27] although it is possible to conduct Py-MS on nanograms of material. Samples typically comprise dried biological tissue: they are usually analysed with little or no pre-processing. Irwin[28] has written an excellent book on analytical pyrolysis and Aries and Gutteridge[29] have reviewed the use of Py-MS in food analysis. Applications of pyrolysis mass spectrometry in food analysis are covered in Chapter 11.

Inorganic Mass Spectrometers

Types of inorganic mass spectrometry most relevant to food and nutrition research have already been discussed above. Further details can be found in the ionisation section, under the headings Thermal Ionisation and Inductively Coupled Plasma Ionisation.

6 Scanning Modes

Only the main scan modes are described briefly below. The interested reader can investigate this topic more thoroughly in the texts cited above.[3,4]

Conventional Scan Modes

Several different types of scan mode (data acquisition) can be used in mass spectrometry, depending on the nature of the information sought. If the maximum amount of information is desired, *full scan mode* is used and *m/z* data are collected over the entire mass range between preselected mass points. For example, EI and CI spectra are often scanned from mass 20 to 1000. Scan speeds are selected according to the type of sample introduction method used. If capillary GC/MS data are collected, the full spectrum might typically be scanned every 0.5 s with a 0.1 s interscan time. Slower scan speeds are generally used if samples are introduced by heated probe or if the chromatographic peak width is broader (often the case in LC/MS).

An alternative to full scan mode is *selected ion monitoring* (SIM, also known as selected ion recording, SIR). Instead of scanning the full spectrum, the mass spectrometer (under computer control) is used to measure a small number of masses, typically the molecular ion and a number of characteristic masses. This yields a very large gain in sensitivity, typically about two or three orders of magnitude, because the dwell time on each peak is increased greatly. To illustrate, a conventional scan acquired during GC/MS analysis will typically sweep the mass range from m/z 20–620 in 1 s. Thus the dwell time at each mass in the spectrum is only 1.67 ms if the instrument is operated at unit mass resolution. However, if the instrument is set to monitor two masses for (say) 0.2 s at a time, switching cyclically from one mass to the other, the gain in sensitivity *per scan* is 120. In practice, several scans are collected as GC or LC peaks elute into the mass spectrometer and are integrated to yield (when compared with an internal reference standard) a quantitative response. Internal standards used are typically an analogue or homologue of the analyte, or (ideally) isotopically labelled analogues of the target compound. Isotopically labelled internal standards are ideal because their physico-chemical properties are almost identically to the analyte. However, they are shifted in mass (according to the number of isotope labels incorporated) and can therefore be distinguished from the target compound. Spiking the analyte with an isotope labelled standard yields the reference quantitative method of isotope dilution mass spectrometry (IDMS). The major drawback of the SIM method is the loss of selectivity consequent on monitoring a small number of masses instead of the full spectrum. This makes the technique prone to interference from components that elute at the same retention time if these components contain masses in the mass channels monitored. Although these masses may be of low abundance, they can still exert a deleterious effect because of the extreme sensitivity of SIM. These problems can be overcome by (i) judicious selection of mass channels monitored, (ii) use of ionisation modes that are selective for the target analyte, (iii) increasing the resolution of the mass spectrometer (where possible) to resolve interferences that have the same nominal mass (isobars) but may have different accurate masses or (iv) using tandem MS techniques such as selected reaction monitoring (see below) that increase selectivity.

Very high sensitivity can be attained even in full scan mode on certain types of instrument. Mass spectrometers equipped with array detectors (in which all or part of the mass spectrum is recorded simultaneously) fall into this category. ToF and FTMS instruments, by the nature of their ion optical systems, are able to record the entire mass spectrum virtually simultaneously and are also capable of high sensitivity in full scan mode.

Tandem Scan Modes

The main tandem mass spectrometry (MS/MS) scan modes are *product ion, precursor ion, constant neutral loss* and MS^n scans. In addition, the special case of *selected reaction monitoring* (SRM) is occasionally used to enhance selectivity in quantitative mass spectrometry. MS/MS methods generally involve activation of selected ions, typically by collision with an inert gas, sufficient to induce fragmentation (collision induced dissociation, CID).

The precursor ion scan involves selection of the ion of interest, activation of that ion and mass analysis of the product ions. This is a widely used technique and is particularly appropriate for aiding structure determination and for biopolymer sequencing.

The product ion scan takes this process in the opposite direction by selecting a fragment (product) ion and scanning the mass spectrometer in a manner that allows the precursor of that product ion to be identified. The scan mode has a number of applications, for example in identifying compounds that contain a common structural feature (yielding a common fragment ion) in a complex chromatogram.

The constant neutral loss scan involves scanning for a fragmentation (neutral loss of fixed, predetermined mass). It is particularly useful for rapid screening in metabolic studies. For example, a molecule of general structure X–Y might lose the neutral fragment Y upon CID ($X–Y^{+\bullet} \rightarrow X^+ + Y^-$). If the X moiety of the molecule is modified during metabolism to yield a different mass, the constant neutral loss scan can be used to identify the metabolites. This approach succeeds because the loss of Y still occurs, so any molecules that fragment by loss of Y can be identified rapidly.

MS^n is commonly applied on particular types of tandem instrument. For example, ion traps and FTMS, that are based on temporal rather than spatial selection of precursor and product ions. These instruments are preferred because of the prohibitive expense of linking more than two mass spectrometers in series. A precursor ion is selected and isolated by selectively ejecting all other masses from the mass spectrometer. CID of the precursor ion yields product ions that may have different masses (MS/MS). A product mass of particular interest is selected and other fragment ions are ejected from the cell. This product ion is, in turn, subjected to CID, generating more product ions that are mass analysed (MS/MS/MS). This process can be repeated several times. MS^n experiments are used to investigate the physical chemistry of gas-phase reactions or as an aid to structural analysis.

Selected reaction monitoring (SRM) is a special case of SIM in which the power of a tandem instrument is used to enhance the selectivity of SIM. For example, a molecule A–B that fragments after CID to yield $A^+ + B^\bullet$. Instead of monitoring the mass of the molecular ion, $A–B^{+\bullet}$ in a simple SIM experiment, the tandem mass spectrometer can be tuned to detect only ions that undergo the fragmentation $A–B^{+\bullet} \rightarrow A^+ + B^\bullet$. This automatically filters out any interfering ions of the same mass as $A–B^{+\bullet}$ that do not exhibit the same fragmentation. As the chances of an interfering mass undergoing the same fragmentation are remote, selectivity is greatly enhanced. This can be at the expense of some loss of sensitivity (typically 5–10-fold) in the more popular types of tandem instrument. If several different reactions are monitored the term *multiple reaction monitoring* (MRM) is sometimes used.

7 References

1 J.F.J. Todd, *Int. J. Mass Spectrom. Ion Proc.*, 1995, **142**, 211.
2 B. Munson and F.H. Field, *J. Am. Chem. Soc.*, 1966, **88**, 2621.
3 J. Throck Watson, *Introduction to Mass Spectrometry*, Raven Press, New York, 1985.
4 R.A.W. Johnstone and M.E. Rose, *Mass Spectrometry for Chemists and Biochemists*, Second Edition, Cambridge University Press, 1996.
5 J.H. Beynon and A.G. Brenton, *An Introduction to Mass Spectrometry*, University of Wales Press, Cardiff, 1982.
6 F.W. McLafferty and F. Turecek, *Interpretation of Mass Spectra*, Fourth Edition, United Science Books, California, 1993.
7 E.C. Horning, D.I. Carroll, I. Dzidic, K.D. Hagele, M.D. Horning and R.N. Stillwell, *J. Chromatogr.*, 1974, **99**, 13.
8 K.G. Heumann, in *Inorganic Mass Spectrometry*, ed. F. Adams, R. Gijbels and R. Van Grieken, Wiley-Interscience, New York, 1988, pp. 319–331.
9 A.L. Gray, in *Inorganic Mass Spectrometry*, ed. F. Adams, R. Gijbels and R. Van Grieken, Wiley-Interscience, New York, 1988, pp. 257–300.
10 A.R. Date, *Spectrochim. Acta Rev.*, 1991, **14**, 3.
11 G. Horlick, *Spectroscopy (Oregon)*, 1992, **7**, 22.
12 A.L. Gray, in *Applications of Inductively Coupled Plasma Mass Spectrometry*, eds. A.L. Gray and A.R. Date, Blackie, Glasgow and London, 1987, pp. 257–300.
13 K.E Jarvis, A.L. Gray and R.S. Houk (eds.), *Handbook of Inductively Coupled Plasma Mass Spectrometry*, Blackie, Glasgow and London, 1992.
14 J.M. Carey, F.A. Byrdy and J.A. Caruso, *J. Chromatogr. Sci.*, 1993, **31**, 330.
15 N. Bradshaw, E.F.H. Hall and N.E. Sanderson, *J. Anal. At. Spectrom.*, 1989, **4**, 801.
16 A.J. Walder and P.A. Freedman, *J. Anal. At. Spectrom.*, 1992, **7**, 571.
17 U. Gießmann and U. Greb, *Fresenius' J. Anal. Chem.*, 1994, **350**, 186.
18 G.C. Eiden, C.J. Barinaga and D.W. Koppenaal, *Rapid Commun. Mass Spectrom.*, 1997, **11**, 37.
19 S.D. Tanner and V.I. Baramov, *J. Am. Soc. Mass Spectrom.*, 1999, **10**, 1083.
20 E. De Hoffman, J. Charette and V. Stroobant, *Mass Spectrometry: Principles and Applications*, John Wiley and Sons, Chichester, 1996, Chapter 2, pp. 39–98.
21 R.G. Cooks and R.E. Kaiser, Jr., *Acc. Chem. Res.*, 1990, **23**, 213.
22 E. De Hoffman, J. Charette and V. Stroobant, *Mass Spectrometry: Principles and Applications*, John Wiley and Sons, Chichester, 1996, pp. 73–84.
23 G.L. Glish and D.E. Goeringer, *Anal. Chem.*, 1984, **56**, 2291.
24 H.R. Morris, T. Paxton, A. Dell, J. Langhorne, M. Berg, R.S. Bordoli, J. Hoyes and R.H. Bateman, *Rapid Commun. Mass Spectrom.*, 1996, **10**, 889.
25 J.T. Brenna, *Acc. Chem. Res.*, 1994, **27**, 340.
26 D.E. Mathews and J.M. Hayes, *Anal. Chem.*, 1978, **50**, 1465.
27 J.J. Boon, *Int. J. Mass Spectrom. Ion Proc.*, 1992, **118/119**, 755.
28 W.J. Irwin, *Analytical Pyrolysis*, Marcel Dekker Inc., New York, 1982.
29 R.E. Aries and C.S. Gutteridge, in *Applications of Mass Spectrometry in Food Science*, ed. J. Gilbert, Elsevier Applied Science, Amsterdam, 1987, pp. 377–431.

CHAPTER 2

Interpretation of Organic Mass Spectrometric Data

1 Introduction

Mass spectral data can contain information relevant to the determination of the mass and/or chemical structure of a sample molecule. Using the conventions defined in Chapter 1, several ionic structures can be defined. These include positively and negatively charged radical ions, protonated and deprotonated molecules, multiply charged ions, and adduct ions of various types. These ion types can be created by different ionisation methods and are, in some way, representative of the molecular weight of the analyte. If the sample is an unknown compound, it is essential to understand the method used to ionise the sample. Furthermore, deduction of the molecular weight of the analyte should be approached cautiously. Table 2.1 shows the mass range and some of the typical ions produced by the common ionisation methods for particular sample types. Once the molecular weight has been determined (or it has been deduced that the mass spectrum does not contain molecular weight information), the remaining ions, assumed to be fragment ions, are examined as part of the interpretation process. Table 2.2 shows the likelihood of obtaining useful fragmentation from the common ionisation agents in use.

Table 2.1 *Comparison of ionisation methods with respect to molecular weight information*

Method	Sample	Range (Th)	Mol. Ions
EI	Gas	600	$M^{+\cdot}$
CI	Gas	600	MH^+, $(M + Na)^+$
APCI	Gas or liquid	1500	MH^+
ESI	Liquid	100 000	$(M + nH)^{n+}$, polymers
FAB	Liquid or solid	15 000	MH^+, $(M + Na)^+$
MALDI	Liquid or solid	300 000	MH^+, polymers, $(M + nH)^{n+}$

Table 2.2 *Comparison of ionisation methods with respect to the abundance of fragment ions in a typical mass spectrum*

Method	Agent	Fragmentation	Press./temp.
EI	70 V electrons	Abundant	Vac./300 °C
CI	500 V electrons	Very little	Vac./300 °C
APCI	Corona discharge	None	Atmos./ambient
ESI	Electric field, evaporation	None	Atmos./ambient
FAB	Equilibrated desorption	Useful	Vac./⩾ambient
MALDI	Heating or equilibrated desorption	Very little	Vac./⩾ambient

At this point it is appropriate to consider all the evidence available from other sources. This includes the record of the extraction, isolation and clean-up processes used, the retention time data from previous separation steps and any other spectral or physical data, *e.g.* the UV spectrum. Often, although the mass spectrum will be the first item of spectroscopic information obtained, especially on trace amounts of sample, it is highly unlikely that the mass spectrum will enable a complete structure to be drawn. Of particular supplementary value is the chromatographic retention time or the electrophoretic migration rate. These are related to physical properties of the sample, such as its polarity or hydrophobicity. Inclusion of these data permits the elimination of certain chemical classes, and of homologues within a class that behave differently.

There are many different methods to acquire mass spectrometric data. It is therefore essential to record the conditions under which the spectra were obtained, and to keep this information with the spectrum during the interpretation. Variables such as ion source temperature, mass range scanned, mass resolution chosen, *etc.*, all affect the qualitative and quantitative nature of the information. These data are often automatically stored with the mass spectra on modern mass spectrometer computer systems.

2 Formation of the Electron Ionisation Mass Spectrum

Molecular Ion Production

The electron is too small to impact a molecule directly (except by very remote chance). It is better to consider an electron wave interacting with the electric field of the molecule, with negligible translational energy being transferred to the molecule. This interaction causes molecular electrons to become excited, moving them to higher orbitals, in common with all other forms of electromagnetic excitation spectroscopy.

$$h\nu \longrightarrow M \longrightarrow M^* \qquad (2.1)$$

At the pressure of the EI source (10^{-3} Pa) only unimolecular interactions need be considered since the long mean free paths exclude ion–molecule reactions. Unlike other forms of electronic excitation spectroscopy, sufficient energy is supplied during electron ionisation to remove an electron completely from the molecule, and therefore the process is not reversible. Electron capture and ion pairing are alternative ionisation processes, but it is the production of positive ions by the ejection of an electron which we will consider at this stage.

The electronic excitation will induce a range of vibrational energy states within the population of molecular ions formed. Morse diagrams and potential energy surfaces can be used to visualise the various transitions resulting in the eventual formation of ground-state ions with an excess of vibrational energy. Since the bond vibrational frequency is of the order of 10^{-13} s, compared to the lifetime of the excited ion (10^{-8} s), there is plenty of time for the energy to be completely dispersed through the molecular bond structure. Fragmentation occurs at different rates depending on the electronic state and excess vibrational energy of each individual ion. The calculation of the rate of ion decomposition due to the fragmentation of a particular bond can be conducted using quasi-equilibrium or RRKM theory. These topics are discussed more fully by Rose and Johnstone[1] and will not be elaborated here.

The distribution of excess internal energy is shown in Figure 2.1. Molecular ions with energies in the lightly-shaded area, *i.e.* above the appearance energy of a fragment ion, dissociate rapidly. Those in the unshaded area at lower energies remain as molecular ions, stabilising small amounts of excess energy within their structure. The higher the residual internal energy of the fragment ion the greater the likelihood that it will be unable to stabilise the excess energy and that further fragmentation will occur. There is an intermediate area (black) in which ions with excess energy have lifetimes of 10^{-6}–10^{-5} s which means in practice they can decompose after leaving the source (10^{-6} s) but before reaching the detector (10^{-5} s). These are the metastable ions (m*).

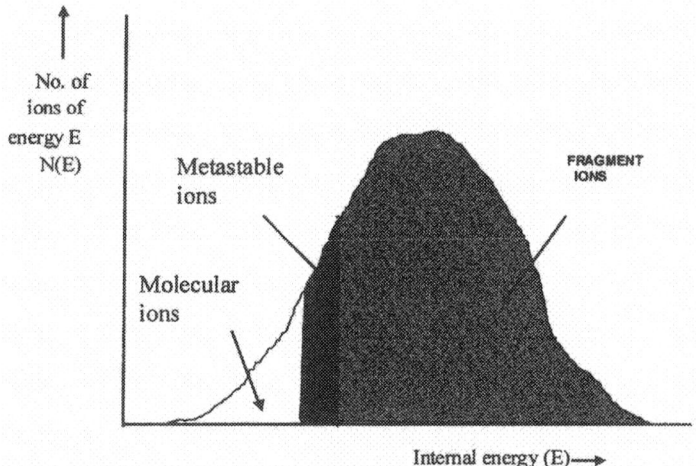

Figure 2.1 *Typical distribution of excess internal energy in a population of ions*

Molecular ions can be formed by the removal of an electron from a lone pair orbital associated with a heteroatom, or from a π- or σ-bonding orbital. Stable organic molecules have an even number of electrons in bonding or non-bonding orbitals and so if the symbol $M^{..}$ represents two non-bonded electrons in a molecule containing a heteroatom subjected to electron ionisation, then

$$e \longrightarrow M^{..} \longrightarrow M^{+\cdot} + 2e \qquad (2.2)$$

showing that the resulting molecular ion $M^{+\cdot}$ has lost an electron. Thus, molecular ions formed by EI possess an odd number of electrons (OE).

It is important to remember that the molecular ion continues to vibrate after formation and may have a different structure from the original molecule.

Fragmentation of the Molecular Ion

From the description of the formation of the molecular ion using 70 eV electrons, it is obvious that many of these will decompose. A summary of the factors influencing their decomposition will include the internal energy of the precursor ion, the bond dissociation energy, the time between formation and detection and the stability of the ionic and neutral products. Large values of vibrational energy will exceed the elastic limit of some of the bonds leading to fragmentation

$$M^{+\cdot} \longrightarrow A^{+} + {}^{\cdot}B \qquad (2.3)$$

where A^{+} is a stable cation, ${}^{\cdot}B$ is a stable neutral radical, and A has a lower IE than B, or

$$M^{+\cdot} \longrightarrow C^{+\cdot} + D \qquad (2.4)$$

where $C^{+\cdot}$ is a stable odd-electron radical ion and D is a stable neutral molecule. Again, C will have a lower IE than D. If the ions formed still possess excess energy, they will continue to decompose,

$$A^{+} \longrightarrow E^{+} + F \qquad (2.5)$$

and
$$C^{+\cdot} \longrightarrow G^{+} + {}^{\cdot}H \qquad (2.6)$$

Alternatively, $M^{+\cdot}$ can decompose directly to any intermediate ion.

Eventually, all the energy has been dissipated. This cascade of ion formation is called the fragmentation pattern. The overall shape of the pattern, *i.e.* the mass (strictly m/z) and intensity of each fragment ion in the cascade – the mass spectrum – is dependent upon the molecular structure. The masses and intensities of ions in the spectrum are characteristic of those structures which can retain the charge and stabilise any excess energy when labile bonds break.

The product ion can be stabilised in several ways. Electron sharing, resonance stabilisation, and distonic radical ions are discussed later. The stability of the neutral counterpart is also important. Neutral molecules are more stable than neutral radicals. Small, multiple bond molecules, such as HCN, with high ionisation energies are particularly stable, and can remove relatively large quantities of internal energy from the precursor ion, speeding the energy dissipation process.

Although there would appear to be the potential for 70 eV electrons to supply sufficient energy to cause every bond in the molecule to cleave, this rarely is the case. Mass spectra often do not contain an ion corresponding to cleavage of the bond with the lowest bond dissociation energy. This is presumably because the prospective fragment cannot stabilise the excess energy, or conversely a suitably stable neutral fragment cannot form to remove enough energy to allow the ion to stabilise. In general, the cleavage of weaker bonds, and loss of energy when associated neutral fragments are expelled, accounts for much of the ionisation energy expended. Ions formed from cleavage of the more stable bonds are absent or present only in trace quantities.

Ultimately, the balance between bond lability and fragment ion stability mediates the mass spectrum. Unsubstituted aromatic molecules can absorb large amounts of energy without exceeding the dissociation energy limit. Such compounds produce intense molecular ions (see below). Other chemical classes contain weaker bonds that reduce the probability of the survival of the molecular ion. Conversely, they enhance the likelihood of fragmentation and the formation of stable product ions characteristic of structural features in the precursor ion. Bond lability alone does not explain the character of the spectrum and fragment ion stability determines which of the many feasible processes actually produce the end-products featured in the mass spectrum.

Some structures are sufficiently stable to withstand fragmentation, even under 70 eV EI conditions. In these cases energetic CID of the molecular ion, usually in conjunction with MS/MS, may be used to produce fragmentation for structure elucidation.

3 Calibration of the Mass Scale and Determination of Molecular Mass

The measurement of molecular mass is a fundamental requirement in chemistry and modern mass spectrometers can provide this property for a wide range of chemical classes including biopolymers up to around 300 000 Da. Determination of molecular mass requires knowledge of the sample preparation protocol and the ionisation method employed in order to establish the chemical nature of the representative ions appearing in the spectrum. Positively charged ions can be radicals or adducts with a cation. They may be singly- or multiply-charged, sample molecules may polymerise before ionisation, or they may be formed by hydride abstraction. Negatively charged ions can be formed by electron capture, deprotonation, or by reaction with anions.

The polarity selected for mass spectrometric detection differentiates between anions and cations, but knowledge of the chemically preferred cationisation or anionisation processes occurring under different operating conditions is a matter of experience. Knowledge of the proton affinity of the sample and any reactant gas involved can be used to predict the production of protonated or hydride abstracted ions. In some experiments, conditions can be selected so that, for example, by using ammonia gas in CI, the cation likely to be formed will be $[M + NH_4]^+$. The formation of these adducts *versus* protonated molecules $[M + H]^+$ depends on the proton affinity of the analyte. If unit mass resolution is maintained in the molecular or adduct ion mass region, the nominal molecular weight of the compound may be deduced. If high-resolution mass spectrometry is available then an accurate measurement of the mass of the peak will allow an atomic composition to be proposed.

The Practice of Mass Measurement

In low-resolution mass spectrometry, with its nominal mass separation, it is usual to use the whole number to indicate molecular weight, based on summing the monoisotopic masses of the elements present ($H = 1$, $O = 16$, $N = 14$, *etc.*). For example, the molecular weight of α-ionone ($C_{13}H_{20}O$) is 192 Da.

Accurate masses are calculated using ^{12}C as the agreed mass scale reference of 12.000000 Da. On this scale, $^1H = 1.007825$, $^{16}O = 15.994914$ and $^{14}N = 14.003074$ due to the fractional mass defects (packing fractions) of the elements (see Table 2.3).

Unknown ion masses are measured relative to an internal standard compound (for convenience, often chosen to contain mass deficient ions if the sample ions are mainly mass sufficient). High accuracy (typically four decimal

Table 2.3 *Accurate masses of the common elements*

Isotope	Mass	Isotope	Mass
^1H	1.007825	^{28}Si	27.976929
^2H	2.014102	^{29}Si	28.976492
^4He	4.002601	^{30}Si	29.973758
^{10}B	10.012940	^{31}P	30.973764
^{11}B	11.009307	^{32}S	31.972073
^{12}C	12.000000	^{33}S	32.971464
^{13}C	13.003355	^{34}S	33.967863
^{14}N	14.003074	^{35}Cl	34.968853
^{15}N	15.000109	^{37}Cl	36.965903
^{16}O	15.994914	^{36}Ar	35.967546
^{17}O	16.999132	^{40}Ar	39.962386
^{18}O	17.999162	^{79}Br	78.918390
^{19}F	18.998402	^{81}Br	80.916420
^{20}Ne	19.992441	^{127}I	126.904660
^{22}Ne	21.991385		

places) is obtained by interpolating between standard peaks close to the unknown. Computed tables of the elemental composition of putative ionic masses can be generated for comparison (see Table 2.3 for the accurate masses of some common elements). Ionic masses of low molecular weight compounds provide useful information on the likely atomic composition of the ion, but as the mass increases, so do the number of possible elemental combinations (isobars) with similar mass values. Using this notation, the molecular weight of α-ionone is calculated as 192.1514 Da. Isomers, by definition, cannot be resolved by molecular mass measurements and other methods have to be employed.

As the molecular mass increases a point is reached where the fractional value for the sum of the lower isotopic masses reaches 1.0, *i.e.* the mass is apparently one unit higher. Similarly, as the mass of the molecule increases, the higher isotopes begin to make significant contributions and first the M + 1 and then the M + 2 isotope peaks become more abundant than the monoisotopic peak M in the spectrum. At still higher masses, where even high resolving power cannot separate adjacent isotopic species, it becomes necessary to adopt the convention of using the average mass of all the unresolved isotopes. This approach is recommended when measuring the molecular weights of proteins by, for example, MALDI-MS.

4 Interpretation of High- and Low-resolution Data

Resolving power is defined as:

$$R = \frac{M_1}{\Delta M} \tag{2.7}$$

where R = resolution required to separate two masses M_1 and M_2 having a mass difference between them of ΔM. Alternative definitions based on the peak width at half-height (or full-width at half-maximum, FWHM) are used for some instrument types. For example, ToF mass spectrometers.

Single sector low-resolution mass spectrometry – typically below a resolving power of 1000 – resolves peaks in the mass range differing by one mass unit, with a specified minimum overlap between adjacent peaks of, typically, a 10% valley. Most instruments operate in this mode to provide nominal molecular weight and structural information. On Fourier transform, double focusing and reflectron ToF instruments the resolution can be increased to the point where isobaric peaks can be separated and their individual masses calculated using the elemental values shown in Table 2.3. Typical high-resolution data allow atomic masses to be measured to three or four decimal places. Table 2.4 shows M_1, ΔM and R required to separate six isobars at mass 100.

A complete list of the accurate masses of all the major ions in a mass spectrum is extremely useful in structure elucidation of an unknown compound. Computer programs provided with most mass spectral data processing packages can be used to generate accurate mass values for alternative empirical formulae.

Table 2.4 *Comparison of the accurate masses, mass differences and resolution required for the separation of a group of ions of mass 100*

Composition	Accurate mass (M)	Mass difference from 100	Resolution required (R)
$C_3H_6N_3O$	100.0511	–	–
$C_4H_8N_2O$	100.0637	+0.0126	7940
$C_5H_{10}NO$	100.0762	+0.0251	3986
$C_6H_{14}N$	100.1126	+0.0615	1627
C_2F_4	99.9936	−0.0575	1741
$C_2H_4OSi_2$	99.9700	−0.0811	1234

5 An Empirical Approach to the Elucidation of Organic Structures

Comparison with Standard Spectra in a Library of Mass Spectra

If the EI mass spectrum of an unknown compound has been recorded under standard conditions then it can be matched against a library of known spectra. The degree of similarity between the unknown and the best fits to spectra in the database provides evidence of the possible identity or structure of the sample molecule. Computers facilitate on-line acquisition and processing of low-resolution mass spectral data. Most GC/MS analyses of unknown mixtures may be processed to give a list of the chemical names of compounds whose mass spectra resemble those of the components of the chromatogram, together with an index of similarity. It would be unwise to accept these data without further scrutiny, but the fact that modern instruments can produce these possible solutions quickly makes the task of the interpreter that much easier. The strategies for interpreting spectra given in most textbooks can be modified to start from this point.

A Strategy for the Interpretation of the EI Mass Spectrum

Firstly, one should check that no previous record exists in the library and that the data were obtained from a single pure compound. If no matching records are found, it is necessary to use all other sources such as spectroscopic, chromatographic and chemical information to reduce the number of alternative candidate structures. It is relatively simple to ascertain the presence of a CO group from the IR spectrum. Furthermore, knowledge of the sample preparation steps can often help to identify the chemical classes that will have been removed before the sample was taken for analysis. The retention time or retention index recorded in the GC/MS analysis will help to eliminate other possible structures simply on their different affinity for the stationary phase employed. Depending on the resources available, the next step might be to design new MS protocols in order to eliminate other compound types. Additional evidence may be generated from specially prepared derivatives. Finally, different decomposition pathways should be investigated in the hope of adding new data. Rules and aids to interpretation assist the process.

Categories of Mass Spectra

The overall nature of the spectrum may take the form shown in Figure 2.2:

(A.1) an intense $M^{+\bullet}$ ion with few weak fragment ions, *e.g.* unsubstituted aromatics;

(A.2) an intense $M^{+\bullet}$ ion with few strong fragment ions, *e.g.* substituted aromatics;

(B.1) a weak $M^{+\bullet}$ ion with many fragment ions, *e.g.* saturated aliphatics;

(B.2) a weak $M^{+\bullet}$ ion with a few strong fragment ions, *e.g.* alkanes with 'in-chain' functional groups.

The Nitrogen Rule

For compounds containing C, H, N and O, the peak corresponding to the molecular ion will have an even mass except when an odd number of nitrogen atoms are present. This is the first test to make and, for many low molecular weight compounds, it is a major item of information.

Stevenson's Rule and the Elimination of the Longest Side Chain

Stevenson's rule states that when a single bond in an odd-electron ion cleaves, the charge resides on the fragment of lowest IE. If we consider a molecule of

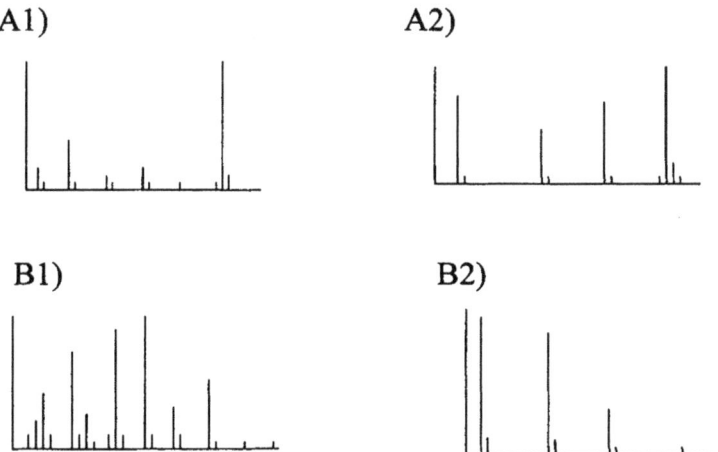

A1)

A2)

B1)

B2)

Figure 2.2   *Categories of mass spectra*

ethanolamine [CH$_2$(OH)CH$_2$NH$_2$], the IE of ethylamine is 9.5 eV whereas that of ethyl alcohol is 10.7 eV. Thus, it is easier to lose an electron from nitrogen than from oxygen, and we would expect the charge to be predominantly situated on the amino group, and for this to determine the course of the fragmentation (Scheme 2.1). In practice, the ammonium ion is ten times more intense than the oxonium ion in the EI spectrum of ethanolamine. If the IEs of the possible fragments are similar, then both ions will be present at approximately the same intensity. The two fragment ions are useful in larger structures because their sum represents the molecular weight. The complementarity of charge migration and simple cleavage processes is a useful aid to interpretation.

$$
\begin{array}{cccccc}
\overset{+}{C}H_2 & \dot{C}H_2 & CH_2\!\!-\!\!CH_2 & CH_2\!\!-\!\!CH_2 & \dot{C}H_2 & CH_2 \\
\| & | & | \,\,\diagup\,\, | & | \,\,\diagdown\,\, | & | & \| \\
OH & NH_2 & OH \quad NH_2 & OH \quad NH_2 & OH & \overset{+}{N}H_2 \\
+ & & +\cdot & +\cdot & &
\end{array}
$$

Scheme 2.1   *Illustration of Stevenson's rule*

An exception to Stevenson's rule is the preferential loss of the largest alkyl side chain in branched-chain compounds. In the compound 2-hydroxybutane, the loss of the ethyl radical to give m/z 45 yields the base peak. The loss of the methyl radical to form m/z 59 has a relative intensity (RI) of 19% while the loss of the hydrogen radical to form m/z 73 has a RI of only 1.2% (Scheme 2.2).

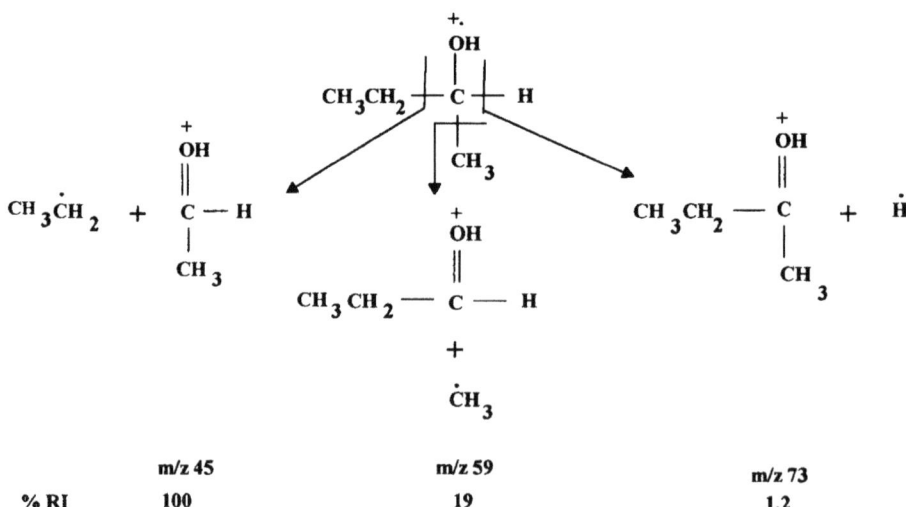

	m/z 45	m/z 59	m/z 73
% RI	100	19	1.2

Scheme 2.2   *An exception to Stevenson's rule*

Field's Rule

In even-electron decompositions forming the same even-electron product, the tendency of a neutral to leave without the charge is greater for molecules of lower proton affinity (PA).

The 'Even Electron' Rule

The molecular ion is an odd-electron species. Simple cleavage will produce an even-electron ion and a neutral radical. A rearrangement process will produce an odd-electron ion and a neutral molecule. Even-electron ions can continue to decompose, producing further even-electron species. The rule states, 'odd electron ions can eliminate either radicals or molecules, whereas even-electron ions can only lose molecules'.

For further reading about the mechanisms of fragmentation, see McLafferty and Turacek.[2]

Fragmentation Characteristic of Chemical Class

Low Resolution Data. Textbooks provide tables of data to serve as an aid to the rapid recognition of the possible structure of the unknown ion. Most useful in this respect are the lists of fragment ions commonly produced by certain chemical classes and the neutral fragments lost from the molecular ion.[3] Other tables present the molecular ion intensities by class.[2] The relative intensities of the molecular ion isotope pattern for molecules containing different numbers of chlorine and/or bromine atoms are a useful aid to interpretation. Although ion clusters can be complex where several halogen atoms are present in the molecule, the pattern accurately reflects the numbers of these atoms present. Clearly, it is prudent to use such experiential data to reduce the options that need to be considered further, but a pedantic approach to their use can be counterproductive. Modern data processing systems have much of this information in user-friendly form, so that this stage in the interpretation process can be rapid and routine.

High Resolution Data. Lists of likely ion structures at nominal masses and tables of accurate masses of these isobars provide useful information. This is needed to determine the elemental composition of ions that may make an unresolved contribution to the peak observed at a given resolving power. It is helpful to calculate the resolving power required to separate isobars at different masses. At mass 28, the resolving power (R) needed to separate CO and N_2 is $R = M_1/\Delta M = 28/0.01124 = 2490$. If the sample mass spectrum contains the ions C_2H_4 and $C^{13}CH_3$ then the resolving power required will be 6264. At mass 43 a common isobaric pair of ions, CH_3CO and $CH_3CH_2CH_2$, require a resolution of 1182 for complete separation. Although low resolution is adopted for most mass spectrometric measurements, there may still be sufficient separation to recognise (in the raw data) the proportional contribution of each isobaric ion present, *e.g.* in the spectrum of methyl isobutyl ketone.

6   Additional Information from the Mass Spectrum

If the empirical approach has not produced a result a more rigorous study is necessary. It is not possible here to provide full details, but the philosophy can be discussed. In brief, it is necessary to utilise all the information contained in the mass spectrum, not only that easily extracted by the automated data processing system. The neutral losses from the proposed molecular ion should be tested to ensure that they are all feasible, *e.g.* an M − 15 ion is feasible for the loss of methyl but M − 14 is most unlikely. The latter 'loss' would suggest that the putative molecular ion is actually a fragment ion (or an homologous mixture component), and that the molecular ion is not present in the mass spectrum. If there is any doubt, softer ionisation techniques should be used. An additional useful item of information is the pattern of isotope abundances associated with the major ions in the spectrum: this can sometimes provide direct information about the elemental composition of the molecule (Table 2.5).

For small molecules the molecular isotope ion cluster may be sufficient to identify the compound (Figure 2.3) but it may not be easy to recognise the monoisotopic ion, for example in Figure 2.3A.

The presence of chlorine, bromine or sulfur is often evident from the characteristic pattern of the molecular ion cluster (see Figure 2.3, B and D). The approximate number of carbon atoms in the molecule is given by the relative intensity of the M + 1 peak: 1.1% of the monoisotopic peak per carbon atom (see Figure 2.3A and C).

Metastable Ions

The presence of metastable ions in the normal mass spectrum may be over-looked because of the nature of modern data processing systems. This is not only because these peaks occur at low intensities, but also because they appear at non-integral masses and are considerably broader than normal mass peaks. These conditions usually result in the rejection of these 'anomalous' peaks by

Table 2.5 *Isotopic abundances of common elements*

Element	Mass	%	Mass	%	Mass	%
H	1	100	2	0.016	–	–
C	12	100	13	1.10	–	–
N	14	100	15	0.36	–	–
O	16	100	17	0.04	18	0.20
F	19	100	–	–	–	–
Si	28	100	29	5.07	30	3.31
P	31	100	–	–	–	–
S	32	100	33	0.78	34	4.39
cL	35	100	–	–	37	32.7
Br	79	100	–	–	81	97.5
I	127	100	–	–	–	–

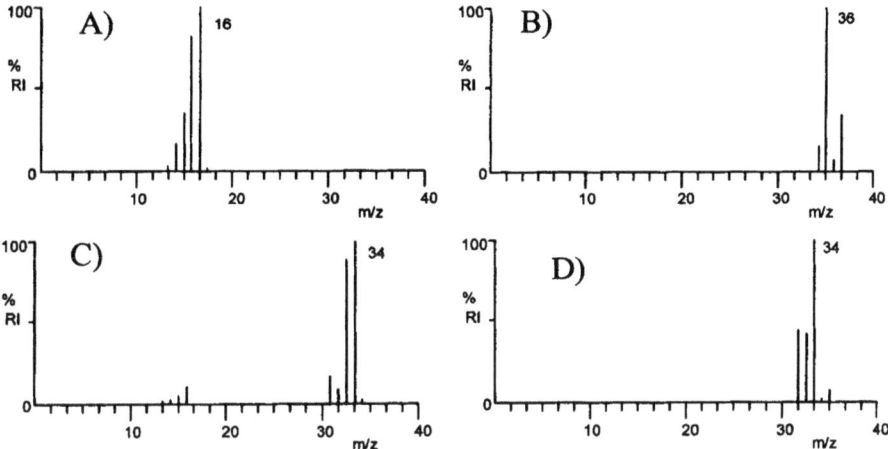

Figure 2.3 *Mass spectra of:* (A) *methane;* (B) *hydrochloric acid;* (C) *fluoromethane;* (D) *hydrogen sulfide*

data systems. However, metastable ions can often give valuable insights into the pathways taken during fragmentation since they link together precursor and product ions (see Chapter 1).

Tandem Mass Spectrometry

MS/MS provides an effective method of studying metastable processes. Precursor ions selected in MS1 under preset conditions of resolving power can be deliberately decomposed in a collision chamber before the resulting product ions are directed to MS2 for mass analysis. The interaction between ions and molecules in collision cells is the key to the use of MS/MS for structure elucidation. The ions of a chosen m/z value are collided with gas molecules, imparting additional internal energy to some ions and causing them to fragment. The gas pressure in the collision cell must not be so high that the ion beam is scattered, but must be sufficient to produce significant numbers of fragment ions. Up to approximately 10 eV of excess energy are required to produce fragment ions by CID; consequently CID and EI mass spectra can often present a similar appearance.

Different Automated Scanning Procedures

The Product Ion Scan. CID decomposes the precursor ion selected from MS1 and the product ions are separated in MS2. This mode is useful for mixture analysis and for structural studies. Under EIMS conditions, by choosing from MS1 each fragment ion as a precursor ion, the whole family of MS2 product ions, and hence the complete fragmentation pattern, can be constructed quickly.

The Precursor Ion Scan. MS2 is tuned to a chosen product ion and the potential precursor ions are scanned in MS1. This mode is useful for recognising homologues that have a common fragment ion in their spectra.

The Constant Neutral Loss Scan. The two analysers are 'ganged' together so that a constant mass difference separates the precursor and product ion. This mode is useful for detecting, say, all compounds losing COOMe, thus recognising possible methyl esters in a mixture of compounds with different organic functional groups.

7 Classical Fragmentation Processes Occurring in EIMS

The Common Fragmentation Processes

A number of fragmentation processes occur under EIMS conditions that are dependent on the chemical structure of the molecule. It is useful to consider the elementary process of cleavage of a covalent bond between two atoms (Scheme 2.3). It is usual for the electrons in a heterolytic cleavage to reside on the more electronegative atom. Most fragmentation of aliphatic compounds occurs within one or two bond lengths of the site of the charge centre. Stevenson's rule proposes that the charge will most likely be situated on the functional group with the lowest ionisation energy. Therefore, if the charge can be localised in the proposed structure, simple homolytic and heterolytic fissions of the α- and β-bonds will be responsible for much of the fragmentation. Both fast and slow processes can occur. Simple cleavage of a single bond requires only the correct amount of energy to be available (fast). However, if bond making and bond breaking are involved in the same process, *e.g.* a hydrogen rearrangement, the correct transition state for the reaction also has to be formed (slow). Slow processes will not be observed in instruments where the precursor ion reaches the detector before the event occurs.

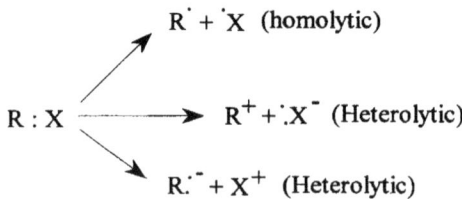

Scheme 2.3 *Homolytic and heterolytic bond cleavage*

Ions travelling through the mass spectrometer's vacuum system have long mean free paths and, in the absence of collisions, retain the energy imparted to them at ionisation. This situation differs from that of ions in solution, where the closer proximity of neighbouring ions causes energy equilibration to occur.

Ion Stabilisation

Systematic strategies have been developed to explain the most likely dissociation. 'The most important general factor affecting the abundance of a product ion is its stability'.[2] Ion stabilisation can be effected through:

(a) electron sharing:

$$CH_3\text{---}\overset{+}{C}\text{=}O \longleftrightarrow CH_3\text{---}C\text{≡}\overset{+}{O} \qquad (2.8)$$

(b) resonance stabilisation:

$$CH_2\text{=}CH\text{---}\overset{+}{C}H_2 \longleftrightarrow \overset{+}{C}H_2\text{---}CH\text{=}CH_2 \qquad (2.9)$$

(c) formation of distonic radical ions which are more stable than their classical counterparts:

$$CH_3\text{---}CH_2\text{---}CH_2\text{---}CH\text{---}O^{+\bullet} \longleftrightarrow {}^{\bullet}CH_2\text{---}CH_2\text{---}CH_2\text{---}CH\text{=}O^+H \quad (2.10)$$

The stability of the neutral product is less important, but radical stabilisation provided by electronegative sites such as O can be important. Certainly, the loss of small stable molecules with high IE and low PA (in parentheses), such as H_2, 15.4 (4.4); CH_4, 12.5 (5.7); H_2O, 12.6 (7.2); CO, 14.0 (6.2); HCN, 13.6 (7.4); H_2S, 10.5 (7.4); HCOOH, 11.3 (7.8), are favoured processes.

Types of Bond Cleavage

The nomenclature adopted in this section is that recommended by McLafferty and Turecek.[2] In general, it is assumed that electrons occupying orbitals where the IE is low are removed preferentially.
    The order of IEs is

$$\sigma\text{-} > \pi\text{-} > \text{n-electrons} > \text{aromatics} \qquad (2.11)$$

Sigma Bond Fragmentation (σ)

Fragmentation of saturated hydrocarbons is a special case since only σ-orbitals are available for ionisation.

$$R^{+}{}^{\bullet}CR_3 \overset{\sigma}{\longrightarrow} R^{\bullet} + {}^{+}CR_3 \qquad (2.12)$$

The radical ion charge centre can be situated at any bond in the chain. Single bond cleavages give rise to radicals and even-electron ions as above. The longer-chain primary products are likely to be unstable and further dissociation occurs with the loss of neutral molecules, normally ethylene or higher homologues, since the loss of methylene is energetically much less likely. Thus the cascade of heterolytic (charge migration) cleavages leaves a distribution of the C_nH_{2n+2} ion series at m/z 15, 29, 43, 57, *etc.* weighted in favour of the shorter-chain length C_3, C_4 and C_5 region.

The presence of a substitution in the chain provides favoured sites for charge localisation and cleavage, producing stabilised secondary and tertiary carbocations (tertiary > secondary). These ions are always present in higher concentrations than the equivalent ions in their primary isomer counterparts and the loss of the largest side chain rule will often determine their relative abundances in the mass spectrum.

One problem with σ-orbital ionisation of unsaturated hydrocarbons is that the even-electron product ions have a tendency to isomerise and undergo random hydrogen rearrangements.

Radical-site Initiation (α-Cleavage)

These are exemplified by homolytic cleavages where single electron transfers are denoted by fishhook arrows

$$R\!-\!CR_2\!-\!\overset{+}{Y}R \xrightarrow{\alpha} {}^{\bullet}R + CR_2\!=\!\overset{+}{Y}R \tag{2.13}$$

Charge-site Initiation (inductive cleavage, i)

These heterolytic cleavages are characterised by electron pair transfers and are denoted by doubly barbed arrows.

$$R\!-\!\overset{+}{Y}\!-\!R \xrightarrow{i} R^+ + {}^{\bullet}YR \tag{2.14}$$

Decomposition of Cyclic Structures

Two bonds must be cleaved to produce a fragment ion from a cyclic system. The first bond breaks to give a distonic radical ion that then donates the radical site for the cleavage of the second bond to release a neutral molecule. The retro-Diels–Alder (RDA) reaction is an important example, often found in the mass spectra of terpenes (Figure 2.4).

Rearrangement Processes

Radical-site Rearrangements

While the random rearrangements following σ-orbital ionisation occurring in unsaturated hydrocarbons are sometimes misleading, the specific rearrangement of a hydrogen atom to a radical site can be of value in structure elucidation. Two steps are required in the well-known 'McLafferty rearrangement' that involves a six-membered ring transition state of an odd-electron radical cation. The first step is formation of the distonic radical ion. This is

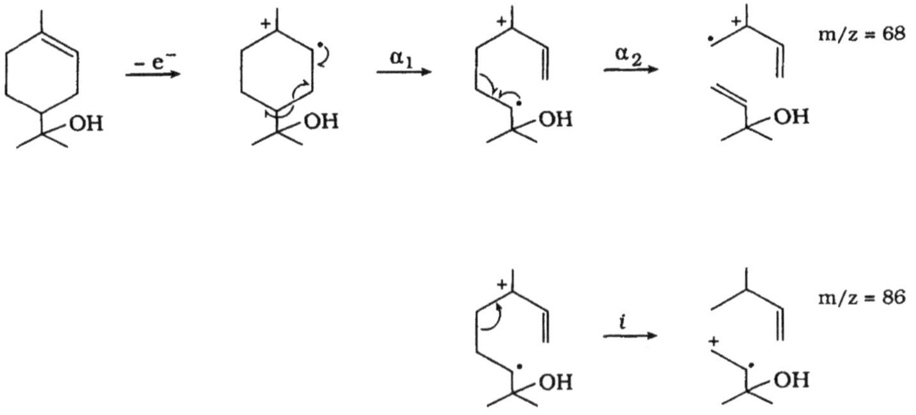

Figure 2.4   *The Retro-Diels–Alder fragmentation process*

followed by a second step, α-cleavage, that usually releases an olefin molecule. The essential pre-requisites for this rearrangement are a hydrogen atom at the γ-carbon and a multiple bond linking the α-carbon to the charge centre. Many of the classes of chemicals constituting food flavours, *e.g.* aldehydes, esters, ketones, *etc.*, undergo this rearrangement.

Other hydrogen rearrangement processes occur where similar steric factors predominate: for example, the dehydration of alcohols with chain lengths

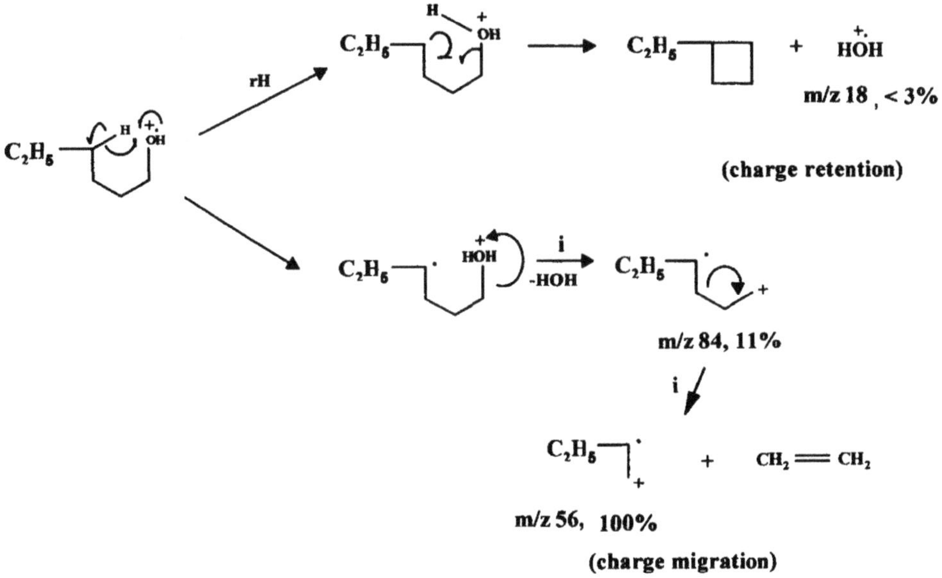

Scheme 2.4   *Dehydration of longer chain alcohols*
      (Reproduced from F.W. McLafferty and F. Turecek, *Interpretation of Mass Spectra*, 4th Edn., United Science Books, Sausolito, Ca., 1993, p. 76 with permission of the authors and publisher)

longer than four carbons (Scheme 2.4). The process is initiated by the formation of a stable -OH bond by donation of an unpaired electron to a nearby atom. Optional homolytic or heterolytic cleavages subsequently lead to charge migration or retention (*i.e.* the formation of the m/z 18 ion or the reciprocal $[M - 18]^+$ carbonium ion).

Displacement reactions involving non-hydrogen rearrangements occur occasionally and should be studied, *e.g.* in relation to the loss of methyl from the ionone structure (see Scheme 2.5 below).

Charge-site Rearrangements

The 'even-electron' rule states that only even-electron ions and neutral molecules can be formed by dissociation of the even-electron fragment ions generated by EI, or protonated (or otherwise cationised) molecules formed under CI conditions.

Occasionally, double rearrangement processes, where a second hydrogen atom is transferred after the initial McLafferty rearrangement, can occur. This produces a resonance stabilised protonated dihydroxy ion and is valuable in the interpretation of the spectra of propyl and higher esters, thioesters, amides, sulfones, *etc.*

Displacement Reactions

Unless there are more energetically-favourable reactions – such as a RDA in α-ionone – displacement reactions can produce intense ions through the creation of stable (usually cyclic) structures – such as in β-ionone (see the general case illustrated in Scheme 2.5).

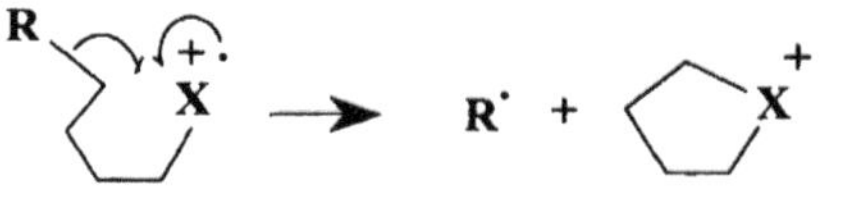

$(M\text{-Alkyl})^+$

Scheme 2.5   *General illustration of a displacement reaction*

General case for long-chain alkyl compounds:

when   X = Cl $[M - Alkyl]^+ = 91 \, [100\%]$
        X = Br $[M - Alkyl]^+ = 135 \, [98\%]$
        X = SH $[M - Alkyl]^+ = 89 \, [12\%]$

Elimination Reactions

Non-hydrogen rearrangements that eliminate small stable radicals and molecules can occur, especially if the resulting product ion has greater stability than the precursor ion, *e.g.* anthraquinone (Scheme 2.6).

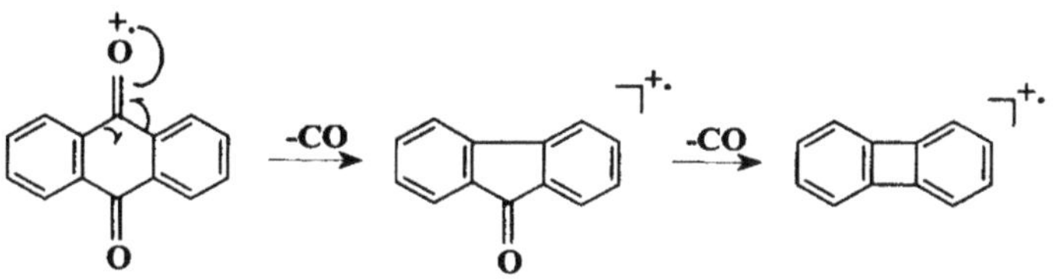

Scheme 2.6   *Elimination of CO molecules from anthraquinone*
(Reproduced from F.W. McLafferty and F. Turecek, *Interpretation of Mass Spectra*, 4th Edn., United Science Books, Sausolito, Ca., 1993, p. 220 with permission of the authors and publisher)

Keto-enol Transformations

The elimination of water from the molecular ion of ketones is common and is best explained by invoking the enol form to understand the process.

The Mass Spectrometric Shift Technique

Often the insertion of a small functional group into a larger molecule does not seriously affect the relative abundances of the fragment ions, so that the mass spectrum merely indicates the increase in mass of any ion containing the functionality. This is called the mass spectrometric shift technique and is especially useful in interpreting an unknown spectrum when certain members of the same chemical class are available as standards.

Remote-site Fragmentation

For long-chain compounds like fatty acids, it is possible to induce a charge site, such as a lithiated cation, at a terminal functional group, *e.g.* carboxyl. This cation may be sufficiently stable to remain the charge centre while remote fragmentation occurs along the chain, with reduced isomerisation.[4] Alternatively, the terminal group may be derivatised to form a charge-stabilising function, such as the nitrogen atom in a pyridinoline group.[5] The sequential cleavage of the carbon chain produces stable fragment ions, enabling structural changes such as the presence of a double bond to be recognised by comparison with the spectrum of the saturated compound. The interpretation of the spectra follows the philosophy of the 'shift technique'.[6] These processes should be compared to the local-charge fragmentation of a carbonium ion.

Summary

Rigorous treatments of the above outlines can be found in advanced texts. Over the years, organic mass spectrometrists have carried out systematic studies covering most of the common chemical classes,[2] bringing a logic to what seemed, in the 1950s, to be an impossibly complex area of gas-phase reaction mechanisms. In practice, it is not always possible to locate the charge because no obvious centres exist in carbonium ions. Furthermore, aromatic systems allow the charge to move about the system freely. However, it is possible to make informed deductions of structures for subsequent validation, despite accepting the occurrence of delocalisation.

8 Some Useful Examples of Features of the Mass Spectrum by Chemical Class

Aliphatic Compounds

Hydrocarbons

σ-Bond cleavages in the series C_nH_{2n+1} with intensity maxima at $n = 3, 4, 5$. Even-mass rearrangement ions at m/z 42, 56, ... occur for olefins depending on chain length. Abundant ions at are formed at secondary and tertiary sites.

Alcohols and Thiols

These yield $[M - H_2O]$ or $[M - H_2S]$ peaks, giving a spectrum similar to the corresponding olefin. GC retention indices (e.g. Kovat's Index) provide evidence of the higher polarity expected for these classes compared to the olefin. The thiols will contain ^{34}S isotopes at 4% of the ^{32}S peak intensity and may contain peaks at 32, 33 and 34 Th.

Ethers and Thioethers

α-Fission and rearrangements in the fragment ions of ethers yield ions in the series $m/z = 31 + n \times 14$, and fission of the C–heteroatom bond creates ions in the homologous series $m/z = 29 + n \times 14$. Molecular ion peaks of thioethers include the ^{34}S isotope at 4% of the ^{32}S isotope intensity. α-Fission gives the m/z 61, 75, 89.... series, analogous to those in the corresponding ether spectrum.

Disulfides and Trisulfides

Molecular ion pattern indicates $[M + 2]$ peaks of 8% and 12% ($2 \times {}^{34}S$ and $3 \times {}^{34}S$) of the ^{32}S isotope (M).

Amines

The dominating interpretative model here is the nitrogen rule – odd numbers of nitrogen atoms yield odd masses, $M^{+\cdot}$ under EI conditions (even masses of $[M + H]^+$ ions under CI). α-Fission gives ions in the series m/z 30, 44, 56....

Aldehydes

α-Fission produces $M - 1$ and m/z 29 ions. H rearrangement (rH) gives the series $M - (44, 58, \ldots)$ [neutral molecule loss] and m/z 44, 58, 72 ... [charge retention]. McLafferty rearrangement (McL rH) yields m/z 44 if there is a γ-hydrogen.

Ketones

A molecular ion is usually present. Ion series for, $[M - R_1]$, $[M - R_2]$, R_1 and R_2 occur. McLafferty products in the series $m/z = 58 + n \times 14$ are also common features.

Esters

α-Fission produces the O-alkyl ion in the series m/z 31, 43, 57 ... characteristic of the alcohol moiety, and the acidic chain length is indicated by the other product of α-fission in the m/z 29, 43, 57 ... series. Esters greater than methyl undergo double rH in the series m/z 47, 61, 75 ... confirming the acidic moiety. McLafferty rH in the alcohol chain can produce ions in the m/z 46, 60, 74 series.

Carboxylic Acids

$M - 45$ and m/z 45 ions and rH ions at $M - (46 + n \times 14)$ or ions in the series $m/z = 29, 43, 57. \ldots$

Amides

The nitrogen rule again applies. α-Fission (to carbonyl, β to the nitrogen) gives series $m/z = 44 + n \times 14$. rH ions in the homologous series $m/z = 59 + n \times 14$. Ions of $m/z = 30 + n \times 14$ appear in the spectra of secondary and tertiary amides.

Isothiocyanates

An ion at m/z 72 corresponds to the CH_2NCS^+ ion. Below pentyl, m/z 59 $(NCSH)^+$ ion and above pentyl, m/z 115 ion and $M - 33$ (loss of SH) are common.

Aromatic Compounds

An intense $M^{+\cdot}$ ion and the presence of ions at $m/z = 39, 50, 51, 52$ and 65 due to benzene ring fragmentation are general signs of aromaticity.

Aldehydes

M, M $-$ 1 and M $-$ CHO ions.

Alkylbenzenes

β-Fission gives m/z 91 $+$ n $\times$ 14 which is usually base peak. rH ions of $m/z = 92 + n \times 14$.

9 References

1 M.E. Rose and R.A.W. Johnstone, *Mass Spectrometry for Chemists and Biochemists*, Cambridge University Press, Cambridge, 1996, pp. 310–312.
2 F.W. McLafferty and F. Turecek, *Interpretation of Mass Spectra*, Fourth Edition, United Science Books, Sausalito, California, 1993.
3 F. Davis and M. Frearson, *Mass Spectrometry, ACOL*, ed. F.E. Prichard, Wiley and Sons, Chichester, 1994, p. 202.
4 K.A. Caldwell and M.L. Gross, in *Mass Spectrometry in Biological Sciences: A Tutorial*, Kluwer Academic Publishers, Dordrecht, 1992, p. 413.
5 J. Eagles, G.R. Fenwick and R. Self, *Biomed. Mass Spectrom.*, 1979, **6**, 462.
6 K. Biemann, *Mass Spectrometry: Organic Chemical Applications*, McGraw-Hill, 1962, p. 370.

Food Flavourings and Taints

1 Introduction

Soon after organic mass spectrometry emerged from the exclusive confines of petroleum research in the 1950s, it was applied to problems in food science.[1] The invention of gas chromatography (GC) in 1952[2] provided the means of separating the complex mixtures of volatile substances comprising food flavours. By the early 1960s, the combination of GC with mass spectrometry[3] was being exploited fully by flavour chemists[4] using ToF mass analysers. The tolerance of these instruments for higher source pressure allowed them to cope better with the output from embryonic interface 'separators' than their high voltage, magnetic sector counterparts.

The rapid development of GC/MS interfaces was possibly the most important factor in the advancement of flavour research. In the course of five years, separation efficiency, detection sensitivity, specificity, and ease of sample handling had been revolutionised. The state of the art was reviewed in 1979.[5] A host of refinements appeared in the 1980s. These included: fused silica columns with over 100 000 theoretical plates, bonded stationary phases, faster scanning speeds, high throughput vacuum pumps, faster and more intelligent data acquisition and processing systems. These allowed the techniques of qualitative and quantitative GC/MS to reach the point where most volatile flavour compounds, including compounds that could be rendered volatile through derivatisation, could be separated, recognised, identified and assayed using on-line automated instrumental methodology.[6] Modern high-temperature stationary phases continue to extend the range of analysable compounds. Such has been the progress in the application of GC/MS to flavour research that identification of volatile flavour components and GC/MS are synonymous. Because of these advances, the remainder of this chapter has been ordered around the general principles and the analytical practice of GC/MS. With the literature growing at the rate of several hundred articles each year, only a small selection of applications that exemplify the newer capabilities of the technique can be mentioned here. A review of food related subjects contains a section on flavour which contains references up to 1993/4 showing the wide variety of applications, many of which use MS in some form or other.[7]

Sampling and Sample Preparation

It is axiomatic that the sample taken for analysis must represent the bulk source material.[8] The basic principles adopted by analytical chemists provide a good academic background for flavour analysis,[9] but additional precautions are important to protect flavour samples from transient changes.[10] Most food flavours change in character with time. This is especially noticeable with flavours generated either by the action of plant enzymes on involatile precursors or during processing. At best, flavour samples should be considered as 'snap-shots' of the aroma at the time of collection.

A recent review of enzymes in flavour generation in food technology illustrates the extent of their use and serves to emphasise the need for analytical methods for the measurement of flavoured products.[11] The relatively low sensitivity of the mass spectrometer as a detector for certain aroma compounds, compared with the sensitivity of the nose, necessitates long collection times. This largely obscures rapid temporal changes. The odour profile of the food changes again in the mouth, as the food is chewed in preparation for swallow-ing, because further enzyme action in the saliva, and other 'mouth effects', alter the composition of the perceived aroma (see below). Thus, standards selected for flavour chemistry should be chosen to minimise the potential for on-going chemical reactions in food flavour components.

Traditional methods for the preparation of essences and flavourings are based on the production of extracts that, although different from the original, unconcentrated aroma (even when diluted back to the original volume) yield commercially acceptable compromises. Modern processes of low temperature pressing, supercritical fluid extraction, *etc.* can reduce decomposition. For example, one group has evaluated different extraction techniques for the GC/MS analysis of blackcurrant volatiles, and then identified almost a hundred compounds including mono- and sesqui-terpenes, carbonyls, esters, and alco-hols, from their Kovats indices and mass spectra.[12] The novel use of ultrasound for the extraction of aroma compounds from wine is claimed to be rapid and simple and to yield good recoveries, linearity and reproducibility for most of the compounds tested.[13] The addition of an internal standard at the outset of the experiment ensured that the assay reliably reflected the original concentration of the target compound.

Sample handling, storage and sub-sampling before insertion into the GC inlet also need attention if reliable quantitative results are to be achieved. The importance of preparative and storage techniques in ensuring the success of GC/IRMS analyses of the origins of chiral furanones has been pointed out.[14] As the sensitivity of detection increases, the amount of sample pre-concentration, and hence the amount of sample handling, may be reduced. Headspace sampling into adsorption tubes eliminates the need for distillation or solvent extraction techniques, which are more susceptible to non-stoichiometric samp-ling errors. Headspace sampling is often sufficient to provide information representative of the bulk material in terms of its aroma. However, alternative headspace solid phase microextraction (HS-SPME) techniques have been developed for use with GC/ITMS in the identification of organic sulfur compounds present in the aromas of two species of truffle.[15] It was concluded that SPME was less suited to quantitative analysis than standard Tenax trapping methods (see below).

If chemical derivatisation forms part of the sample preparation process, extra care is needed to avoid introducing 'interference' errors, *etc.* However, if the internal standard chosen to calibrate the extraction accurately mimics the behaviour of analyte through this stage, and if good laboratory practice (GLP) prevails, the effects of the chemical reaction are minimised.

The Total Perception of Flavour

It is generally agreed that flavour is a complex sensation. It includes, amongst other factors, aroma, taste, texture and mouthfeel. These are stimulated by chemicals from foods interacting with sensors in the nose and mouth. Since the only form of mass spectrometry available until recently employed gas-phase ionisation methods (but see below), its role in flavour research was confined to the estimation of the volatile substances emanating from food. For many years, researchers have been gathering evidence about the composition of the aroma of foods, comparing long lists of chemical constituents identified by GC/MS and aligning them with the biochemical precursors known to be present in the food matrices, *e.g.* amino acids, sugars, *etc.* It was realised as early as 1966 that there was a qualitative similarity among the odour profiles from foods with very different flavours.[16] Consequently, only a quantitative estimation would provide the necessary differentiation between food aromas. Thus GC/MS in the SIM mode, together with some form of stoichiometric sample concentration, was used to measure the relative abundance of the individual components in mixtures of volatile substances constituting the aroma of foods. It may be useful here to recommend a brief review of the principles and guidelines for sensory research.[17]

GC/MS Analysis

The identification of volatile compounds by GC/MS has been extremely successful. Approaching 1000 volatile substances have been recognised from extracts of coffee, not all of which are odorous and many of which make no contribution to the overall aroma of the source material. As an exercise in highly efficient separation and structure elucidation, this serves to illustrate the point. Because of the complexity of most flavour extracts, volatile compound 'trophy hunting' is now considered to be largely unrewarding and attention is focusing on the sensory significance of components in the mixture. One hundred and twenty volatiles in 13 strawberry cultivars and one wild type have been identified in Freon extracts of fresh and frozen fruits.[18] Using a sniffing technique they located 17 key compounds that were studied for further varietal differences. There is still an important role for GC/MS in the SIM mode for quantitative analysis of target compounds in flavour mixtures. For example, a quantitative assay for diacetyl in wine and beer has been described.[19]

Improvements in technique are sought and the identification of off-odours was facilitated by a data processing technique. Ions produced in the mass spectrum were summed into 14 homologous ion series, from which convenient comparisons between contaminated and reference samples could be made.[20]

GC/MS and Consecutive Odour Port 'Sniffing'

Odorous compounds eluting from the GC column can be detected and often characterised by their smell. If the odour description is made at the same time as the GC retention data and mass spectrum are recorded, the three pieces of information can be used to confirm or validate tentative structures.[21] Recent references to the use of consecutive odour detection (OD) are to be found in several of the papers discussed in this chapter. A particularly good example is a study of tequila flavour.[22] GC with FID, sulfur chemiluminescence detection and GC/MS/OD were used in the identification of 175 volatile components. Sensory testing *via* the odour port allowed 60 of these volatiles to be recognised as odorants, five of which, isovaleraldehyde, isoamyl alcohol, β-damascenone, 2-phenylethanol and vanillin, were judged important in the overall flavour. The critical analytical view of a study even as thorough as this would be to question the problem areas where unresolved odours cancel out, or where minor but intense odours are related to major but non-odorous constituents.

Modern Mass Spectrometric Techniques

In Relation to the Odour Intensity

Once aroma sampling is accurate and reproducible, and adsorption and chemical conversion during the chromatographic step minimised, even if the results are not totally compatible with the aroma as perceived by the nose, attention can be turned to the reliability of the mass spectrometric assay. Statistical methods have been applied (see below) to the rationalisation of information from GC/MS data obtained on samples from natural sources. In particular, quantitative GC/MS data have been correlated with the intensity of odour-impact chemicals, or related to changes in the strength of flavour. However, there are often several orders of magnitude difference between the absolute values expressed.

Some researchers have conducted multi-assay comparisons. However, the validation process seldom continued to the point where interlaboratory comparisons were made, yet this is necessary if unacceptable discrepancies are to be revealed.

Stable Isotope Studies

Isotope Dilution Mass Spectrometry (IDMS). If stable isotopes are added in known quantities to mixtures under investigation they act as internal standards for the quantitative estimation of the 'natural' analogue. Odour impact components of stewed beef juice have been measured in this way.[23] Preliminary studies were carried out using aroma extract dilution analysis and static headspace techniques, and by detecting odorous compounds as they eluted from the GC column. Twelve of the components were judged to be odour impact chemicals.

Isotope Ratio Mass Spectrometry (IRMS). The application of GC/MS to more difficult problems in flavour analysis is expanding. Of particular relevance is the use of GC/IRMS for differentiating between natural and synthetic substances found in food extracts and processed products. This is achieved by measuring the different isotopic fractionation of substances generated by bio- and chemico-synthetic processes. One group of researchers developed pyrolysis (Py)-GC/MS and Py-GC/combustion (C)-IRMS, but preferred Py-continuous-flow (CF)-IRMS for studies on the site-specific labelling of vanillin.[24] This was because the generation of CO allowed the carbon isotope content of oxygen bearing carbon atoms to be measured. PY-CF-IRMS was shown to provide a rapid method for the authentication of vanillin and other food flavours.

CI-MS and APCI-MS

While EI-MS has sufficient sensitivity for some experiments on single compounds, there has been a movement toward the use of CI-MS and APCI-MS in SIM mode. These techniques can provide additional sensitivity in the recognition of known compounds in mixtures of volatiles, and include such fundamental investigations as the detection of aromas released in the mouth (see below).

Ion Trap Mass Spectrometry (ITMS)

Ion trap GC/MS has been used to identify over 70 volatiles from coffee,[25] and other researchers[15] used ITMS for work on organic sulfur compounds in truffles. A detailed report on the use of GC/ITMS for the analysis of microextracts of red wine has also appeared.[26] 25 compounds were assayed quantitatively at levels of 0.1–1000 μg l^{-1} with RSDs of 3–7%. LOD ranged from 20 to 1000 ng l^{-1}, which meant that all analytes were detected at levels at or below their flavour thresholds.

ESI and MALDI Mass Spectrometry

Compounds that are potential flavour precursors, such as the glycosides, and that are ideally suited to analysis by desorption and desolvation techniques are dealt with later. ESI and/or MALDI mass spectrometry has been used in the study of odorant binding proteins.[27]

Resonance-Enhanced Multi-photon Ionisation Time-of-Flight Mass Spectrometry (REMPI/TOFMS)

The application of laser ionisation mass spectrometry to on-line monitoring of volatiles in the headspace during the roasting and brewing of coffee has been described.[28] A frequency quadrupled Nd:YAG laser (266 nm) was used for REMPI ionisation of the volatiles, which were sampled in an effusive molecular beam inside the ion source of a linear ToF mass spectrometer. REMPI at 266 nm is highly selective for ionisation of phenolic compounds, *e.g.* 4-vinylguaiacol. Indole and caffeine were also detected.

Tandem Mass Spectrometry

Although tandem mass spectrometry (MS/MS) has enormous capability in structure elucidation, especially in complex chromatography/MS applications where components are often unresolved, its use in flavour chemistry has not been extensive. This may be because, until recently, it was very expensive. The potential of MS/MS in flavour research was illustrated as early as 1985.[29] An interesting recent application described LC/MS/MS in the characterisation of a quinone-trapping substance in papaya preparations as part of a continuing investigation of enzymatic browning and the oxidation of endogenous phenols.[30] The mass spectrum of a 4-methylcatechol conjugate was shown (Figure 3.1) but the tandem MS experiment was not described in detail.

In a detailed report the use of GC/MS/MS, FABMS/MS and Thermospray LC/MS/MS in flavour science was shown to provide valuable additional selectivity compared with SIM in GC/EI-MS.[31] A triple quadrupole mass spectrometer was used to record product ion spectra of the Maillard reaction species, and the oxidation of caffeine was also studied. The CID mass spectrum of the molecular ion (EI) of authentic 4-hydroxy-2,5-dimethyl-3(2H)-furanone was compared with a mass spectrum derived from a GC/MS analysis of reaction products. This study demonstrated the value of the uncontaminated CID spectrum as an aid to interpretation. The recognition of 8-oxocaffeine at m/z 211 (protonated molecular ion) by LC/MS and selected reaction monitoring of the transition m/z 211 → 196 in LC/MS/MS are compared in Figure 3.2. This example dramatically illustrates the improvement in selectivity yielded by MS/MS in the detection of the target compound. The use of tandem mass spectrometry in the analysis of both flavour products and bioactive peptide precursors has also been demonstrated.[32]

An array of modern techniques, including ESI-MS/MS, has been used to identify an important flavour component of strawberry fruit, 2,5-dimethyl-4-hydroxy-3(2H)-furanone (DMHF) as its β-D-glucuronide, in the urine of humans who were fed strawberries.[33] 2D-NMR was used to verify the structure of the synthetic DMHF glucuronide. Solid-phase extraction and HPLC

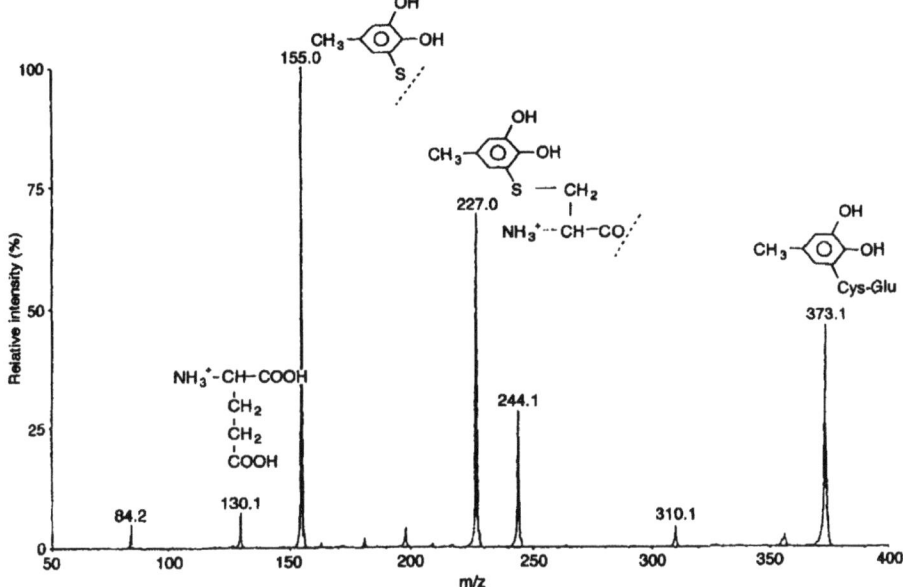

Figure 3.1 *The tandem mass spectrum of 4-methylcatechol thiol conjugate observed in the presence of papaya crude extract*
(Reproduced from F. Richard-Forget, M. Cerny, N. Fayad, T. Saunier and P. Varoquaux, *J. Food Sci. Technol.*, 1998, **33**, 285, with the permission of the authors and publishers)

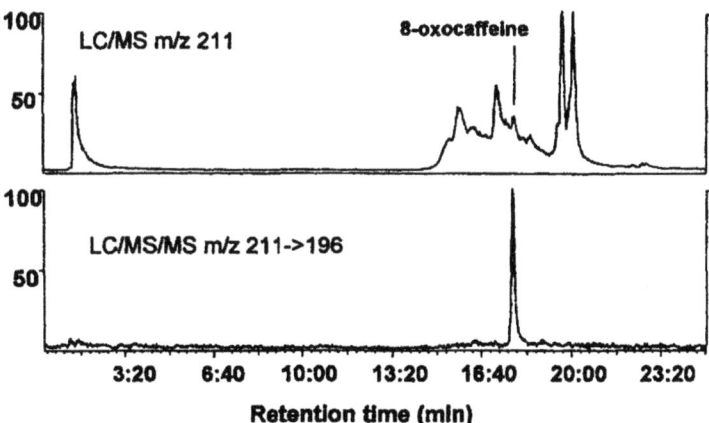

Figure 3.2 *Comparison of LC/MS and LC/MS/MS in the detection of 8-oxocaffeine from a solid-phase treated instant coffee*
(Reproduced from L.B. Fay, I. Blank and C. Cerny, in *Flavour Science: Recent Developments*, Special Publication No. 197, ed. D.S. Mottram and A.J. Taylor, Royal Society of Chemistry, Cambridge, 1997, p. 271, with the permission of the authors and publishers)

separation, with either on line UV-Vis or ESI-MS/MS detection, were then used to measure the amount of DMHF glucuronide excreted over 24 hours. There is a real future for ESI-MS/MS methods in food analysis of flavour precursors. The high mass range – up to 100 000 Da – achievable through the production of multiply charged ions will enable the study of intact biopolymers carrying the active molecule.

Summary of Modern Mass Spectrometric Objectives

The sub-divisions of modern flavour analysis include:

1. Identification of the components in a volatile mixture;
   (a) from a new biological source,
   (b) as a result of a new extraction process,
   (c) after different experimental processing.
2. Quantitative assay of odour impact chemicals;
   (a) to gain more accurate knowledge of the aroma profile,
   (b) to explore the nature of the odours actually perceived by the nose compared to those produced by the food.
3. Checking for the presence of additives and contaminants in natural flavour mixtures.
4. Comparing natural and synthetic flavours.
5. Authentication of natural and synthetic products.
6. Characterisation of flavour precursors.

The experimental design should take account of the combined resolution and specificity of the chromatographic and mass spectrometric methods. The synergism of high-resolution capillary column chromatography and multidimensional gas chromatography (MDGC), MDGC/IR/MS[34] and high-resolution EI/CI mass spectrometry will reduce significantly the need for multi-step pre-separation routines, providing the detection sensitivity is adequate.

Reviews in Related Areas

Several excellent recent reviews of the advances in specific areas of flavour research are noted. Over one hundred references are cited in a review of fish flavours.[35] Over 200 references were cited to update work on potato flavour, mainly quoting GC/MS experiments for the identification of the volatile constituents. Developments in fruit flavour research have been described under several headings. These include analytical and sensory measurements, flavour precursors, authentication of natural flavours, *etc.* The role of lipids in relation to the production of aldehydes, ketones and furans in food flavour has also been discussed.[36] The authors comment on autoxidation, singlet oxygen oxidation and lipoxygenase reactions, but most importantly they discuss the interaction between the Maillard reaction and lipid degradation in the formation of desirable odours of foods (see also refs 37–40).

In a special issue on flavour and flavour perception, the *International Journal of Food Science and Technology* marked the retirement of our colleague Dr D.G. Land from the editorship after 12 years. His contribution to flavour research was described[41] and several leading flavour researchers published reviews of their subject areas, including sensory analysis,[42] chemical tainting of foods,[43] microbiology of food taints,[44] physical chemistry of flavour,[45] and the molecular biology of olfactory perception.[46]

2 Flavour Analyses

Within the remit for this chapter on flavours and taints, it is convenient to include studies relating to the perception of odours.

New Methods of Analysis of Aromas and Their Precursors

The use of dynamic headspace-mass spectrometry (DHS-MS) has been described for the rapid recognition of food products.[47] Volatiles collected for 20 minutes by purge and trap techniques were desorbed in 1 minute and transferred directly to the mass spectrometer. Statistical discriminant analysis allowed four different textured French cheeses to be classified on the basis of 6 m/z values: 93 (terpenes), 47, 59 (various, including lipid oxidation products), 106 (benzenes), 18 (water content) and 21 (unknown). Direct mass spectrometric detection of simple mixtures of flavours generated (micro)biologically has been advocated.[48]

The use of *tert*-butyldimethylsilyl substituted cyclodextrin derivatives as versatile chiral stationary phases in capillary column GC has allowed important chiral food and flavour ingredients to be studied. The technique is of interest in the design of high performance GC/MS assays.[49] Thick films of polydimethylsiloxane stationary phase coated capillary columns gave good results for a standard mixture of 37 volatile flavour components and a sample of Swiss Emmental cheese flavour.[50]

MDGC employing a double oven chromatograph was used to resolve enantiomeric theaspiranes and theaspirones, widely used as flavourings.[51] Reference materials were isolated from quinces and their GC retention data recorded. Separation on an achiral column was followed by analysis on a chiral column and the complexity of the natural flavour isolates required MS identification after MDGC. SIM was used to determine the enantiomeric distribution at low levels. This technology was applied to the analysis of vitispiranes in Riesling wine when LC-MS/MS was used to identify the specific glucopyranoside precursor in the wine and vine leaves.[52] MDGC and GC/MS with simultaneous sniffing at an odour port were also used in studies of Cheddar cheese flavour[53] and a similar approach was taken in flavour dilution studies to identify the major odour impact chemicals of ginger.[54] Five new compounds were recognised.

Combined, simultaneous IR and MS detection of static headspace volatiles from heated triolein, separated by capillary column GC, has been used to identify and quantify products of oxidative and thermal decomposition.[55]

Progress in the analysis of garlic flavour has been enhanced by the co-operative use of GC/AES and GC/MS.[56] The enzymatic degradation of alliine to allicine and further decompositions produced a very complex mixture of odorous, sulfur-containing molecules. A sulfur-specific chromatogram was visualised by using the sulfur atomic emission line and the structure analysis was performed by simultaneous GC/MS.

Sample Preparation Methods and the GC/MS Analysis of Flavour Volatiles

Before considering the various GC/MS applications using the different sample preparation techniques, it is useful to consult a recent review in which a wide range of sample preparation methods were discussed in relation to the analysis of milk and dairy products.[57] All the methods described below are discussed, plus dialysis and molecular distillation. Mass spectrometry is included in the discussion of detection methods. Consecutive papers from groups working independently have described the GC/MS analysis of the volatiles from fresh and stored mackerel.[58,59] One group used distillation and solvent extraction methods[58] while the other used static and dynamic headspace methods.[59] Similar numbers of compounds were identified, so a general comparison of the different sample preparation methods is of interest.

Static and Dynamic Headspace Analysis

An example of the enormous power of GC/MS in flavour research is demonstrated by a thorough survey of the volatiles produced during the baking of white bread.[60] Over 375 compounds were identified from crust and crumb. Dynamic and static headspace methods were used to collect samples for GC/MS. New compounds were identified as 5-methyl-4,5-dihydro-3-[*H*]-thiophenone and 1-[*H*]-pyrrolo-[2,1-*c*]-1,4-thiazine.

Adequate precision was reported when a purge and trap method was devised for measuring VOCs in table ready foods.[61] A Vocarb 3000 trap was used prior to quantitative GC/MS when 45 U.S. EPA Method 524.2 and analytes were recovered from nine fortified food samples in duplicate. Two hundred and thirty-four table ready foods of the FDA Total Diet Program were analysed. 77 showed residues > 50 ppb, 43 > 100 ppb and 47 contained no residues.

γ-Ray irradiation of fresh onion is an accepted method of preventing sprouting during storage. Changes in the aroma character of irradiated stored onion have been studied by headspace analysis with a thermal desorption cold trap injection system for GC.[62] It was found that the recommended WHO/FAO dose of 0.15 kGy caused no change in aroma.

An investigation of lipid oxidation products, aimed at elucidating the stoichiometry of adsorbent traps used for dynamic headspace analysis, has been performed.[63] This resulted in the recommendation that thermal desorption with helium as carrier gas was preferable to the use of hydrogen and reduced the production of partially and fully hydrogenated artefacts.

A comparison of volatiles released by an extrusion cooker and those collected by purge and trap methods from corn flour extrudate has been made.[64] A special sample collection device designed to fit around the extruder die was devised. Ninety-one compounds were identified. Significant differences found between the two profiles were believed to provide additional information on possible mechanisms of flavour generation during extrusion cooking.

Purge and trap and cryofocusing are useful in the concentration of Cheddar cheese flavour volatiles for GC/MS analysis.[53] Purge and trap collection for analysis by GC/MSD-FID and retention indices has also been used in the analysis of 47 grazing plants in cheese producing areas.[65] Fifty-four terpenoid volatiles were identified, allowing a linkage to be made between changes in cheese flavour and differences in terpene composition of the grazing plants. The authors also add an analytical note about the likelihood of polarity discrimination of the purge and trap method, and the relative sensitivity of the MS detector to different chemical classes. Fifty-three volatiles have been identified by GC/MS from ripened cheese using the dynamic headspace technique.[66]

Both headspace concentration and solvent extraction (see below) techniques have been employed in extensive work on the preparation of samples of mushroom volatiles for GC/MS analysis.[67,68] The first paper reported the identification of 27 monoterpenes among 82 wild mushroom species, declaring both sample preparation methods to be appropriate for monoterpene analyses.[67] In contrast, more volatiles were identified by dynamic headspace than the solvent extraction methods in the analysis of *Marasmius alliaceus*.[68]

Solid-phase Extraction

Solid-phase extraction (SPE) is commonly used to separate the analyte(s) from biological matrices. This is achieved by adsorption of either the analyte or the matrix onto a solid phase, in either normal-phase, reversed-phase or ion-exchange liquid chromatography. SPE has been applied extensively to preparation of food samples contaminated with pesticides, fungicides, *etc.*[69] The modern application of SPE is based on the use of commercially available cartridges fitted into multi-sample manifolds. These devices provide automatic processing through the conditioning, loading, rinsing, and elution stages, making SPE a convenient sample preparation method for chromatography–MS assays. There are a small number of examples of its use in flavour research and readers are referred to a useful review.[70] SPE of the minor volatiles of cider preceded the quantitative analysis of alcohols, esters, lactones, phenols and medium- and long-chain fatty acids by GC/MS.[71] The concept of chromatographic selectivity is well understood in the simplification of complex flavour extracts, and automated SPE will increase in popularity as more specific adsorbents are added to the commercial catalogues.

Solid-phase Microextraction

Solid-phase Microextraction (SPME) is emerging as an efficient method of collecting and concentrating VOCs from both headspace and aqueous liquid phase environments.[72] It has been applied to the analysis of wine bouquet,[73] apple fruit,[74] strawberry aroma[75] and volatile trapping in spray-dried whey protein.[76] Maillard product artefacts formed during liquid sampling using a polyacrylate-coated fibre were observed during a study of strawberry flavours.[77] Surface contamination of the probe with high concentrations of sample-derived carbohydrates and amines was thought to be responsible. The authors recommended simply washing the probe with water before thermal desorption of the volatiles. Absorption times of a number of compounds on a 100 μm PDMS-coated probe have been studied.[75] A wide range – from 2 to 30 minutes – was found. However, the response of the system was linear in the ppb to ppm range when absorption time was standardised.

The success of SPME in analytical chemistry is due to its simplicity of operation, in particular the convenience of direct injections for gas chromatographic analysis. Most authors compare SPME to classical methods of concentration such as purge and trap. Experience with cooked vegetable aroma analysis[78] and by the originators of the technique[79] has shown that SPME is not as effective as static headspace collection for very low boiling volatiles. A modification to enable the boiling point range of VOCs to be extended has been described.[79] This extended range modification provides sensitivity in the low ppt region. SPME for beverages and SPME with microwave assisted extraction of solid foods has been used with SIM mode of GC/MS to determine Veltol® and Veltol-plus® flavour ingredients in a range of food products.[80]

A GC/MS survey of volatiles collectable by SPME from the static headspace above boiling potatoes[79] showed that, apart from the absence of the low boiling volatiles known to be present,[81] the VOCs were similar to those reported previously. However, the quantitative efficiency of SPME probes requires further study. The need for mild extraction methods has been emphasised.[39] Furthermore, SPME-GC/MS with a 100 μm polydimethylsiloxane (PDMS) coating has extended the range of compounds identified to include the C_{10}–C_{20} free fatty acids (normally analysed as their methyl ester derivatives) and other semi-volatiles such as the substituted naphthalenes. This enhancement will provide a convenient analytical pathway for further flavour precursor studies. However, the ability to sample directly from the aqueous phase has created a new concept in flavour research; that of gas/liquid phase partition analysis (GLPA). The GLPA of broccoli volatiles collected using PDMS on a SPME probe and analysed by GC/MS are shown in Figure 3.3.

Regions in which examples of the four types of gas/liquid partition distribution ($D_{g/l}$) which can occur are highlighted, and the detail of each region is shown in Figure 3.4.

GLPA has proved to be particularly useful for the analysis of volatile and semi-volatile glucosinolate breakdown products.[82] The retention times and

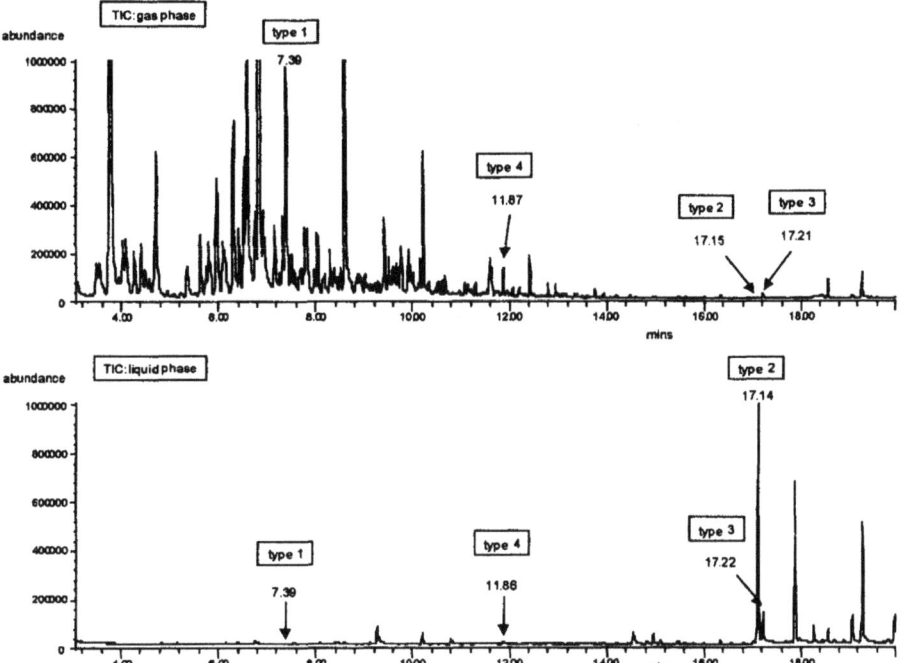

Figure 3.3 *The GLPA of broccoli volatiles with examples of the four different types of partition distribution that can occur. For detail of the four types see Figure 3.4a–3.4d*

identification of glucosinolate derived isothiocyanates and nitriles have also been studied.[83]

A small section around the elution time of caffeine of the GPLA of tea volatiles is shown in Figure 3.5. A cluster of type 2 compounds that would have required solvent extraction methods prior to GC/MS analysis is seen in this region. These are readily visualised 'among' the gas-phase volatiles and an approximate idea of the partition coefficient for type 2 compounds, such as the long-chain fatty acids, is also available in a single diagram.

Figure 3.6 shows a section of the GLPA of potato volatiles. The unsaturated aldehydes are non-condensable type 1 compounds while the saturated aldehydes and the aromatic aldehyde are partitioned between the two phases (type 3 compounds). The hydroxypyranone is a type 2 compound.

Conventional analytical methodology uses two different collection techniques for the two phases, making data correlation difficult. SPME-GC/MS with GLPA data presentation provides a more complete audit of the total volatile contribution to flavour as perceived by the nose (gas phase – orthonasal) and in the mouth (liquid phase – retronasal). Simultaneously, it provides valuable distribution data for research into the partitioning of chemical classes and its effect on the perceived flavour.

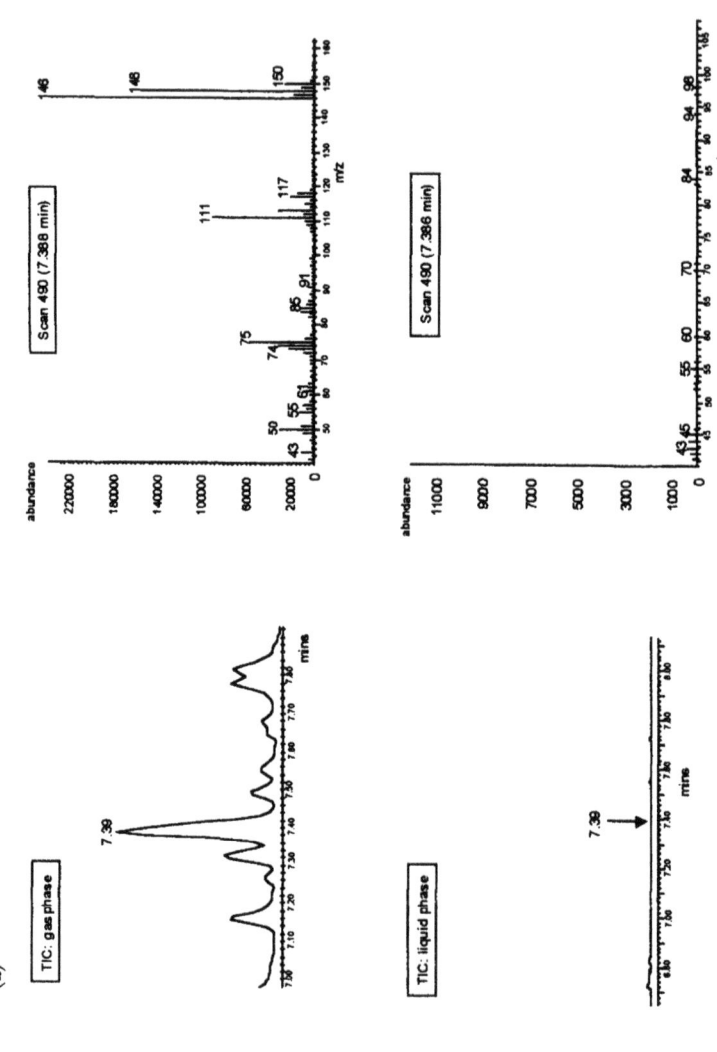

Figure 3.4   *Close up of the TIC and mass spectrum of each example marked in Figure 3. The GC retention time is normally reproducible to within two seconds, and the related mass spectral scans are shown for the equivalent gas and liquid phases*
(a) The compound at retention time = 7.39 minutes is classed as Type 1 – a single hydrophobic substance undetectable in the liquid phase. At the sensitivity of the experiment, the peak area measurement suggests that $D_{g/l} > 100$

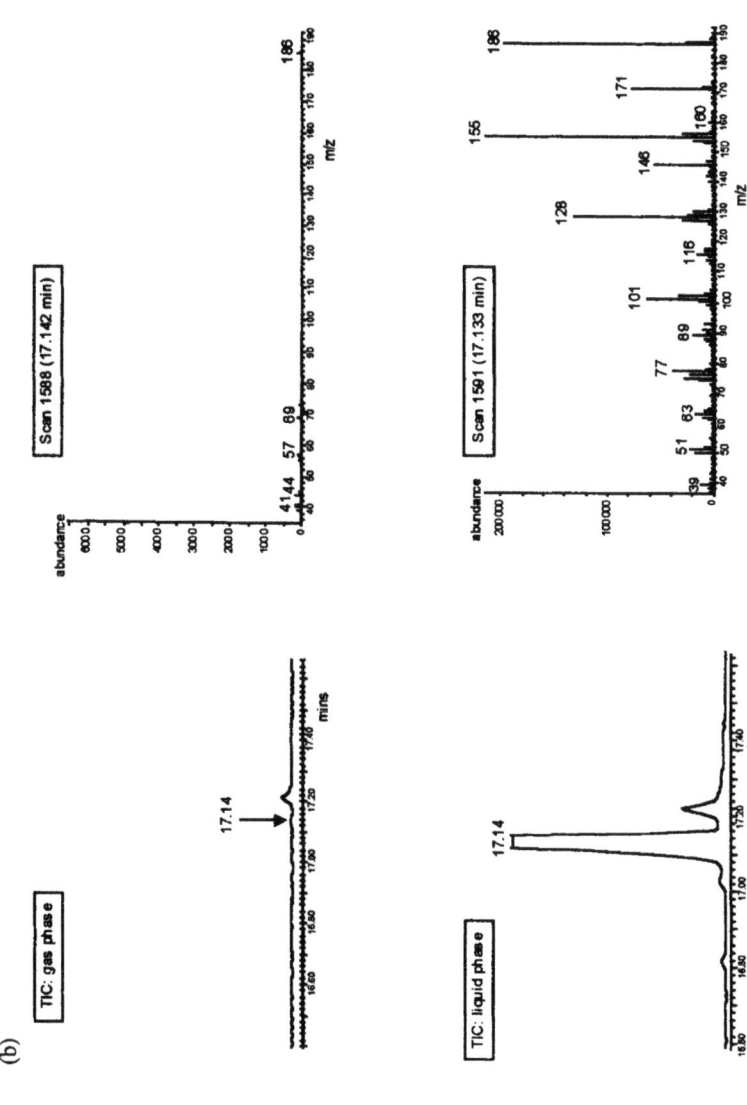

Figure 3.4 (b) *The compound at retention time 17.14 minutes is classed as Type 2 – a single hydrophilic substance undetectable in the gas phase. At the sensitivity of the experiment, $D_{g/l} < 0.01$*

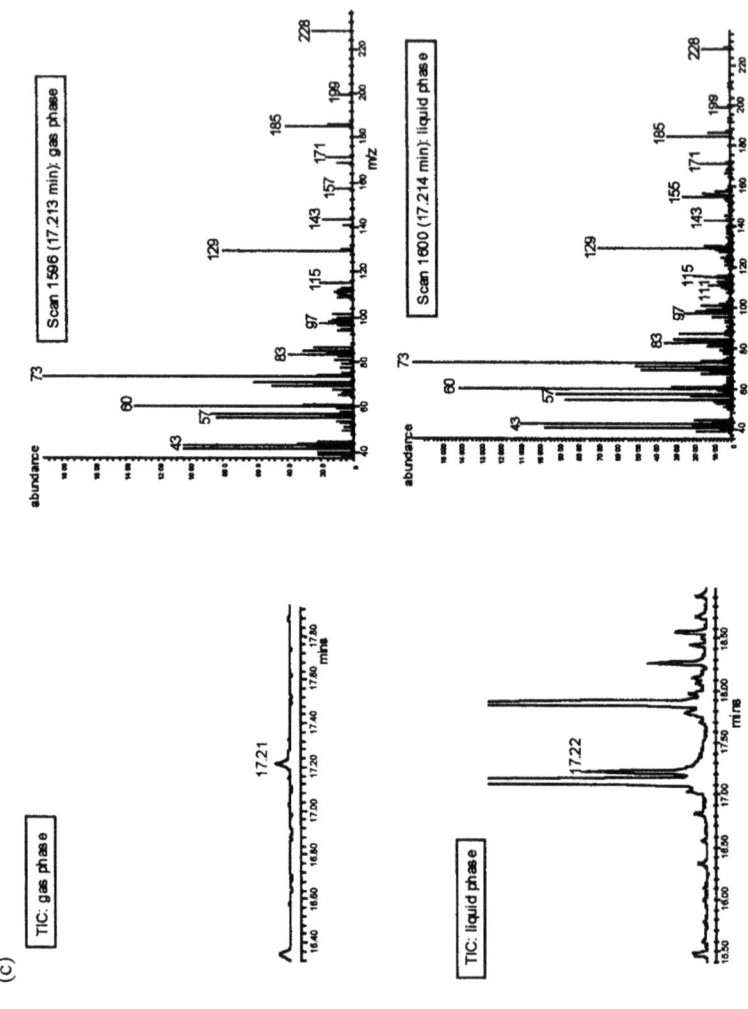

Figure 3.4 (c) *The compound at retention time 17.21 minutes is classed as Type 3 – a single volatile substance partitioned between the gas and liquid phases. The areas under the peaks in the TIC chromatogram give an approximate distribution. In this case, approximately 80% of the substance is in the condensed phase, i.e.* $D_{g/l} = 0.21$

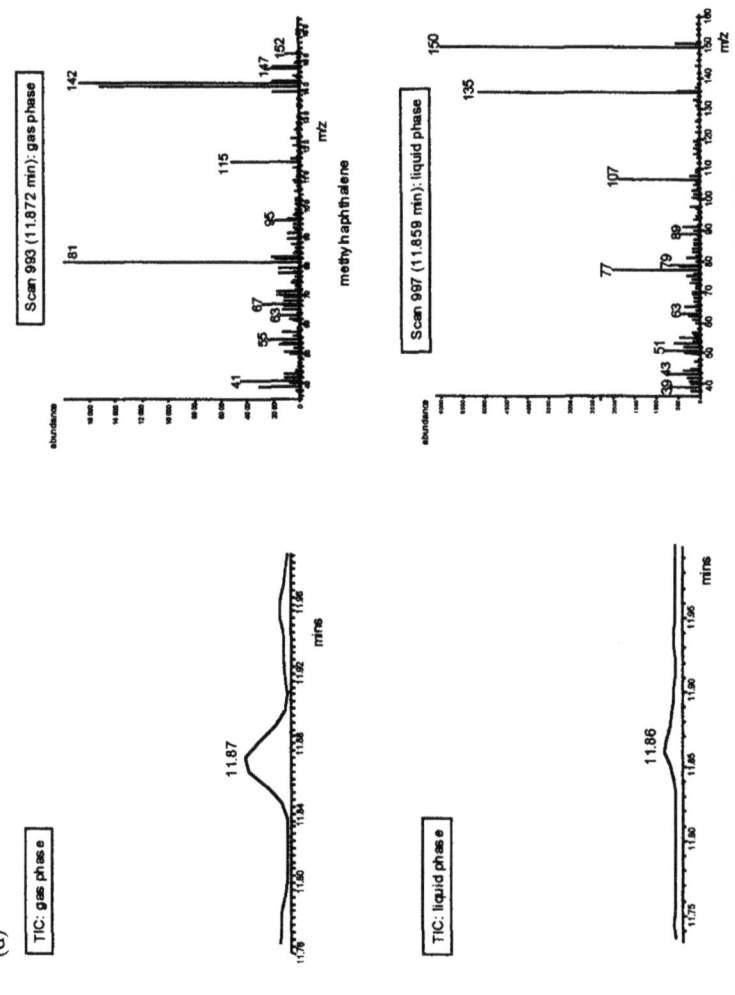

Figure 3.4 (d) *At the chosen retention time (11.87 minutes), the two different mass spectra indicate that unrelated substances, one Type 1 ($D_{g/l} > 1$) and one Type 2 ($D_{g/l} < 1$) elute at this time. Because the analytes are in trace amounts only, at the sensitivity of the experiment it is not possible to estimate the upper and lower limits respectively of the value of $D_{g/l}$*

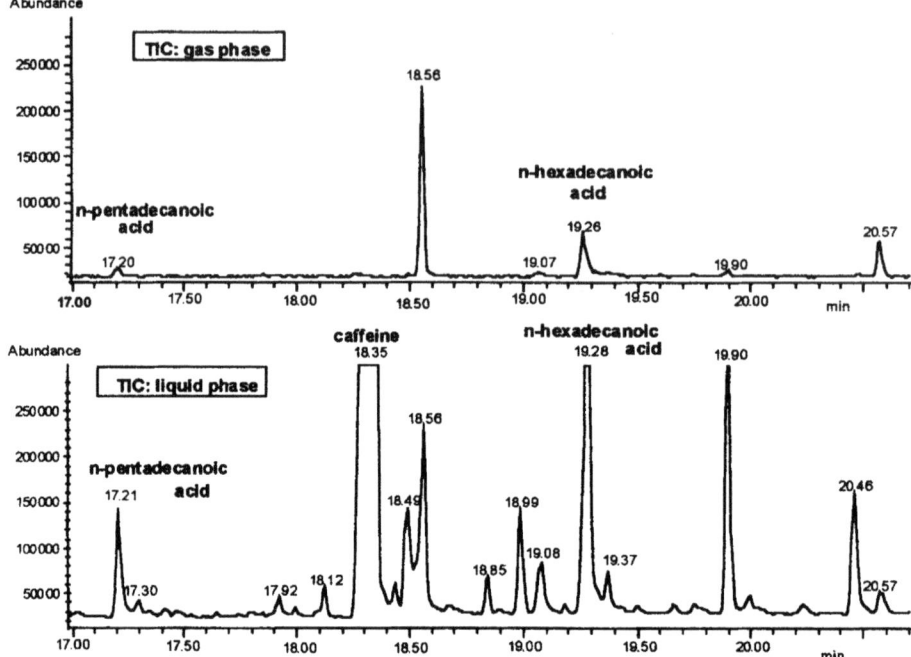

Figure 3.5 *Detail from the GLPA of tea. The region around the elution of caffeine, a condensed phase volatile (type 2, $D_{g/l} < 0.01$), retention time 18.35 minutes. Long-chain fatty acids appear as type 3 compounds ($D_{g/l} < 1$) at 17.20 and 19.26 minutes*

Distillation and Solvent Extraction Methods

Six new compounds occurring in bread flavour were reported from a study of the crust and crumb of anise white bread extracted with dichloromethane–diethyl ether and analysed by GC/MS.[84] Twenty-three compounds were found in the crust and 34 in the crumb, but only six of them were common to both. The authors also discussed the origins of the volatiles in relation to anise, dough fermentation, Maillard reaction and the bread making process.

An example of the effective use of GC/MS in volatile flavour analysis is given in the identification of 122 compounds in the simultaneous steam distillation – solvent extraction of snow crab cooker effluent and effluent concentrate.[85] Comparisons were made of the fate of N- and S-containing compounds in the effluents. A later contribution by the same team applied these methods to the analysis of the flavour of other fish products.[86] One hundred and fifty-five volatile compounds were detected; including 111 positively identified compounds comprising aldehydes, alcohols, esters, aromatics and nitrogen- or sulfur-containing compounds. The majority of volatiles from fish pastes were lipid-derived components such as aldehydes, alcohols, and esters, while hetero-cyclic nitrogen-containing compounds such as pyrazines were predominant in shrimp pastes.

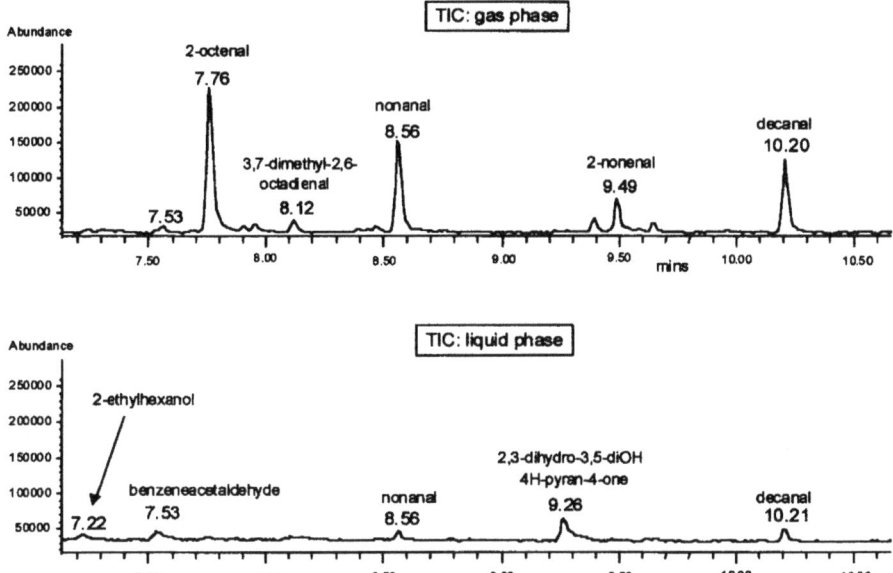

Figure 3.6 *Detail from the GLPA of potato. The unsaturated aldehydes behave as type 1 partition compounds whereas the saturated aldehydes are type 2 partition compounds, and the substituted pyranone, e.g. is a type 3 partition compound*

The importance of allium species as flavour precursor sources has been recognised since early times and modern food processing companies use large quantities of dried products as flavouring agents. The chemistry of alliums has been reviewed:[87] mass spectrometry has been used extensively to characterise the chemical composition of various members of the family.

GC/MS has been used to compare the effect of different extraction methods on the composition of sulfur compounds from *Allium cepa* L.[88] Steam distillation followed by solvent extraction destroyed methyl propyl disulfide (MPDS). In contrast, direct extraction of macerated plant tissue with diethyl ether yielded MPDS as the major component. The same allium variety was the subject of a steam distillation extraction studied by GC and GC/MS when 49 compounds were identified.[89] These included two branched dithiols, three dithiazines and a trithiane (reported in the oil for the first time). All were thought to be reaction products of fatty aldehydes, hydrogen sulfide, thiols and ammonia, induced by the distillation process.

The therapeutic effect of allium spp., especially garlic, has been known since 3000 BC.[87] The identification of anticarcinogens in garlic employed both instrumental and biological assay techniques to analyse, qualitatively and quantitatively, nine organosulfur compounds.[90] The instability of the extract prevented the development of a thoroughgoing GC/MS assay. A bioassay designed to quantify the same compounds in the serum of test animals, that had been fed either the garlic or liquorice extract, recognised only one of the nine.

The same nine organosulfur compounds were candidate marker compounds in a later study to monitor dietary compliance in free-living humans participating in dietary intervention trials for cancer research. Again, the GC/MS data on the garlic extract were not reproducible.[91]

Headspace techniques were used in the 1960s for the collection of volatiles formed during the cooking of potatoes, with identification by GC/MS.[81] More recently, volatiles from the baking of sweet potatoes were collected in methylene chloride and the solvent was reduced in a Kuderna-Danish concentrator prior to characterisation by GC/MS.[92] Quantitative data were presented and the eight major components identified. The state-of-the-art was reviewed in an article comparing the lipid-derived aroma compounds in cooked potatoes and reconstituted dehydrated potato granules.[93] The authors also observed that both distillation and absorption methods yielded highly reproducible qualitative and quantitative results. Nevertheless, distillation/extraction produced higher quantities of the less volatile components, while the purge and trap methods enhanced the lower-boiling constituents.

Methylene chloride extraction was also used to concentrate the flavour components from processed grapefruit juices.[94] Fifty-two aroma and five taste components were analysed by GC/MS and HPLC.

Allyl isothiocyanate, allegedly an important flavour compound in chopped cabbage, was not found in the methylene chloride extracts of juices from fresh cabbages.[95] Instead the primary volatile found in all extracts was an isomer, 1-cyano-2,3-epithiopropane. This raises questions about the suitability of the extraction processes involved and the stability and odour intensity of the isomers concerned. Several aroma-active compounds, derived from pressure-cooked hen meat by simultaneous distillation–extraction, have been identified.[96] These were assumed to be artefacts because they were not present when extraction–high vacuum distillation was used as a preliminary to GC/MS and GC/OD.

An internal standard, 2-heptanol, was added to the water used in a reduced-pressure steam distillation of non-fermented sausage.[97] The distillate was then passed through Porapak Q and the adsorbed volatiles were eluted with diethyl ether. GC and GC/MS analyses revealed 108 peaks, of which 99 were identified on the basis of the mass spectrum and Kovats index. Although the authors used the internal standard to calibrate the extraction and assay stages, they did not quantify the analytes, concentrating instead on the origin of components and possible changes occurring during processing.

The same group investigated the volatiles of spray-dried skim milk powder using similar methods.[98] They separated 196 GC peaks and identified 187 compounds including 48 hydrocarbons, 18 aldehydes, 20 ketones, 21 alcohols, 29 fatty acids, 8 esters, 2 furans, 7 phenolics, 10 lactones and 14 nitrogenous compounds. Subsequently, storage experiments were extended to the off-flavour developed in commercially processed spray-dried skim milk powder.[99] Compounds present in the 'off-odour' fraction, in greatly increased concentrations compared to the original milk extract, were thought to be responsible for 'sickening', 'hay-like', 'sulfuric' and quinoline-like odours.

It is symptomatic of the ready availability and analytical versatility of GC/MS that exotic food/pharmaceuticals ('nutraceuticals'), such as white desert truffles have been studied for their volatile content.[100] Eleven compounds were identified: the major components were unsaturated fatty acids with hexadecanoic acid comprising 49% of the total volatile isolate.

Supercritical Fluid Extraction (SFE)

In research into the flavour of olive oil, a higher proportion of semivolatile compounds was extracted by SFE than by the dynamic headspace method.[101] The SFE volatiles were concentrated on Tenax TA traps, thermally desorbed and cryofocused for analysis by high resolution GC/MS. Supercritical fluid extraction and GC/MS have also been used to characterise hydrocarbons and alkylcyclobutanones (putative irradiation markers) in irradiated fatty foods.[102]

Membrane Extraction with Sorbent Interface (MESI)

MESI evolved from investigations of membrane separators for coupling GC to MS. It is related to the use of membranes in microporous filtration, ultrafiltration, reverse osmosis, microdialysis, electrodialysis, *etc.*[103] An integrated approach to the use of membranes and sorbent traps for extraction of volatiles above aqueous systems, and trapping for subsequent desorption and analysis by GC and GC/MS, will undoubtedly be of interest to flavour analysts.[104]

Pervaporation

Pervaporation has progressed rapidly in the past five years. Its place in the armoury of transport processes such as dialysis, reverse osmosis, gel permeation, and its transport equations have been secured using the solution–diffusion model.[105] The use of pervaporation for the removal of organics from dilute aqueous solutions has been reviewed.[106] The four model processes involved in the transfer of volatiles across the membrane were described in relation to aroma recovery. Subsequently the methodology was developed for the food processing industry.[107] Pervaporation of commercial grape essence has been described as producing 'highly enriched flavours'.[108] Flavour analysts will no doubt see the potential of developing narrow-bore micropervaporation tubes for the transfer of volatiles from aqueous (or high humidity headspace environments) to a low moisture gas line for cryotrapping, *etc.*, with subsequent MDGC and GC/MS analysis.

Fruit Juice Concentration

The volatiles of carambola and yellow passion fruit juices have been examined by GC/MS after thermally accelerated short-time evaporation (TASTE).[109] Esters and aliphatic and terpinyl alcohols were identified.

Interlaboratory Trials

An interlaboratory trial has been used to validate methods, required by Control Agencies, for the measurement of irradiation of meat.[110] A GC method, validated by GC/MS, for chicken carcass, pork and beef, supplied by two different producers, was submitted to an interlaboratory trial involving 17 laboratories. Four different radiation-induced volatile hydrocarbons in the fat fraction of coded specimens were measured after three and six months storage. The method gave the correct answer in 98.3% of the 864 samples of irradiated and non-irradiated meat tested.

3   Taints and Off-flavours

Work in the 1960s implicated various naturally occurring hormones as the source of the characteristic taint of boar pork (boar taint).[111] More recent work has described a comparison of a rapid HPLC method with radio-immunoassay (RIA) and two GC/MS methods in the analysis of boar taint compounds, androstenone, skatole and indole in pig back fat.[112] The limit for quantification for indole and skatole was 30 ng g^{-1} and for androstenone, 200 ng g^{-1}. Correlations for the four methods were in the range 0.946–0.993. GC/MS headspace analysis of adipose tissue correlated four volatiles with fat androstene content.[113] A recent paper on the subject describes an analytical method for the measurement of androstenone (5-α-androst-16-en-3-one) in boar fat.[114] Supercritical fluid extraction and SIM GC/MS form the basis of the method. The authors gave valuable and detailed instructions for setting up the SFE and optimising the selectivity of the method (essential information if the 0.05 μg g^{-1} detection level is to be attained). Similar accuracies were observed for RIA and LC methods.

'Iodoform' taint in Australian prawns was attributed to the presence of 2,6-dibromophenol at the 60 ppb level.[115] Its origins were investigated by examining likely sources among a variety of marine algae, sponges, *etc.* 2,6-Dibromophenol and several related bromophenols were identified by high resolution GC/MID-MS in all species tested.

The rancidity of hazelnuts has been measured using organoleptic, GC/MS, total fatty acid and iodine value methods.[116] The authors suggest that the GC/MS assay of hexanal and octanal could be used to assess rancidity, instead of non-specific peroxide and iodine value assays. Flavour fade and off-flavours in ground, roasted peanuts have been studied.[38] GC/MS experiments followed the development of pyrazines from Maillard browning and aldehydes from lipid oxidation during storage. It was concluded that the production of large quantities of aldehydes masked the 'roast peanut' pyrazine aroma, not loss of pyrazines.

A review on the use of hexanal as an indicator of flavour deterioration of meat and meat products has 83 references.[117] The authors compared the use of hexanal with the classical 2-thiobarbituric acid test and with sensory analysis.

They recommended caution in the use of hexanal as a marker for lipid oxidation and meat flavour deterioration.

Many reports describe the use of GC/MS to solve problems associated with the contamination of foods from wrapping material, for example, the identification of (mainly) carbonyls from high-density polythene.[118] GC/MS/OD and GC/FTIR/OD were employed to recognise compounds responsible for six odour and nine taste attributes.

4   Flavour Chemistry

The convenience of separation and identification of volatile compounds in a single procedure has been invaluable in advancing understanding of the mechanisms of odour production from involatile precursors. The thermal degradation of essential vitamins is an obvious area of interest to food chemists and nutritionists and a recent study utilised many of the techniques now available for this kind of experiment.[119] The study was designed to observe reactions between thiamin and sulfur-containing amino acids in an autoclave. The Likens and Nickerson distillation and extraction method produced concentrates that were separated by liquid chromatography, with subsequent analysis by capillary column GC and GC/MS identification. Preparative capillary column GC provided sufficient material for IR and NMR spectroscopy and olfactory analysis. Some of the identified compounds were synthesised, including new thiophenes, resulting in an explanation of one of the main degradation pathways of thermally treated thiamin. The degradation of thiamin produced many products of organoleptic interest. These compounds contained molecules with one or more sulfur and/or nitrogen atoms, many of which were heterocycles. GC/MS has also been used to study the oxidation of thiols to disulfides.[120.]

GC/MS analysis of the thermal degradation of alliin and deoxyalliin revealed mainly diallyl sulfides (mono-, di-, tri-, and tetrasulfides) and allyl alcohol.[121] Other complex cyclic sulfides (mainly appearing at higher temperatures) were discussed in relation to the non-enzymic formation of garlic aroma.

GC/MS featured in a multi-step biomimetic synthesis of solerone (5-oxo-4-hexanolide) from L-glutamic acid 5-ethyl ester.[122] Seven intermediates were identified from sherry for the first time. The effect of pH on the formation of degradation products from glucosamine also employed GC/MS.[123] Furfurals were the major products at acidic pH values. The formation of pyrazine and methylpyrazine at pH 8.5 was attributed to the self-condensation of the retro-aldol products α-aminoacetaldehyde and α-aminopropanal. The presence of 3-hydroxypyridines and pyrrole-2-carboxaldehyde was also reported.

Flavour Precursors

As flavour chemists increasingly turn their attention to the characterisation of precursors it is gratifying to observe that mass spectrometry has kept pace with their needs.

In an exemplary study, the decomposition products formed in used frying oils were determined by using a number of sample preparation techniques.[124] These included dynamic headspace, steam distillation–extraction, adsorption chromatography, and preparative GC. 4-, and 5-oxoaldehydes (thought to be the precursors of alkylfurans that display anticancer effects), were synthesised and characterised by GC (Kovats Indices), GC/MS, GC/FTIR and NMR spectroscopy. Only trace amounts of 2-pentylfuran were found after a long reflux in hexane (40 days). However, frying oils from commercial food processors contained 4-oxohexanal, 4-oxooctanal, 4-oxononanal and 4-oxodecanal, as expected. The EI mass spectrum of 4-oxononanal is shown in Figure 3.7.

Interest is growing in the reaction of lipids with classical Maillard precursors (see below). The Maillard cysteine/ribose reaction has been studied, with and without the interaction of phosphatidylcholine.[125] Three methylthio-substituted furans and thiophenes were formed in the phospholipid reaction but were not present in the lipid-free reaction. The reaction between (*E*)-4,5-epoxy-(*E*)-2-heptenal and lysine or bovine serum albumin has been studied to investigate the production of pyrroles.[126] It was concluded, from GC/MS identifications of, amongst other compounds, alkyl- and acyl-pyrroles, that this new mechanism may work in conditions unsuitable for the classical Maillard reaction.

The origins of aromas associated with the partial hydrogenation of soybean oil in air, such as the 'melon' odour, have been studied using a synthetic model compound, tri-(*cis,cis*-9,15-linoleoyl)glycerol.[127] Allylic radicals reacted to form monohydroxyperoxide precursors of a range of alkenals and alkadienals, of which 6-nonenal was reported to have the most intense odour associated with

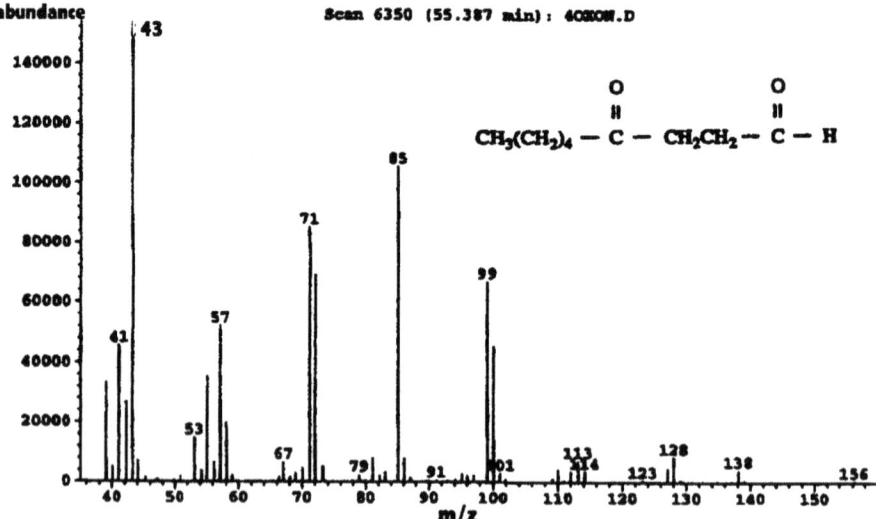

Figure 3.7 *The EI mass spectrum of 4-oxononanal*
(Reproduced with permission from G.R. Takeoka, R.G. Buttery and C.T. Perrino, *J. Agric. Food Chem.*, 1995, **43**, 22, © 1995 American Chemical Society)

deteriorated, partially-hydrogenated soybean oil. Further work on the thermal and oxidative decomposition products of methyl arachidonate has been reported.[128] GC/MS/OD and Kovats indices have successfully identified 29 compounds (6 odour impact chemicals and 12 with flavours characteristic of Korean soybean paste) from the neutral fraction of *Bacillus subtilis* PM3 isolated from Korean soybean paste.[129]

In work on the volatiles of chorizo, 115 of the 126 GC peaks were identified using GC/MS.[130] The authors observed that, out of the long list of chemical classes found, typical lipid autoxidation products were virtually negligible in chorizo.

Cheese-like flavours can be developed from amino acid mixtures containing free methionine or by enzyme action on mixtures of peptides containing methionine. Volatile sulfur compounds formed from methionine precursors[131] have been identified by GC/MS.[132]

GC retention time and GC/MS analysis have revealed the identity of a new compound extracted from wine by ethyl acetate.[133] While the compound bis (2-hydroxyethyl) disulfide had no specific odour, the precursor of 2-mercapto-ethanol had an unpleasant off-odour. Free and bound fractions of aroma of grape cultivars were separated on nonionic resin (Amberlite XAD-2). GC and GC/MS of the free fraction and enzyme hydrolysis and TFA derivatisation of the bound fraction allowed 60 compounds to be identified.[134]

GC/CIMS with methane as reactant gas has been used in the analysis of TFA derivatives of monoterpene glycosides. In common with other soft ionisation techniques used for analysis of plant glycosides,[135] the CI spectra contained protonated molecules and/or adduct ions from which the molecular weight could be determined.[136] Characteristic fragmentation reactions indicated the nature of the sugar moiety. The method was applied to the analysis of passion fruit and Muscat wine.

Identification of Maillard Reaction Products

GC/MS is used routinely to monitor the products of chemical reactions between amino acids and sugars in model systems, and there are many examples of this in the literature.[137] A classic example of the use of mass spectrometry includes the use of high resolution in the identification of products of the glycine/xylose reaction.[138] A range of modern methods, including MDGC, GC and GC/MS, have been used in the investigation of the thermal degradation of ^{13}C-labelled and unlabelled 2,3-dihydro-3,5-dihydroxy-6-methyl-(4*H*)-pyran-4-one in the Maillard reaction.[139] GC/MS and GC/MS/MS were employed to characterise caramel-like flavour producing hydroxdimethylfuranones from Maillard reactions involving various pentoses and their mechanisms of formation were proposed. The production of flavour compounds during Maillard reactions has been reviewed.[140] Fundamental studies have been applied to the production of meat-like aromas in the reaction of alkanediols with H_2S and furanthiols.[37] GC/MS and GC/odour port analyses were used to establish that disulfides containing the 2-methyl-3-furyl group were 'meat-like' in aroma while other disulfides were 'sulfurous' or 'onion-like' in character. An investigation of the effect of processing variables on the aroma of extrusion-cooked maize flour identified 80 compounds using GC/MS and GC/OD.[40] For example, at low temperature, (120 °C) and high moisture content (22%), extrusion cooking generated lipid degradation compounds, rather than Maillard products.

A model system of aqueous L-cysteine and D-glucose has established the differences between the flavour of foods cooked in microwave and conventional ovens.[141] GC/MS analysis showed that qualitative and quantitative differences in the heterocycles, such as the substituted pyridines, oxazoles, pyrones, furanones, thiazoles and pyrazines, were responsible for the differences in flavour of foods cooked by the two methods. The lack of flavour from microwave-cooked food was thought to be due to the absence of, or a decrease in, the quantity of pyrazines and furans. The different flavour profiles from the two methods suggested that different mechanisms of formation were involved. A cysteine/ribose system has been monitored by GC/MS to follow the formation of furfuryl mercaptan, a critical component of coffee flavour.[142]

Solutions of histidine and glucose, heated at different temperatures from 100 to 220 °C, yielded an incredible total of 231 identified volatiles.[143] Nine new pyrido[3,4-d]imidazoles were described for the first time. High-resolution mass spectrometry (and one- and two-dimensional NMR) has enabled elucidation of a new popcorn-like aroma from a cysteine/ribose reaction, 5-acetyl-2,3-dihydro-1,4-thiazine.[144]

The relative rates of production of pyrazines from glycine and another amino acid (Glu, Gln, Asp, Asn, Lys, Arg, Phe, and Ile) were measured. Fifty-six pyrazines were recognised and while lysine was able to increase the reactivity of glycine, arginine actually decreased it.[145] Thirteen pyrazines, 6 dioxolanes, 2 aldehydes, 2 acids and 2 pyridines have been identified by GC/MS from L-leucine, L-valine and D-glucose mixtures refluxed at 140 °C for 2 h in propylene glycol.[146]

The Effect of Diet on Meat Quality

The importance of methyl-branched long-chain fatty acids (formed through feeding barley to lambs) in the softening of carcass was recognised some time ago. Recent studies, using headspace analysis and GC/MS combined with sensory analysis, described 'strong', 'grassy' and 'lamby' odours quantitatively associated with the finishing diet.[147] Animals fed on the various diets, forage, radish top, ryegrass and clover, blackeye (cowpea) forage, alfalfa and grass, could be separated into classes by discriminant and canonical statistical analyses of the volatile compound data from GC/MS analyses. It has also been found that sheep meat with an induced high final pH has a significantly lower cooking odour and flavour than meat with a more normal pH.[148] Purge and trap concentration and GC/MS analysis identified 57 of the total 325 volatiles.

Oxidative decomposition of membrane polyunsaturated fatty acids has been suggested as the first step in the formation of aldehydes and alcohols produced by distillation–extraction and GC/MS of the volatile components of duck fillets (smoked, smoked and dry-cured, brined or raw).[149] Sixty-two compounds were identified by GC/MS, seven for the first time.

The quantitative GC/MS analysis of nine wild and two cultivated species of prawns (shrimps) for five bromophenols known to be important flavour components has been described.[150] The total bromophenol content of three commercially important wild species was between 9.5 and 1114 ng g^{-1}, while that of the major cultivated species was < 1 ng g^{-1}. Comparative total bromophenol analysis of the heads (including the gut) and tails of wild and cultivated prawns 'support the opinion that the bromophenols are derived from components of the diet' of the wild species.

5 Aroma Release from Foods

The on-column cryogenic trapping of headspace samples taken from above odorous food was used extensively for the pre-concentration of aroma volatiles for gas chromatography from an early stage[151,152] and was subsequently adapted for measuring the odour of cooked foods by GC/MS.[21] Details of a more recent example of this technique for collecting headspace volatiles from cheese have been given.[153] In general, headspace analysis is a good first approximation to the smell of the food perceived by the nose and formed the basis of early attempts to correlate odour and chemical composition by simultaneously recording odour description and mass spectra.[21] An elegant example is the analysis of bell peppers, in which chemical compounds were identified while recording their odour description:[154] '2,3-butanedione (caramel), 1-penten-3-one (chemical/pungent, spicy), hexanal (grassy), 3-carene (red bell peppers, rubbery), (Z)-β-ocimene (rancid, sweaty), octanal (fruity) and (Z)-isobutyl-3-methoxypyrazine (green bell peppers)'.

Improvements in the instrumental measurement of trace compounds in air and water triggered a renewed interest in the measurement of aroma from food in the mouth (commonly called the taste of food) as actually perceived by the nose. This has led to renewed attempts to develop sampling methods that avoid degradation before measurement takes place. It is necessary to adjust the overall flavour of processed foods when added components, such as gelling agents, affect the balance between adsorption and volatilisation. Therefore mass spectrometry is being reconfigured to assist in these modern olfactory studies, at the point of creation from the food and perception in the mouth. Low-temperature concentration (cryotrapping), or specific trapping materials, have been used to provide sufficient sample for analysis. However, doubt has always existed concerning the validity of the data acquired.

As mass spectrometric detection becomes more sensitive, it is now possible to undertake direct analyses of the aroma generated from the food while it is being chewed in the mouth. Care is still required to avoid losses due to adsorption or decomposition while the volatiles are being transported to the mass spectrometer. When natural changes are taking place, *e.g.* when enzymes are active, it is necessary to record qualitative and quantitative status against time, and it is this dynamic approach that modern mass spectrometry is adept at measuring. The changing composition of breath whilst food is chewed for swallowing has been compared with breath sampled some time before eating (see below).

GC/MS has been used to measure breath taken directly and continuously from the nose and mouth[155–157] and from model simulations of the processes occurring in the mouth.[158–162] Purge and trap collection from the headspace above synthetic saliva solutions at 23 °C and 37 °C, followed by GC/FID and GC/MS analyses, yielded < 5% CV and $mg\,l^{-1}$ sensitivity.[159] State-of-the-art technology encountered problems in sampling volatiles directly from the mouth and nose.[163] The high concentration of air reduced the overall sensitivity to the point where the signal-to-noise level was unacceptable. On-column cryogenic trapping followed by GC/MS provided different profiles for headspace, 'mouthspace' and 'nosespace'. The presence of water vapour and inefficient trapping limited the usefulness of the method. Tenax traps were better, but only menthone and menthol could be measured during the eating of 'extra strong' mints. Further development is required before the methods can be applied to other foods. Other workers were using Tenax traps for a 1 hour collection prior to GC/MS and observed differences depending on the sampling method used.[164] The same authors went on to use GC/OD of volatiles released in a 'mouth model system', and GC/MS was used to identify volatiles from rehydrated french beans, bell peppers, and leeks.[165]

Direct introduction of an air stream into an EI-MS in SIM mode was not successful, presumably because the pumping systems were optimised for helium/hydrogen carrier gases.[155] Further studies by the same team with Tenax and GC/MS, using limonene, menthone and menthol as markers, when four subjects chewed mint-flavoured sweets, sampled at 10 s intervals over 5 minute periods, yielded very different time–release curves for the three compounds.

This phase of research was summarised recently.[166,167] Although EI-MS was capable of detecting components at the 1 ppb level, this left no room for losses that would occur during the removal of air and water from 'mouth' samples and sub-ppb detection levels would be required in future. A shorter periodic sampling time, preferably a continuous measurement, and higher flow rates were also suggested. These conditions are more compatible with the flow of expired air ($350\ ml\ min^{-1}$), an important criterion for 'in mouth' measurement.[166]

A membrane separator has been used to sample a stream of air containing odour volatiles in order to achieve the continuous-flow condition.[168] This facilitated continuous sampling by EI-MS in the SIM mode. Sensitivity was somewhat limited and the stoichiometry of the mixture might have been changed due to different solubilities of the volatile components. However, 2-butanone in air or water solutions in the mouth at 25–100 ppm have been measured by similar methods.[169] Breath-by-breath analyses were performed showing the pulses of volatiles and the intermediate blanks.

The notion of using classical GC/MS molecular separators as concentrators for volatiles in moist air, capable of handling flow rates from the mouth of around $30\ ml\ min^{-1}$, was taken further by using a jet separator coupled to a mass spectrometer used in SIM GC/MS mode.[162] The technique was applied to the measurement of dynamic flavour release from liquid foods in a novel flavour release vessel. The authors identified several operational problems, such as frequent source cleaning, even when high purity isobutane was used as reagent gas, and short multiplier lifetimes when working with air. Nevertheless, they went on to model the effect of changing the texture/composition of processed food, containing carboxymethylcellulose, sucrose and guar gum, on the release of volatile flavour compounds, monitoring the dynamic conditions by GC/MS.[170]

The most recent application of mass spectrometry to flavour research utilises APCI for the breath by breath analysis of air expired during eating.[171] This enables researchers 'to sample volatiles in the most appropriate place' – in the mouth. Furthermore, it was possible to differentiate between orthonasal and retronasal expiration by monitoring background acetone levels. This excellent review explains the principles and practice of these instrumental methods in investigating the relationship between flavour stimuli and flavour perception, giving details of experiments conducted over the past three years.[171] By working with model systems and foods with different fat contents (such as biscuits) it was demonstrated that the rate of release of aroma volatiles from different matrices could be measured.

For those compounds that form protonated molecules in the APCI source, the method permits continuous monitoring of the changing profile in breath of a chosen flavour component over the time required to masticate and swallow a food (Figure 3.8).

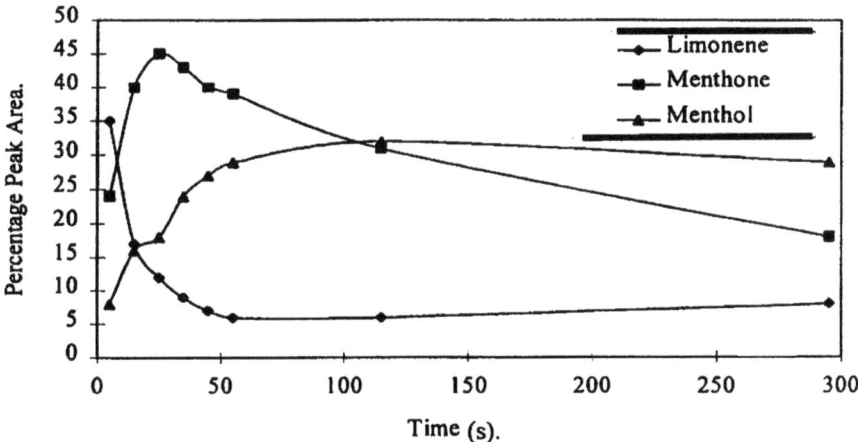

Figure 3.8 *Average volatile release curves for menthol, menthone and limonene released from mint flavoured sweets during eating*
(Reproduced from R.S.T. Linforth, K.E. Ingham and A.J. Taylor, *Flavour Science: Recent Developments*, ed. A.J. Taylor and D.S. Mottram, Royal Society of Chemistry, Cambridge, 1996, pp. 361–368 with permission of the authors)

In contrast to this approach, an alternative method uses a fibre interface to a conventional, bench-top EI/CI mass spectrometer.[172] This simple but very effective 'separator' method essentially increases the concentration of volatiles flowing into the EI mass spectrometer, simultaneously removing troublesome water peaks. The device, shown schematically in Figure 3.9, introduced a time delay of 43 s from sample vessel to mass spectrometer.

Mass Spectrometry in Support of Sensory Assessment of Flavour

Ultimately, the interaction of the total volatile aroma population with solid phases governs the nature of the perceived aroma. The science of sensory assessment[173] has evolved slowly in the wake of information about the chemical structure of the volatile constituents of the aroma, largely provided by GC and GC/MS assays. Mass spectrometry continues to provide evidence of the identity of chemicals contributing to the flavour of foods. Once identified, the value of each chemical as an odour impact substance has to be determined from odour threshold values. Total ion current chromatograms obtained in the GC/MS mode provide the chemical evidence and the chromatographic framework for the presentation of data as an aromagram, where the peak intensities are proportional to the intensity of the odour.[174]

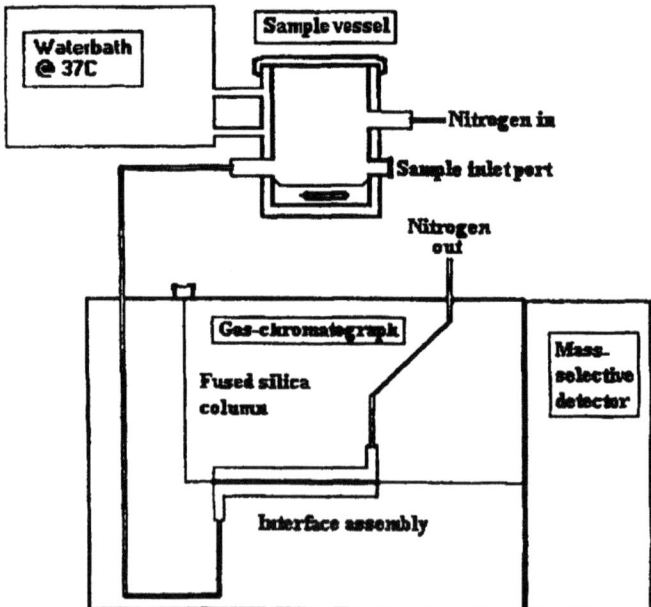

Figure 3.9 *Schematic diagram of semi-permeable membrane fibre apparatus for sampling aroma volatiles directly into the ion source of a conventional EI mass spectrometer*
(Reproduced with permission from M.B. Springett, V. Rosier and J. Bakker, *J. Agric Food Chem.*, 1999, **47**, 1125, © 1999 American Chemical Society)

An excellent review of the estimation of the contribution and interactions of volatile components to the overall flavour of solid foods has been presented, using hard cheese as an example.[175] The references cited cover the GC/MS identifications (largely made in the 1960s and 70s); the physicochemical detection in a two-phase system, *e.g.* binding onto protein, or a three-phase system involving the air/lipid in water interface and instrumental and sensory analyses, including aroma release in the mouth.

6   Breath and Body Odours

An interesting mass fragmentography experiment carried out on the alveolar breath of 12 normal volunteers involved measuring the concentrations of the 24 most abundant VOCs therein, and also in the air they inspired, in order to ascertain whether they originated from metabolic processes or from the environment. Two hundred or more VOCs have been identified in human alveolar breath, but their origins are still largely unknown.[176]

The use of GC/MS to analyse hydrolysates from individual proteins found in apocrine secretions has been described.[177] 3-Methyl-2-hexenoic acid (3M2H) was liberated from two proteins of 26 and 45 kDa molecular weight. Three of these body fluids, tears, nasal secretions and saliva, were separated into aqueous and organic soluble fractions and hydrolysed to show that 3M2H could be liberated from the aqueous soluble fraction. These results contribute towards an understanding of the role of human axillary odours as pheromones in relation to non-human mammalian odours used as chemical signals.

Methods have been described for the non-invasive measurement of low levels of volatile food-borne phytochemicals (*e.g.* those induced by eating garlic) in time-course experiments using thermal desorption GC/MS.[178]

GC/Ion trap MS has been used to identify volatiles from human breath collected by SPME. Several fibre coatings were compared for sensitivity, linear range, precision and detection limits, and the effects of temperature and humidity evaluated.[179]

7   Authentication of Natural and Synthetic Food Components

A 1984 paper and the references therein form an excellent starting point to a discussion of gas isotope ratio mass spectrometry (GIRMS) in relation to food analysis.[180] Small but distinct differences in the pattern of natural isotope abundances can be measured accurately by MS. Differences in the ratio of isotopes present in organic compounds occur as a result of, *e.g.* the different biosynthetic pathways followed during their creation. For example, the $^{13}C/^{12}C$ ratio is different for products synthesised by C_3 or C_4 plants. Biosynthetic pathways discriminate against the heavier isotope to a different extent to synthetic chemical equivalents, consequently natural and synthetic flavourings can be differentiated. This has resulted in applications of MS of interest to food control agencies and biosynthesis research groups. By way of example in flavour chemistry, the $\delta^{13}C$ values of natural sources of vanillin, *V. paniflora* and *V. tahitiensis* were -20 and $-17‰$, whereas synthetic sources, lignin, eugenol and guaiacol yielded values of -27, -31 and $-33‰$ respectively, making the detection of adulteration a simple matter.[180]

GC/IRMS measurement of $^{13}C/^{12}C$ ratio yields an additional dimension of selectivity. It has been used to differentiate between naturally grown, fermentatively produced, and synthetic (nature-identical) aldehydes from different orange oils and commercial products: synthetic nonanal, decanal, and dodecanal were distinguishable from their natural counterparts.[181] A further example is the use of multidimensional GC on cyclodextrin derivatives to separate chiral compounds, is combined with GC/IRMS to authenticate fruit and plant sources for the food and perfume industries.[182]

GC/IRMS in experiments formed the basis of tests to ascertain whether isotope fractionation occurs during liquid–liquid extraction of volatile components form Italia grapes and Primofiori lemons.[183] Using 10 different ^{13}C-labelled compounds (six added to grapes and four to lemons) no significant isotope fractionation was found for the extraction procedures employed.

High resolution GC/IRMS for the measurement of enantioselectivity and isotope discrimination in the authentication of flavours and essential oils has been reviewed and its scope and limitations discussed.[184]

8 Data Processing and Statistical Analysis

As the emphasis of flavour research moves towards the quantitative measurement of odour impact chemicals, the need for statistical analysis increases. Publications are increasingly including correlation between the chemical composition and the odour intensity of each constituent, in an attempt to reconstruct the aroma of the starting material. Thus mass spectrometric data are becoming of secondary importance, although still required to present an accurate chemical description of flavours. A reflection of this trend can be seen in recent papers: these increasingly refer to the smaller number of components that are considered for further study.

Chemical analyses, including MS identification, of 31 components separated by GC from steam distillates of 12 commercial Cheddar cheeses, were related to sensory attributes using multivariate calibration.[185] Cheese maturity and intensity of Cheddar cheese flavour were modelled with a high degree of fit.

GC/MS is seen in its proper, modern context in the work on multivariate statistical analysis of 37 peak areas in gas chromatograms chosen to differentiate cocoa masses from 13 geographical origins and four different roasting conditions by five manufacturers.[186] It is essential that the compounds be accurately described, either qualitatively or quantitatively. However, the important parameters of any flavour experiment are those relating to the odour imparting properties of some of the constituent volatiles which need to be uniquely identified by bioassay (the nose?). Only nine out of the 37 peak areas studied contained compounds that affected the variability of the response influenced by origin, roast or manufacturer. These peaks, identified by GC/MS, were used as variables in principal component analysis, hierarchical clustering and discriminant analysis. Two components, hexane and 2-methoxy-4-methylphenol, were thought to be markers of origin.

Three teas made from the same plant were analysed by GC and GC/MS.[187] Pattern recognition techniques, applied to the analysis of the GC data, showed that partial least squares and linear discriminant analysis correctly discriminated tea samples. The aldehyde (*E*)-2-hexenal was the most efficient discriminant. Principal component analysis produced distinct clusters for the three teas but a sub-cluster of one variety appeared close to the main cluster of one of the others.

9 References

1 G.R. Hercus and J.D. Morrison, *Austral. J. Sci. Res.*, 1951, **4**, 290.
2 A.T James and A.J.P. Martin, *Biochem. J.* 1952, **50**, 679.
3 J.C. Holmes and F.A. Morrell, *Appl. Spectrosc.*, 1957, **11**, 86.
4 W.H. McFadden and R. Teranishi, *Nature*, 1963, **200**, 329.
5 R. Self, in *Progress in Flavour Research*, ed. D.G. Land and H.E. Nursten, Applied Science Publishers, 1979, p. 135.
6 D.A. Cronin and P.J. Caplan, in *Applications of Mass Spectrometry in Food Science*, ed. J. Gilbert, Elsevier Applied Science, London, 1987, p. 1.
7 S.K.C. Chang, E. Holm, J. Schwarz and P. Rayas-Duarte, *Anal. Chem.*, 1995, **67**, 127.
8 S.R. Smith and G.V. James, *The sampling of bulk materials*, Analytical Sciences Monographs, Royal Society of Chemistry, London, 1981.
9 B.W. Woodget and D. Cooper, *Samples and standards*, ACOL series, Wiley, 1991.
10 M.Gy. Pierre, *Sampling of Heterogeneous and Dynamic Material Systems*, Data Handling in Science and Technology, Elsevier, Amsterdam, 1992, Vol. 10.
11 P. Christen and A. Lopezmunguia, *Food Biotechnol.*, 1994, **8**, 167.
12 J. Piry, A. Pribela, J. Durcanska and P. Farkas, *Food Chem.*, 1995, **54**, 73.
13 C. Cocito, G. Gaetano and C. Delfini, *Food Chem.*, 1995, **52**, 311.
14 G. Bruche, A. Dietrich and A. Mosandl, *Z. Lebensm. Unters. Forsch.*, 1995, **201**, 249.
15 F. Pelusio, T. Nilsson, L. Montanarella, R. Tilio, B. Larsen, S. Facchetti and J.O. Madsen, *J. Agric. Food Chem.*, 1995, **43**, 2138.
16 R.W. Moncrieff, in *Symposium on Foods: the Chemistry and Physiology of Flavors*, ed. H.W. Schultz, E.A. Day and L.M. Libby, AVI Publishing Co, Westport, CT., 1967, p. 542.
17 C.W. Naswari, *ACS Symp. Ser.*, 1993, **536**, 32.
18 D. Ulrich, S. Eunert, E. Hoberg and A. Rapp, *Deutsche Lebens. Rundschau*, 1995, **91**, 349.
19 B. Martineau, T. Acree and T. Henick-Kling, *Biotechnol. Tech.*, 1994, **8**, 7.
20 L.B. Fay and A.A. Staempfli, *J. AOAC Int.*, 1995, **70**, 1429.
21 R. Self, in *Mass Spectrometry: Proceedings of Symposium on Mass Spectrometry*, ed. R. Brymner and J.R. Penny, Butterworths, London, 1968, p. 93.
22 S.M. Benn and T.L. Peppard, *J. Agric. Food Chem.*, 1996, **44**, 557.
23 H. Guth and W. Grosch, *J. Agric. Food Chem.*, 1994, **42**, 2862.
24 M.J. Dennis, P. Wilson, S. Kelly and I. Parker, *J. Anal. Appl. Pyrolysis*, 1998, **47**, 95.
25 X. Yu, C.T. Ho and J.D. Rosen, *Dev. Food Sci.*, 1993, **32** (Food Flavors, Ingredients and Composition), p. 23.
26 V. Ferreira, R. Lopez, A. Escudero and J.F. Cacho, *J. Chromatogr. A*, 1998, **806**, 349.

27 M.F. Herent, S. Collins and P. Pelosi, *Chem. Senses*, 1995, **20**, 601.

28 R. Zimmermann, H.J. Heger, C. Yeretzian, H. Nagel and U. Boesl, *Rapid Commun. Mass Spectrom.*, 1996, **10**, 1975.

29 K.L. Busch and K. Kroha, *ACS Symp. Ser.*, 1985, **289**, 121.

30 F. Richard-Forget, M. Cerny, N. Fayad, T. Saunier and P. Varoquaux, *J. Food Sci. Technol.*, 1998, **33**, 285.

31 L.B. Fay, I. Blank and C. Cerny, in *Flavour Science: Recent Developments*, Special Publication No. 197, ed. D.S. Mottram and A.J. Taylor, Royal Society of Chemistry, Cambridge, 1997, p. 271.

32 H. Jiang, P.A. Grieve, R.J. Marschke, A.F. Wood and D.A. Dionysius, *Aust. J. Dairy Technol.*, 1998, **53**, 119.

33 R. Roscher, H. Koch, M. Herderich, P. Schreier and W. Schwab, *Food Chem. Toxicol.*, 1997, **35**, 777.

34 C.L. Wilkins, *Anal. Chem.*, 1994, **66**, 295A.

35 T. Kawai, *Crit. Rev. Food Sci. and Nutr.*, 1996, **36**, 257.

36 C.T. Ho and Q.Y. Chen, *ACS Symp. Ser.* 1994, **558**, 2.

37 D.S. Mottram and F.B. Whitfield, *J. Agric. Food Chem.*, 1995, **43**, 1302.

38 K.J.H. Warner, P.S. Dimick, G.R. Ziegler, R.O. Mumma and R. Hollender, *J. Food Sci.*, 1996, **61**, 469.

39 M.A. Petersen, L. Poll and L.M. Larsen, *Food Chem.*, 1998, **61**, 461.

40 W.L.P. Bredie, D.S. Mottram and R.C.E. Guy, *J. Agric. Food Chem.*, 1998, **46**, 1479.

41 H. Nursten, *J. Food Sci. Technol.*, 1998, **33**, 1.

42 J.R. Piggott, S.J. Simpson and S.A.R. Williams, *J. Food Sci. Technol.*, 1998, **33**, 7.

43 D.S. Mottram, *J. Food Sci Technol.*, 1998, **33**, 19.

44 F.B. Whitfield, *J. Food Sci Technol.*, 1998, **33**, 31.

45 A.J. Taylor, *J. Food Sci Technol.*, 1998, **33**, 53.

46 P.W. Goodenough, *J. Food Sci Technol.*, 1998, **33**, 63.

47 G. Vernat and J.L. Berdague, in *Bioflavour 95 Dijon*, ed. INRA, 1995 (Les Colloques No. 75), p. 59.

48 E. Heinzle, in *Bioflavour 95 Dijon*, ed. INRA, Paris, 1995 (Les colloques No. 75), p. 67.

49 B. Maas, A. Dietrich, V. Karl, A. Kaunzinger, D. Lehmann, T. Koppe and A. Mosandl, *J. Microcolumn Sep.*, 1993, **5**, 421.

50 R. Imhof and J.O. Bosset, *HRC – J. High Res. Chromatogr.*, 1994, **17**, 25.

51 G. Full and P. Winterhalter, *Vitis*, 1994, **33**, 241.

52 G. Full, P. Winterhalter, G. Schmidt, P. Herion and P. Schreier, *HRC – J. High Resol. Chromatogr.*, 1993, **16**, 642.

53 G. Arora, F. Cormier and B. Lee, *J. Agric. Food Chem.*, 1995, **43**, 748.

54 O. Nishimura, *J. Agric. Food Chem.*, 1995, **43**, 2941.

55 S.M. Mahunga, S.L. Hansen and W.E. Artz, *J. Am. Oil Chem. Soc*, 1994, **71**, 453.

56 D. Deruaz, F. Soussanmarchal, I. Joseph, M. Desage, A. Bannier and J.L. Brazier, *J. Chromatogr. A*, 1994, **677**, 345.

57 R.G. Mariaca and J.O.Bosset, *Lait*, 1997, **77**, 13.

58 H.Z. Zhang and T.C. Lee, *ACS Symp. Ser.*, 1997, **674**, 55.

59 C. Alasalvar, P.C. Quantick and J.M. Grigor, *ACS Symp. Ser.*, 1997, **674**, 39.

60 W. Baltes and C. Song, *ACS Symp. Ser.*, 1994, **543**, 192.

61 D.L. Heikes, S.R. Jensen and M.E. Flemingjones, *J. Agric. Food Chem.*, 1995, **43**, 2869.

62 A. Kobayashi, R. Itagaki, Y. Tokitomo and K. Kubota, *J. Jn. Soc. Food Sci. Technol. – Nippon Shokuhin Kogyo Gakkaishi*, 1994, **41**, 682.

63 P.S.W. Park, *J. Food Sci.*, 1993, **58**, 220.

64 M. Nair, Z.L. Shi, M.V. Karwe, C.T. Ho and H. Daun, *ACS Symp. Ser.*, 1994, **543**, 334.

65 R.G. Mariaca, T.F.H. Berger, R. Gauch, M.I. Imhof, B. Jeangros and J.O. Bosset, *J. Agric. Food Chem.*, 1997, **45**, 4423.
66 W.J.M. Engels, R. Dekker, C. DeJong, R. Neeter and S. Visser, *Int. Dairy J.*, 1997, **7**, 255.
67 S. Breheret, T. Talou, S. Rapior and J.M. Bessiere, *J. Agric. Food Chem.*, 1997, **45**, 831.
68 S. Rapior, S. Breheret, T. Talou and J.M. Bessiere, *J. Agric. Food Chem.*, 1997, **45**, 820.
69 M. Hiemstra, J.A. Joosten and A. Dekok, *J. AOAC Int.*, 1995, **78**, 1267.
70 K. Coulibaly and I.J. Jeon, *Food Rev. Int.*, 1996, **12**, 131.
71 J.J. Mangas, M.P. Gonzalez, R. Rodriguez and D. Blanco, *Chromatographia*, 1996, **42**, 101.
72 C.L. Arthur and J. Pawliszyn, *Anal. Chem.*, 1990, **62**, 2145.
73 D.D. Garcia, S. Magnaghi, M. Reichenbacher and K. Danzer, *HRC–J. High Resol. Chromatogr.*, 1996, **19**, 257.
74 J. Song, B.D. Gardner, J.F. Holland and R.M. Beaudry, *J. Agric. Food Chem.*, 1997, **45**, 1801.
75 D. Ulrich, A. Rapp and E. Hoberg, *Z. Lebensm. Unters. Forsch.*, 1995, **200**, 217.
76 R.J. Stevenson and X.D. Chen, *Food Res. Int.*, 1996, **29**, 495.
77 H. Verhoeven, T. Beuerle and W. Schwab, *Chromatographia*, 1997, **46**, 63.
78 L.N. Surugau, MSc Thesis, University of East Anglia, 1998.
79 Z.Y. Zhang and J. Pawliszyn, *HRC–J. High Resol. Chromatogr.*, 1996, **19**, 155.
80 Y.W. Wang, M. Bonilla, H.M. McNair and M. Khaled, *HRC–J. High Resol. Chromatogr.*, 1997, **20**, 213.
81 R. Self and T. Swain, *Proc. Nut. Soc.*, 1963, **22**, 176.
82 E.A.S. Rosa, R.K. Heaney, G.R. Fenwick and C.A.M. Portas, *Hort. Rev.*, 1997, **19**, 99.
83 L.N. Surugau, G.R. Stephenson and R. Self, 1999, personal communication (manuscript in preparation).
84 M.E. Komaitis and G. Aggelousis, *Sciences Aliments*, 1993, **13**, 585–91.
85 Y.J. Cha, K.R. Cadwallader and H.H. Baek, *J. Food Sci.*, 1993, **58**, 525.
86 Y.J. Cha and K.R. Cadwallader, *J. Food Sci.*, 1995, **60**, 19.
87 G.R. Fenwick and B.H. Hanley, *CRC Crit. Rev. Food Sci. Nutr.*, 1985, **23**, 1.
88 R.A. Martinlagos, M.F.O. Serrano and M.D. Ruizlopez, *Food Chem.*, 1992, **44**, 305.
89 P. Farkas, P. Hradsky and M. Kovac, *Z. Lebensm. Unters. Forschung.*, 1992, **195**, 459.
90 D.S. Weinberg, M.L. Mainer, M.D. Richardson, F.G. Haibach and T.S. Rogers, *HRC–J. High Resol. Chromatogr.*, 1992, **15**, 641.
91 D.S. Weinberg, M.L. Mainer, M.D. Richardson and F.G. Haibach. *J. Agric. Food Chem.*, 1993, **41**, 37.
92 J.B. Sun, R.F. Severson and S.J. Kays, *Hortscience*, 1993, **28**, 1110.
93 J.P. Salinas, T.G. Hartman, K. Karmas, J. Lech and R.T. Rosen, *ACS Symp. Ser.*, 1994, **558**, 108.
94 P. Jella, R. Rouseff, K. Goodner and W. Widmer, *J. Agric. Food Chem.*, 1998, **46**, 242.
95 K.H. Kyung, H.P. Fleming, C.T. Young and C.A. Haney, *J. Food Sci.*, 1995, **60**, 157.
96 P. Farkas, J. Sadecka, M. Kovac, B. Siegmund, E. Leitner and W. Pfannhauser, *Food Chem.*, 1997, **60**, 617.
97 H. Shiratsuchi, M. Shimoda, Y. Minigishi and Y. Osajima, *J. Agric. Food Chem.*, 1993, **41**, 647.
98 H. Shiratsuchi, M. Shimoda, K. Imayoshi, K. Noda and Y. Osajima, *J. Agric. Food Chem.*, 1994, **42**, 984.
99 H. Shiratsuchi, M. Shimoda, K. Imayoshi, K. Noda and Y. Osajima, *J. Agric. Chem.*, 1994, **42**, 1323.

100 E.A. Omer, D.L. Smith, K.V. Wood and B.S. el Menshawi, *Plant Foods Human Nutrition*, 1994, **45**, 247.
101 M.T. Morales, A.J. Berry, P.S. McIntyre and R. Aparicio, *J. Chromatogr. A*, 1998, **819**, 267.
102 P. Lembke, J.Bornert and H. Engelhardt, *J. Agric. Food Chem.*, 1995, **43**, 38.
103 D. Warren, *Anal. Chem.*, 1984, **56**, 1529A.
104 M.J. Yang and J. Pawliszyn, *LC – GC Int.*, May 1996, p. 283.
105 J.G. Wijmans and R.W. Baker, *J. Membrane Sci.*, 1995, **7**, 1.
106 H.O.E. Karlsson and G. Tragardh, *J. Membrane Sci.*, 1993, **76**, 121.
107 H.O.E. Karlsson and G. Tragardh, *Trends Food Sci. Technol.*, 1996, **7**, 78.
108 N. Rajagopalan and M. Cheryan, *J. Membrane Sci.*, 1995, **104**, 243.
109 S. Nagy, S. Barros and C.S. Chen, *ACS Symp. Ser.*, 1995, **596**, 48.
110 G.A. Schreiber, G. Schulzki, A. Spiegelberg, N. Helle and K.W. Bogl, *J. AOAC Int.*, 1994, **77**, 1202.
111 R.L.S. Patterson, *J. Sci. Food Agric.*, 1968, **19**, 31.
112 J. Hansen-Moller, *J. Chromatogr. B.*, 1994, **661**, 219.
113 J.L. Berdargue, C. Viallon, M. Bonneau and M. Le Denmat, in *Measurement of Prevention of Boar Taint in Entire Male Pigs*, Colloq. INRA, 1993, **60**, p. 49.
114 M.A. Magard, M.E.B. Berg, V. Tagesson, M.L.G. Jaremo, L.L.H. Karlsson, L.J.E. Mathiasson, M. Bonneau and J. Hansenmoller, *J. Agric. Food Chem.*, 1995, **43**, 114.
115 F.B. Whitfield, K.J. Shaw and D.J. Walker, *Water Sci. and Technol.*, 1992, **25**, 131.
116 J.L. Kinderlerer and S. Johnson, *J. Sci. Food Agric.*, 1992, **58**, 89.
117 F. Shahidi and R.B. Pegg, *ACS Symp. Ser.*, 1994, **558**, 256.
118 K. Villberg, A. Veijanen, I. Gustafsson and K. Wickstrom, *J. Chromatogr. A*, 1997, **791**, 213.
119 M. Guntert, H.J. Bertram, R. Emberger, R. Hopp, H. Sommer and P. Werkhoff, *ACS Symp. Ser.*, 1994, **564**, 199.
120 T. Hofmann, P. Schieberle and W. Grosch, *J. Agric. Food Chem.*, 1996, **44**, 251.
121 R. Kubec, J. Velicek, M. Dolezal and V. Kubelka, *J. Agric. Food Chem.*, 1997, **45**, 3580.
122 D. Haring, P. Schreier and M. Herderich, *J. Agric. Food Chem.*, 1997, **45**, 369.
123 C.K. Shu, *J. Agric. Food Chem.*, 1998, **46**, 1129.
124 G.R. Takeoka, R.G. Buttery and C.T. Perrino, *J. Agric. Food Chem.*, 1995, **43**, 22.
125 D.S. Mottram, M.S. Madruga and F.B. Whitfield, *J. Agric. Food Chem.*, 1995, **43**, 189.
126 R. Zamora, J.J. Rios and F.J. Hidalgo, *J. Sci. Food Agric.*, 1994, **66**, 543.
127 W.E. Neff and E. Selke, *J. Am. Oil Chem. Soc.*, 1993, **70**, 157.
128 W.E. Artz, E.G. Perkins and L. Salvadorhenson, *J. Am. Oil Chem. Soc.*, 1993, **70**, 377.
129 W.D. Ji, S.H. Yang, M.R. Choi and J.K. Kim, *J. Microbiol. Biotechnol.*, 1995, **5**, 143.
130 J. Mateo and J.M. Zumalacarrequi, *Meat Sci.*, 1996, **44**, 255.
131 R. Self, in *Symposium on Foods: The Chemistry and Physiology of Flavours*, ed. H.W. Schultz, E.A. Day and L.M. Libby, Avi, Westport, CT, 1967, p. 362.
132 W.J.M. Engels and S. Visser, *Neth. Milk Dairy J.*, 1996, **50**, 3.
133 A.A. Beloqui, P.G. Depinho and A. Bertrand, *Am. J. Enology Viticulture*, 1995, **46**, 84.
134 N.C. Marino, E.L. Tamares and C.M.G. Jares, *Food Sci. Technol. Int.*, 1995, **1**, 105.
135 R. Self, in *Applications of Mass Spectrometry in Food Science*, ed. J. Gilbert, Elsevier, London, 1987, p. 239.
136 D. Chassagne, J. Crouzet, R.L. Baumes, J.P. Lepoutre and C.L. Bayonove, *J. Chromatogr. A*, 1995, **694**, 441.
137 J.M. Ames and G. MacLeod, in *The Maillard Reaction in Food Processing, Human Nutrition and Physiology*, ed. P.A. Finot, H.U. Aeschbacher, R.F. Hurrell and R. Liardon, Birkhauser Verlag, Basel, 1990, p. 209.

138 H.E. Nursten and R. O'Reilly, *ACS Symp. Ser.*, 1983, **215**, 103.
139 I. Blank and L.B. Fay, *J. Agric. Food Chem.*, 1996, **44**, 531.
140 D.S. Mottram, *ACS Symp. Ser.*, 1994, **543**, 104.
141 T. Shibamoto and H. Yeo, *ACS Symp. Ser.*, 1994, **543**, 457.
142 T.H. Parliment and H.D. Stahl, *ACS Symp. Ser.*, 1994, **564**, 160.
143 U.S. Gi and W. Baltes, *J. Agric. Food Chem.*, 1995, **43**, 2226.
144 T. Hofmann, R. Hassner and P. Schieberle, *J. Agric. Food Chem.*, 1995, **43**, 2195.
145 H.I. Hwang, T.G. Hartman and C.T. Ho, *J. Agric. Food Chem.*, 1995, **43**, 179.
146 M.W. Samsudin, R.T. Sun and I.M. Said, *J. Agric. Food Chem.*, 1996, **44**, 247.
147 M.E. Bailey, J. Susuki, L.N. Fernando, H.A. Swartz and R.W. Purchas, *ACS Symp. Ser.*, 1994, **558**, 170.
148 T.J. Braggins, *J. Agric. Food Chem.*, 1996, **44**, 2352.
149 S. Lesimple, L. Torres, S. Mitjavila, Y. Fernandez and L. Durand, *J. Food Sci.*, 1995, **60**, 615.
150 F.B. Whitfield, F. Helidoniotis, K.J. Shaw and D. Svoronos, *J. Agric. Food Chem.*, 1997, **45**, 4398.
151 R. Self, *Nature*, 1961, **189**, 223.
152 R. Self, *J. Sci. Food Agric.*, 1963, **14**, 8.
153 A.F. Wood, J.W. Aston and G.K. Douglas, *Aust. J. Dairy Technol.*, 1994, **49**, 42.
154 P.A. Lunning, T. Derijk, H.J. Wichers and J.P. Roozen, *J. Agric. Food Chem.*, 1994, **42**, 977.
155 R.S.T. Linforth and A.J. Taylor, *Food Chem.*, 1993, **48**, 115.
156 C.M. Delahunty, J.R. Piggott, J.M. Conner and A. Paterson, in *Trends in Flavour Research*, ed. H. Maase and D.G. van der Heij, Elsevier, Amsterdam, 1994, p. 47.
157 A.J. Taylor and R.S.T. Linforth, in *Trends in Flavour Research*, ed. H. Maase, and D.G. van der Heij, Elsevier, Amsterdam, 1994, p. 3.
158 W.E. Lee, *J. Food Sci.*, 1986, **51**, 249.
159 D.D. Roberts and T.E. Acree, *J. Agric. Food Chem.*, 1995, **43**, 2179.
160 K.E. Ingham, R.S.T. Linforth and A.J. Taylor, *Food Sci. Technol Lebens.-Wiss. Technol.*, 1995, **28**, 105.
161 K.E. Ingham, R.S.T. Linforth and A.J. Taylor, *Food Chem.*, 1995, **54**, 283.
162 J.S. Elmore and K.R. Langley, *J. Agric. Food Chem.*, 1996, **44**, 3560.
163 A.J. Taylor and R.S.T. Linforth, *Food Chem.*, 1993, **48**, 115.
164 S.M. van Ruth, J.P. Roosen and J.L. Cozijnsen, in *Trends in Flavour Research*, ed. H. Maase and D.G. van der Heij, Elsevier, Amsterdam, 1994, p. 59.
165 S.M. van Ruth, J.P. Roosen, J.L. Cozijnsen and M.A. Posthumus, *Food Chem.*, 1995, **54**, 1.
166 A.J. Taylor, *CRC Rev. Food Sci. Nutr.*, 1996, **36**, 765.
167 A.J. Taylor and R.S.T. Linforth, *Trends in Food Sci. Technol.*, 1996, **7**, 444.
168 W.J. Soeting and J. Heidema, *Chem. Senses*, 1988, **13**, 607.
169 P.G.M. Haring, in *Flavor Science and Technology*, ed. Y. Bessiere and A.F. Thomas, John Wiley, Chichester, 1990, p. 351.
170 D.D. Roberts, J.S. Elmore, K.R. Langley and J Bakker, *J. Agric. Food Chem.*, 1996, **44**, 1321.
171 A.J. Taylor, R.S.T. Linforth, I. Baek, M. Brauss, J. Davidson and D. A. Gray, *Advances in Food Chemistry and Technology*, ed. S. Risch and C.-T. Ho, ACS, in press.
172 M. B Springett, V. Rosier and J. Bakker, *J. Agric Food Chem.*, 1999, **47**, 1125.
173 C.W. Naswari, *ACS Symp. Ser.*, 1993, **536**, 32.
174 C.M. Delahunty, J.R. Piggott, J.M. Conner and A. Paterson, *ACS Symp. Ser.*, 1996, **633**, 202.
175 C.M. Delahunty and J.R. Piggott, *Int. J. Food Sci. Technol.*, 1995, **30**, 555.
176 M. Phillips, J. Greenberg and J. Awad, *J. Clin. Pathol.*, 1994, **47**, 1052.
177 A.I. Speilman, X.N. Zeng, J.J. Leyden and G. Preti, *Experientia.*, 1995, **51**, 40.

178 R. Ruiz, T.G. Hartman, K. Karmas, J. Lech and R.T. Rosen, *ACS Symp. Ser.*, 1994, **546**, 102.
179 C. Grote and J. Pawliszyn, *Anal. Chem.*, 1997, **69**, 587.
180 F.J. Winkler, in *Chromatography and Mass Spectrometry in Nutrition Science and Food Safety*, ed. A. Frigerio and H. Milon, Analytical Chemistry Symposium Series Vol. 21. Elsevier, 1984, p.173.
181 R. Braunsdorf, U. Hener and A. Mosandl, *Z. Lebensm. Unters. Forsch.*, 1992, **194**, 426.
182 H. Casabianca, J.B. Graff, P. Jame, C. Perrucchietti and M. Chastrette, *J. High Resol. Chromatogr.*, 1995, **18**, 279.
183 S. Dautraix, K. Gerola, R. Guilloy, J.L. Brazier, A. Chateau, E. Guichard and P.Etivant, *J. Agric. Food Chem.*, 1995, **43**, 981.
184 A. Mosandl, *Food Rev Int.*, 1995, **11**, 597.
185 J.M. Banks, E.Y. Brechany, W.W. Christie, E.A. Hunter and D.D. Muir, *Food Res. Int.*, 1992, **25**, 365.
186 C.V. Hernandez and D.N. Rutledge, *Analyst*, 1994, **119**, 1171.
187 N. Togari, A. Kobayashi and T. Aishima, *Food Res. Int.*, 1995, **28**, 495.

CHAPTER 4

Bioactive Non-nutrients in Foods

1 Introduction

Bioactive non-nutrients in foods comprise a wide range of natural substances that may have deleterious (*e.g.* toxicants, mutagens and antinutrients) or beneficial (*e.g.* anticarcinogens, antioxidants) effects. Although bioactive compounds in foods have not been as widely studied as anthropogenic compounds, such as synthetic pesticides, their importance should not be underestimated. The concentration of natural pesticides in foods, for example, can be 10 000 times higher than that of synthetic pesticides.[1] It has been suggested that the relative neglect of natural toxicants may be a consequence of the strongly held public belief that 'natural' or 'organic' foods are safer to consume. This contrasts strongly with the negative perception of foods exposed to treatment with pesticides or synthetic fertilisers.[2] However, the attitude to research into all types of bioactive natural substances in foods may now be changing. Bioactive food components that have beneficial effects are currently the subject of increasing study. Strong epidemiological evidence that fruits and vegetables exert an overall protective effect against cancer is now available.[3,4] Protective factors in foods may function through a variety of mechanisms. Examples include the flavonoids, which not only possess antioxidant activity[5] but may also act by stimulating the body's own defensive enzyme systems.[6–8]

Because many natural chemicals confer pest and disease resistance in plants, research programmes to increase their levels by plant breeding or genetic engineering are now quite common. Consequently, it becomes even more important to increase our understanding of their structure, distribution, biological effects and metabolism. Mass spectrometry can aid the exploration of all of these areas by contributing to structure determination, by providing accurate and definitive quantitative data and by defining metabolic pathways with the aid of stable isotope labelling studies.

Undesirable naturally occurring compounds include secondary metabolites, for example the glycoalkaloids found in potatoes and tomatoes. Fungi growing on food that may produce mycotoxins, for example aflatoxins from *Aspergillus flavus* and flavonoids. Some natural toxicants may also be formed by pyrolysis during cooking, for example 2-amino-3,4-dimethylimidazo[4,5-*f*]quinoline

(MeIQ). Besides acute toxins, food plants may contain natural substances that possess longer-term deleterious effects. These include teratogens, antinutrients and natural oestrogens. Some compound classes contain molecules that can have both beneficial and adverse effects. Flavonoids, for example, have more beneficial than adverse health properties.[2] Glucosinolates and their breakdown products may have beneficial properties as inducers of enzymes that protect against cancer, besides detrimental qualities such as antithyroid activity.

Despite the relative neglect of natural toxicants in food as a popular research topic, many publications describe mass spectrometric methods for their identification and determination. Early examples of mass spectrometric analysis of bioactive compounds relied heavily on the traditional ionisation techniques of EI and CI. These methods often required extensive sample preparation and derivatisation, to render the analytes sufficiently pure or volatile for mass spectrometric analysis. These approaches always possess the dangers of artefact formation through loss of important sample components or uncontrolled structural modifications. The development of new, soft ionisation mass spectrometric methods and the more widespread availability of tandem mass spectrometers have allowed researchers to analyse plant material with minimal sample preparation. This reduces considerably the dangers of introducing artefacts. Nevertheless, conventional mass spectrometric methods, particularly GC/MS, still have an important function in the analysis of bioactive non-nutrients. However, the speed and convenience of techniques such as ESI mass spectrometry, especially when enhanced by MS/MS techniques, will undoubtedly lead to their increasing adoption as the methods of choice in this area. Applications described below are organised by compound class, rather than the methodology used. The list of applications is not exhaustive, but is designed to suggest the potential of mass spectrometry in the analysis of natural toxicants in foods. A review of thermospray LC/MS in phytochemistry cites many applications of LC/MS techniques to the analysis of bioactive and other components in food and non-food plants.[9]

2   Alkaloids

Alkaloid-containing plants and plant products are notable sources of human and animal exposure to natural toxins and have been throughout human existence.[10] Most cases of acute poisoning by alkaloids are the consequence of accidental consumption of non-food plants that have a similar appearance to edible species, for example deadly nightshade or *Amanita* species of mushroom. However, some common food plants such as tomatoes (especially when unripe), aubergines (eggplant) and potatoes contain significant quantities of toxic glycoalkaloids (steroidal alkaloids attached to mono- or more commonly oligo-saccharides). Under normal circumstances the concentration of these chemicals is insufficient to be harmful to health: the only known cases of human poisoning by potato tubers have occurred because the plants contained larger quantities of the glycoalkaloids than normal.[10] Other notable potential sources of toxic alkaloids include some types of herbal tea. An example is comfrey, which contains cumulatively hepatotoxic and carcinogenic pyrrolizidine alkaloids. Contamination of food grains with seeds of toxic plants, and ergotamine, produced by *Claviceps pupurea*, a fungal pathogen of grains (especially rye) used in bread-making may also occur. Human ergotism now appears to have been eliminated by increased awareness of the dangers of *Claviceps*-parasitised grain and by modern agricultural practices.

Although alkaloid poisoning by food plants is now a rare occurrence, improved methods of analysis of alkaloids, especially glycoalkaloids, are desirable. These methods are needed to identify and monitor the levels of natural toxicants in new varieties, including those produced by genetic manipulation. Interest has also increased in developing sensitive and specific methods of determining these alkaloids and their metabolites to assess the long-term effects of chronic exposure. Mass spectrometric techniques have an important role to play in these developments.

Combined LC/MS of several different classes of alkaloids has been reviewed.[11] The review is recommended reading for those interested in the advantages and disadvantages of different LC/MS methods for determining various types of alkaloid. A general LC/MS approach, based on generating profiles of mixture components, for chemical identification of botanical components has been described.[12]

The Glycoalkaloids

Glycoalkaloids and Glycoalkaloid Aglycones

The structures of the representative glycoalkaloids α-tomatine (found in green tomatoes) and α-solanine, α-chaconine and β_1-chaconine (all found in potato tubers and/or shoots) are shown in Figure 4.1.

Few reports of mass spectrometric studies of glycoalkaloids have appeared, although papers on this topic generally reflect use of state-of-the-art techniques because of the difficulties attendant on analysing these complex molecules. Conventional mass spectrometric methods for the structural analysis of glycoalkaloids are based on the hydrolysis of the parent molecules to their component steroidal aglycones and sugar groups. Aglycones can then be analysed by EI mass spectrometry. Sugar moieties are analysed by GC/MS of their partially methylated alditol acetates, a well-established method for determining the nature of sugars and their linkage positions in oligosaccharides.[13,14] A recent publication describes techniques for optimising the acid hydrolysis of glycoalkaloids, prior to mass spectrometric analysis of the oligosaccharide side-chain by GC/MS of partially methylated alditol acetate derivatives.[15]

The first mass spectrometric investigation of glycoalkaloids was a detailed report of the EI direct probe mass spectrometry of ten aglycones formed by chemical and/or enzymatic removal of the oligosaccharides.[16] Major fragment ions were rationalised and complete spectra reproduced for most of the aglycones studied.

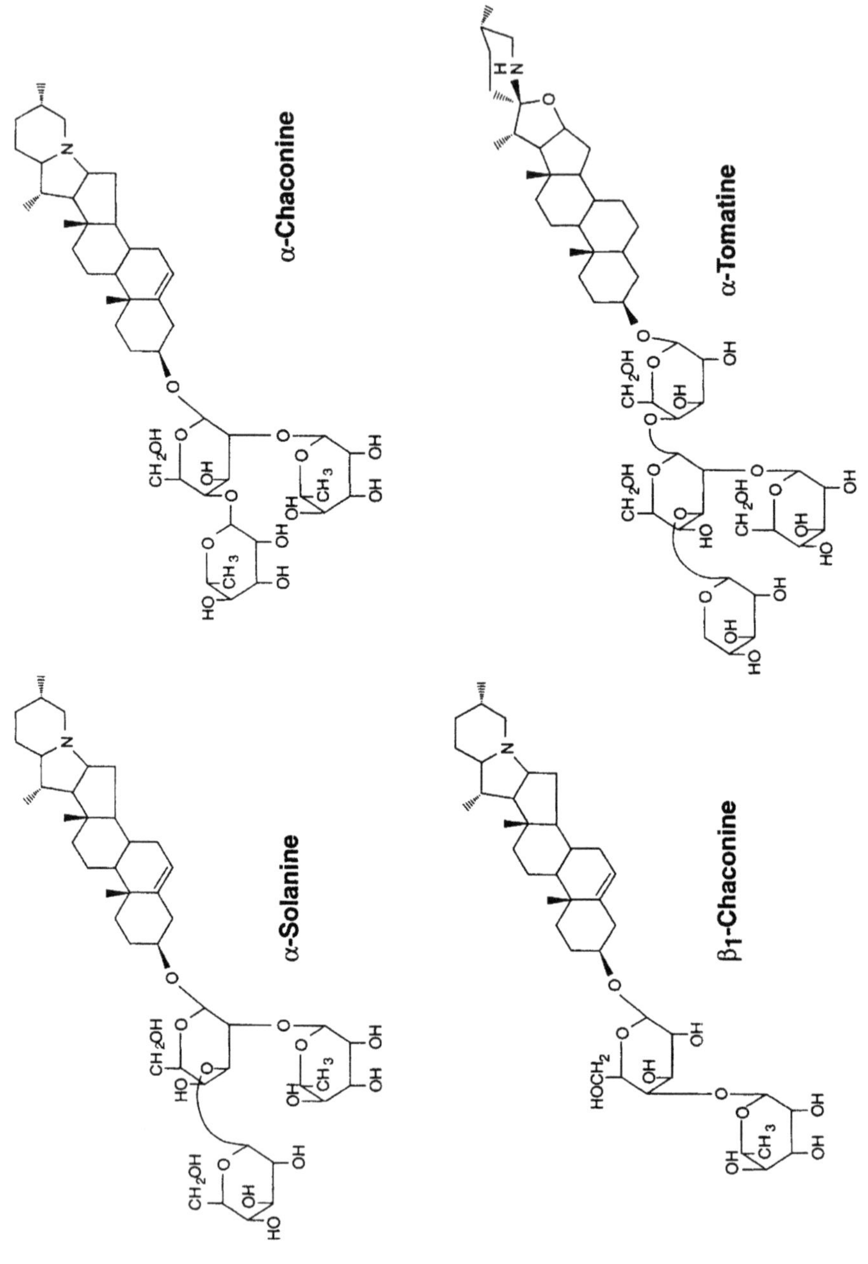

Figure 4.1   *Structures of the representative glycoalkaloids α-solanine, α-chaconine, β₁-chaconine and α-tomatine*

GC/MS of Glycoalkaloid Derivatives

A GC/MS method for the characterisation of novel steroidal alkaloids from tubers of *Solanum* species has been described.[17] Both EI and CI measurements were conducted on the intact aglycones and high-resolution mass spectrometry helped to confirm the empirical formulae of molecular and fragment ions. The presence of novel steroidal alkaloids in the wild *Solanum* species studied was confirmed by these measurements, showing the usefulness of the technique, although complete structural elucidation was not possible.

Intact Glycoalkaloids (Direct Probe Mass Spectrometry)

FABMS of ten glycoalkaloids yielded good quality spectra that contained molecular weight information. Structurally significant fragment ions, formed by cleavage of the individual carbohydrate groups, were also observed.[18,19] Fragment ions confirm the molecular weights of individual sugar units and yield some indication of branching position, but do not establish linkage position or the structures of the sugars. Direct analysis of the juice from etiolated sprouts and outer tuber layers of a commercial potato cultivar was possible by FABMS, after minimal sample processing.[18] The method was not quantitative but provided a very simple qualitative method for obtaining a rapid profile of glycoalkaloids in potatoes. This work led to the development of a quantitative FABMS technique for determining α-tomatine in tomato fruit.[20] The levels found (7 ppm in green fruit and $\leqslant 0.8$ ppm in red fruit) agreed with figures reported using a radiolabelling approach. Structure determination of the carbohydrates formed by hydrolysis of the potato glycoalkaloids α-chaconine and α-solanine has been aided by FABMS.[21]

LC/MS of Glycoalkaloids

A brief report of on- and off-line LC/MS analysis of glycoalkaloids and their aglycones by FAB and thermospray showed that these techniques were suitable for rapid screening of plant extracts.[22] The development of an HPLC method for determining tomato glycoalkaloids included a report of LC/MS studies of tomato extracts using the ion spray technique.[23] In addition to tomatine, other glycoalkaloids were observed in the LC/MS traces, although these were minor components. It was speculated that two of the minor components were a tomatine-like molecule with a modified tomatidine aglycone that contained a double bond, and commersonine.

Tandem Mass Spectrometry of Glycoalkaloids

Four-sector tandem mass spectrometry with scanning array detection has shown that as little as 200 femtomoles of α-tomatine can be characterised structurally.[24] Informative spectra could be obtained from a single 10 s scan and diagnostic fragmentation of the $[M + H]^+$ ion yielded information on sugar residues and the aglycone. Subsequent investigations expanded these data by analysing a wider range of glycoalkaloids by positive and negative ion tandem mass spectrometry of $[M + H]^+$ and $[M - H]^-$ ions generated by LSIMS.[25] The tandem mass spectrum of the $[M + H]^+$ ions of α-tomatine is shown in Figure 4.2.

Peaks formed by glycosidic fragmentation are labelled according to the nomenclature of Domon and Costello.[26] In positive ion mode, ions formed by Z_0, Y_0 and $^{1,5}X_0$ cleavages provide information on the nature of aglycone moieties in the glycoalkaloids. Linkages and positions of sugars in tri- and tetra-saccharide-containing glycoalkaloids are shown by the presence or absence of $Z_{\alpha/\beta}$ and $Y_{\alpha/\beta}$ cleavages and intensity differences of $^{1,5}X_\alpha$ and $^{1,5}X_\beta$ cleavages, respectively. The positive ion tandem mass spectra were more informative than negative ion spectra and had the advantage of lower detection limits. Considerable structural information was also determined from crude plant extracts containing glycoalkaloids. *Definitive* structural assignments are not possible using this technique: only NMR can generate conclusive structural information on the glycoalkaloids. However, the mass spectrometric method is rapid and works on small quantities of material. Furthermore, comparable amounts of information cannot be obtained from crude mixtures by alternative analytical techniques. High- and low-energy collisional activation tandem mass spectrometric techniques have been compared for their ability to generate structural information on glycoalkaloids and their aglycones.[27] Both types of tandem MS yielded informative spectra, with charge-driven reactions dominating the low-

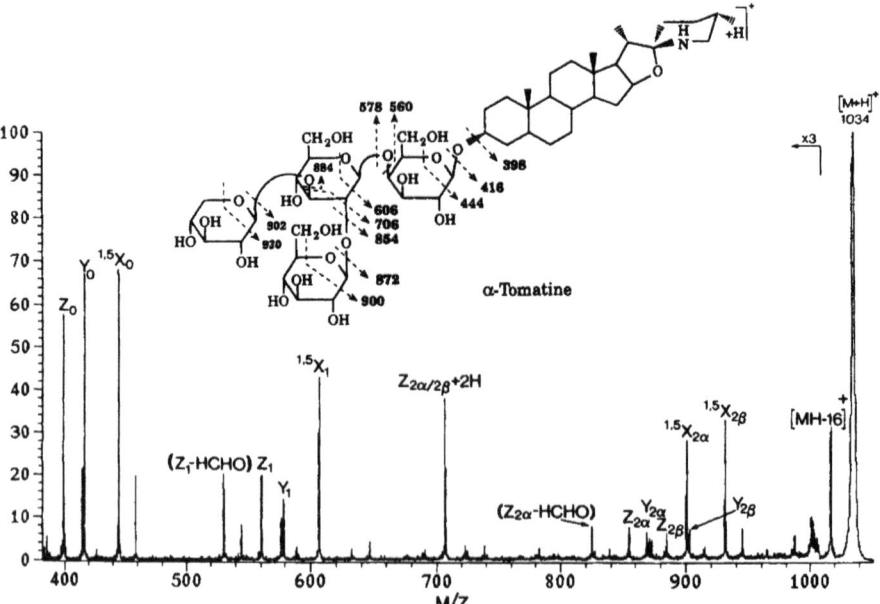

Figure 4.2   *Tandem mass spectrum of the [M + H]⁺ ion of α-tomatine*
(Reproduced from S. Chen, P.J. Derrick, F.A. Mellon and K.R. Price, *Anal. Biochem.*, 1994, **218**, 157, with permission of the authors and publishers)

energy spectra and more complex fragmentation patterns, including charge remote and multiple bond cleavages, in the high-energy spectra.

Recommended Mass Spectrometric Procedures for Glycoalkaloids

Conventional FABMS or LSIMS, especially in positive ion mode, is a useful method for confirming the molecular weights of glycoalkaloids and yielding some structural information. These techniques are also suitable for obtaining a rapid profile of the major components of crude, glycoalkaloid-containing plant extracts. Tandem mass spectrometry is a useful aid to structure determination or confirmation on components of small samples, on mixtures or on possible new glycoalkaloids. Finally, ion spray (or electrospray) LC/MS shows considerable promise in the analysis of glycoalkaloids. If used with tandem mass spectrometry, electrospray has the potential to produce structurally informative data on novel or minor components, besides confirming the presence of known glycoalkaloids.

Pyrrolizidine Alkaloids

More than 250 pyrrolizidine alkaloids have been isolated and characterised.[10] The general structure of the pyrrolizidine alkaloids comprises a basic, necine part and the 'ester' moiety (necic acid) and is shown in Figure 4.3. The ester portion of the molecule may comprise non-esters, monoesters, acyclic diesters or macrocyclic diesters. All hepatotoxic pyrrolizidine alkaloids are esterified and contain an unsaturated (usually in the 1,2 position) heterocyclic ring.[28] Besides direct consumption of, for example, some types of herbal tea, pyrrolizidine alkaloids can enter the food chain indirectly *via* the products (*e.g.* milk and honey) of animals that feed on plants containing these substances.

Intact Pyrrolizidine Alkaloids (Direct Probe Mass Spectrometry)

Several EI mass spectrometric studies of pyrrolizidine alkaloids have been conducted.[29-31] The EI spectra contain molecular and characteristic fragment ions. A typical use of EI mass spectrometry has been to confirm the empirical formulae of new alkaloid species by accurate mass measurement in studies where the main source of structural data is NMR.[32]

Both positive and negative ion CI of pyrrolizidine alkaloids yields molecular weight and some structural information.[33,34] Negative ion CI, with OH⁻

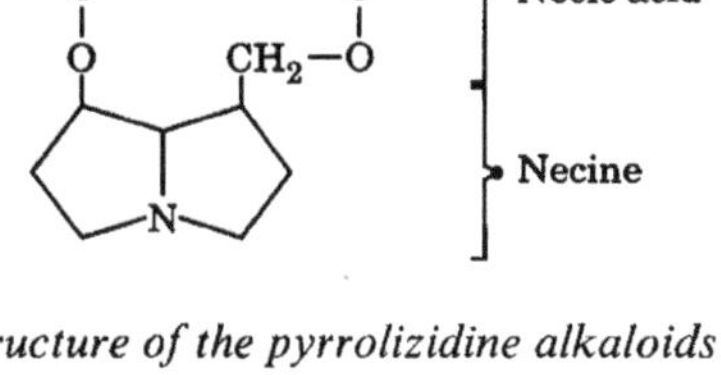

Figure 4.3 *The general structure of the pyrrolizidine alkaloids*

reactant ion yields $[M - H]^-$ ions or, with macrocyclic diesters, $[M + OH]^-$ adducts.[34] In addition, fragment ions often defined the masses of the intact side chain(s) and the basic necine nucleus. Methane positive CI and electron capture negative ion CI mass spectrometry of pyrrolizidine alkaloids yielded molecular weight information on the intact molecules and on ester fragments.[35] Positive CI was more useful for establishing molecular weights as abundant $[MH]^+$ ions were formed; negative CI spectra were often dominated by fragments formed by dissociative electron capture. Esterification at the C-9 position can be distinguished from the C-7 position in the positive CI spectra of non-cyclic retronecine esters because C-9 esters yield a characteristic acyl carbonium ion.

GC/MS of Pyrrolizidine Alkaloids

Several reports of GC/MS analysis of pyrrolizidine alkaloid derivatives have appeared. The first step of these procedures usually requires hydrolysis of the alkaloids to the common base, retronecine, followed by derivatisation. The first reported study described GC/MS of bistrifluoroacetate, bisheptafluoro-butyrate, diacetate and bistrimethylsilyl ether derivatives of retronecine using EI.[36] This demonstrated the feasibility of using GC/MS to determine pyrrolizi-dine bases. The halogenated derivatives were subsequently used in an electron capture GC determination of these alkaloids in goats' milk, following GC/MS confirmation of the residues. Ammonium positive ion CI and OH$^-$ negative ion CI GC/MS of trimethylsilyl derivatives of pyrrolizidine alkaloids offers a potentially rapid method of structure elucidation.[37] The positive CI spectra exhibit abundant $[M + H]^+$ and the negative spectra exhibit ester bond cleavage products. The methodology was used to identify alkaloids in several plant species.[37,38]

EI and ammonia CI GC/MS data can be combined with GC/Matrix Isolation/ FTIR to yield more detailed structural information on pyrrolizidine alkaloids than can be generated by either technique alone.[39] The information from the FTIR spectra was sensitive to subtle structural variations and enhanced the mass spectrometric data. Pyrrolizidine alkaloid metabolites from mouse liver micro-somes have been identified with the aid of GC/MS and tandem mass spec-trometry.[40] GC/MS of trimethylsilyl derivatives revealed the presence of the toxic metabolite dihydropyrrolizidine and the results suggested that the major pathways of pyrrolizidine alkaloid metabolism (*N*-oxidation, hydrolysis and oxidation to pyrrolic compounds) could be assessed using the GC/MS technique. Despite the proliferation of ionisation techniques and derivatisation methods used, straightforward EI GC/MS of trimethylsilyl derivatives is capable of yielding definitive information in many cases. This is shown by a comparative analysis of more than 100 pyrrolizidine alkaloids from natural sources.[41] A combination of retention indices, molecular ion identification and group specific fragmentation allowed unequivocal assignment of the alkaloids in complex mixtures, even in the presence of different geometrical isomers.

Mass spectrometric methods are particularly useful for identifying cases of pyrrolizidine alkaloid poisoning. GC/MS and FABMS identified pyrrolizidine alkaloids from Alpendost (*Adenostyles alliariae*) in a herbal tea preparation in a recent case of hepatic veno-occlusive disease in an 18-month-old child.[42] The patient had been given an herbal tea mixture, erroneously believed to contain coltsfoot.

LC/MS of Pyrrolizidine Alkaloids

LC/MS of a series of *Senecio* alkaloids has been conducted using a thermospray interface.[43] Positive and negative ion thermospray in discharge ionisation mode yielded [M + H]$^+$ and [M − H]$^-$ ions respectively, with most fragmentation occurring in the negative ion spectra. Thermospray LC/MS has also been used to aid the identification of several types of alkaloid, including pyrrolizidines, in *Castanospermum australe*.[44] LC/MS is a particularly useful adjunct to conventional HPLC because of the poor UV absorption of many alkaloids studied. Positive ion APCI LC/MS has successfully detected several pyrollizidine alkaloids in honey produced by bees that have access to Ragwort pollen.[45] The APCI LC/MS reconstructed selected ion chromatograms of a 'ragwort honey' are shown in Figure 4.4, together with the mass spectra of jacobine, jacozine, senecionone and seneciphylline. Monocrotaline (a non-naturally occurring alkaloid) was used as a quantitative internal standard and the detection limit for pyrollizidine alkaloids in honey was 0.002 mg kg^{-1}. This APCI method was robust, rapid, sensitive and specific and is a potential reference method for determining pyrrolizidine alkaloids in selected foods. It was superior, in terms of specificity, sensitivity and speed, to previous approaches but was not suitable for determination of jaconine, a minor alkaloid in ragwort.

An alternative method for determining pyrrolizidine alkaloids is ESI LC/MS and this technique has been exploited in the analysis of these compounds in plants and herbal medicines.[46] Both in-source CID and tandem mass spectrometry were used in combination with HPLC to distinguish the hepatotoxic alkaloids from non-toxic ones. The alkaloids were also detected in the blood of rats dosed with the pyrrolizidine alkaloids.

FABMS of Pyrrolizidine Alkaloids

FABMS has been used to confirm the molecular weights of pyrrolizidine alkaloids in Chinese medicinal herbs.[47] The FABMS data were useful for screening extracts of the plants for the presence of suspected alkaloids. However, the main identification was conducted by EI GC/MS of trimethylsilyl and methylboronate–acetyl derivatives of the alkaloids.

Tandem Mass Spectrometry of Pyrrolizidine Alkaloids

Tandem mass spectrometry of pyrrolizidine alkaloids has been used to confirm decomposition pathways in negative ion CI spectra.[34] Specific metabolites have also been identified by comparing the Collision Induced Decomposition spectra of unknowns with standard compounds[40] and by the LC/MS/MS method mentioned above.[46]

(a)

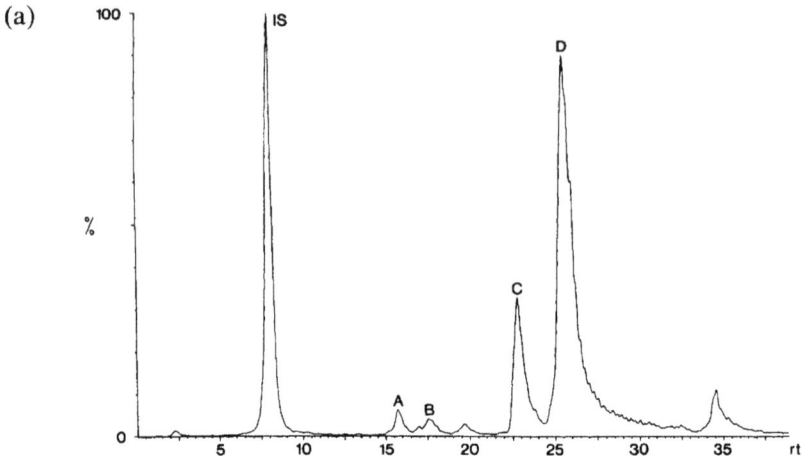

(b)

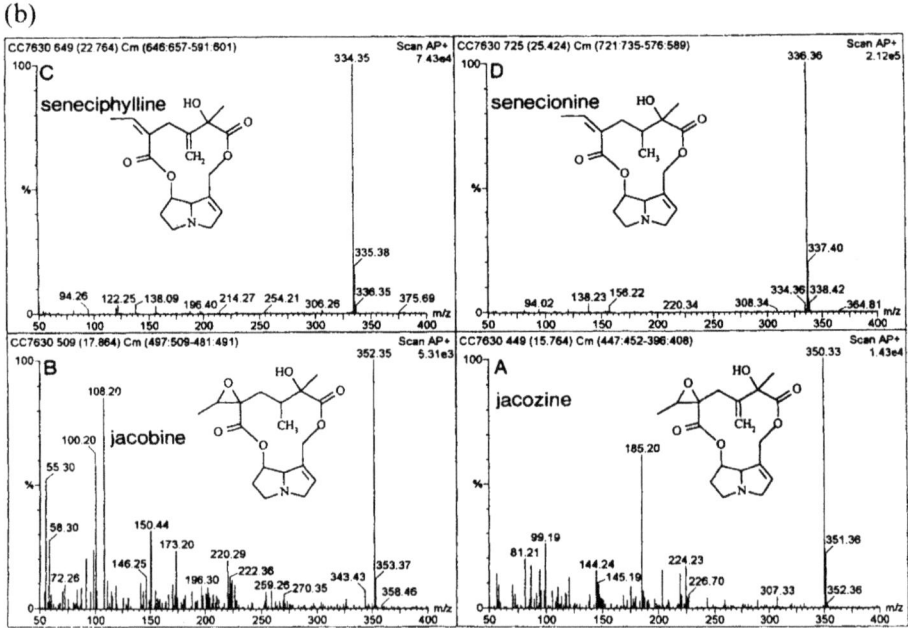

Figure 4.4   *APCI LC/MS reconstructed ion chromatogram* (a) *of a 'Ragwort honey' and the mass spectra* (b) *of jacobine, jacozine, senecionone and seneciphylline* (Reproduced from C. Crews, J.R. Startin and P.A. Clarke, *Food Add. Contam.*, 1997, **14**, 419, with permission of the authors and publishers)

Recommended Mass Spectrometric Procedures for Analysing Pyrrolizidine Alkaloids

EI and CI GC/MS of silylated derivatives using apolar capillary columns (OV-1 or DB-1) is, at the time of writing, the most useful general mass spectrometric method for determining pyrrolizidine alkaloids and their metabolites. However, the success of APCI LC/MS in determining pyrrolizidines in a food matrix[45] clearly demonstrates the considerable potential of this method for analysing these natural products both qualitatively and quantitatively. APCI is generally more robust (less prone to suppression effects from co-eluting compounds or solvent modifiers) than ESI and so may be preferred over the ESI LC/MS method discussed above.[46] However, a definitive comparison of the two techniques will be necessary to establish which is best for 'real world' samples.

Ergot Alkaloids

The virtual worldwide elimination of *Claviceps*-parasitised grain means that the once dreaded 'disease' of ergotism, *i.e.* poisoning by ergot alkaloids, is extremely rare. Consequently, there is little need for advanced techniques for analysing ergot alkaloids and only brief details of mass spectrometric methods for studying these compounds are presented here. The alkaloids present in ergot fungi vary considerably between individual types of fungus and their geographical location, but the major components (Figure 4.5) are often the lysergic acid derivatives ergocristine and ergotamine.[48]

EI mass spectrometric studies of the ergot cyclol and clavine alkaloids have been reported in several publications.[49–53] Although the EI spectra contain abundant diagnostic fragments, molecular ions are absent from the spectra of ergopeptides, making interpretation difficult. Isobutane CI mass spectrometry does not yield molecular weight information on ergot cyclol alkaloids but spectra are simplified and a combination of EI and CI data is helpful in differentiating structures.[54]

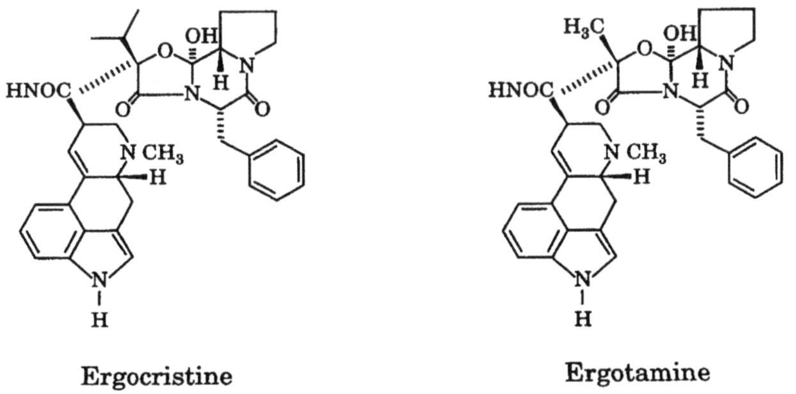

| Ergocristine | Ergotamine |

Figure 4.5 *Structures of the ergot alkaloids ergocristine and ergotamine*

A sensitive and specific GC/MS method for identifying and quantifying ergotamine in human plasma has been developed.[55] Although this methodology was developed for pharmacological studies (ergotamine can be used in the treatment of migraine), it could also be used as a definitive technique for confirming rare cases of ergotism. An alternative method based on tandem mass spectrometry with negative ion CI was rapid and sufficiently sensitive to quantify down to 2 pg of ergotamine in human plasma.[56] EI and CI LC/MS using the (now outdated) moving belt interface has been used to identify ergot alkaloids in fermentation broths.[57,58] Tandem mass spectrometry was also used to analyse the alkaloids but the LC/MS method was superior, in this example, because it enabled isomers to be distinguished. Isobutane CI and tandem mass spectrometry of ergot cyclols can differentiate twelve different alkaloids and distinguish ten out of the fourteen ergots present in a crude extract.[59] FABMS (conventional and tandem) is a useful aid to the structural elucidation of elymoclavine glycosides[60] and positive and negative ion FABMS of intact ergots and dihydroergots appears to be superior to EI for characterising these compounds.[61]

A recent example of more up to date mass spectrometric methods applied to the analysis of ergot alkaloids describes an ESI LC/MS method for their detection in endophyte-infected tall fescue.[62]

Quinolizidine Alkaloids

New varieties of lupin crops have been developed as a possible substitute for soya in human foodstuffs.[63] This has increased interest in the development of sensitive analytical techniques for the qualitative and quantitative analysis of the quinolizidine alkaloids found in these crops. Lupin alkaloids are typified by lupinine and sparteine (Figure 4.6) and related compounds.

Intact Quinolizidine Alkaloids (Direct Probe Mass Spectrometry)

Detailed investigations of the EI mass spectrometry of quinolizidine alkaloids have revealed that ions characteristic of structural and stereoisomers are present in the low and high resolution mass spectra.[64] This paper reappraises earlier work[65-67] in the light of high-resolution and metastable ion measurements and provides useful rules for distinguishing particular structural features.

Lupinine   Sparteine

Figure 4.6   *Structures of the Lupin alkaloids lupinine and sparteine*

GC/MS of Quinolizidine Alkaloids

The technique of choice for identifying quinolizidine alkaloids is EI GC/MS of underivatised material.[68-74] Considerable information can be garnered by EI GC/MS of quinolizidine alkaloids, as shown by the analysis of 100 alkaloids in 90 taxa of the genus *Lupinus*.[75] This paper contains very useful information on retention indices and the principal ions in the mass spectra of a variety of quinolizidine alkaloids. An example of the practical utility of the GC/MS approach is provided by the confirmation of the low quinolizidine alkaloid content of the Washington lupin, confirming its potential as a feed and food crop.[76]

Alternative Mass Spectrometric Techniques for Quinolizidine Alkaloids

The distribution of quinolizidine alkaloids in stem tissue has been determined by Laser Desorption Mass Spectrometry.[77] Thin sections of plant material yielded spectra containing intense ions assigned to sparteine and lupinine. Only relatively high local concentrations (typically > 5 mmol kg^{-1}) of organic compounds were detectable; however, some indication of the spatial distribution of the alkaloids was evident.

3 Coumarins

Coumarins are naturally present in the tissues of umbelliferous plants, including food plants such as carrot (methoxymelleins), celery and parsnip. Celery and parsnip contain high levels of furanocoumarins when healthy (10 mg/100 g) and even higher concentrations in diseased tissues (*ca.* 250 mg/100 g). The furanocoumarins possess a range of biological activities, including anticoagulant, antibacterial, fungicidal, cytostatic and potential anti-cancer properties. The psoralens, a group of naturally occurring furanocoumarins, are noted for their photo-toxicity, an effect thought to be related to their capacity to form photo-induced adducts with pyrimidine bases in DNA.[78,79] The structure of psoralen is shown in Figure 4.7; a number of substituted psoralens are also known to occur naturally.

A useful early description of the EI mass spectrometry of coumarins shows the potential utility of this technique in structure elucidation.[80] The mass spectrometry of furanocoumarins, pyranocoumarins and chromones has been reviewed more recently.[81]

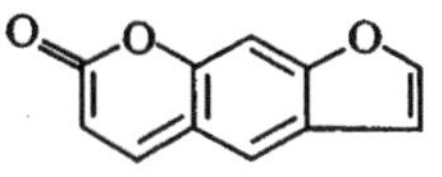

Psoralen

Figure 4.7 *The structure of psoralen*

Direct Probe Mass Spectrometry of Coumarins

In an early example of coumarin analysis by direct probe mass spectrometry, mass spectrometric and NMR data were combined to define the structures of substituted coumarins isolated from dill.[82] EI,[83,84] electron attachment[85] and positive[86] and negative CI[87,88] mass spectrometric analyses of coumarins and furanocoumarins have been reported. All these techniques yielded useful structural information. The intensity of the phenoxide fragment ion in the negative ion spectra of furanocoumarins is an aid to assignment of ring substitution patterns.[87] With positive CI, isomer-specific data were only evident in collision induced dissociation spectra. Conventional EI mass spectrometry can be very useful in characterising furanocoumarins in plants, exemplified by the confirmation of the presence of psoralen, bergapten, xanthotoxin, isoimperatorin, isopimpinellin, oxypeudanin and its hydrate and graveolone in common herb parsley.[89]

GC/MS of Coumarins

GC/MS analysis of coumarins in plant extracts can be conducted successfully if the extracts are first treated with diazomethane to methylate free hydroxy groups. This increases the volatility of the individual components so that they are suitable for GC.[90] GC/MS of coumarins (and some additional natural products) has been used to confirm dietary compliance in free-living human studies in which the volunteers were fed a diet rich in anticarcinogenic, umbelliferous vegetables.[91] GC/MS analysis has also been used to confirm the reduction of coumarin content in Tonka bean extracts by supercritical CO_2 extraction.[92]

LC/MS of Coumarins

The first report of LC/MS of naturally occurring coumarins utilised a moving belt interface to obtain EI and ammonia CI mass spectra of root extracts. Components of *Imperatoria ostruthium* and *Calophyllum inophyllum*, both rich sources of coumarins, were separated by microbore HPLC.[93] Although useful data were obtained, the belt LC/MS interface is now considered obsolete. Subsequent studies have used the newer and more convenient thermospray and particle beam interfaces.[94-96] Although the second of these papers dealt with LC/MS of coumarin oral anticoagulants, the methodology can be extended to the study of natural coumarins. The previously mentioned review of thermospray LC/MS in phytochemical analysis contains discussion of LC/MS of coumarins.[9] The analysis of 18 coumarins in an extract of the plant *Smyrnium perfoliatum* L. was accomplished successfully using a microbore HPLC LC/MS system.[96] Although both thermospray and particle beam LC/MS yielded analytically useful data, the EI spectra obtained using the particle beam systems displayed better reproducibility and were more structurally informative.

More recently, an up to date LC/MS method, APCI, has been used to 'fingerprint' coumarins in plant extracts.[97]

FABMS of Coumarins

Although many coumarins can be analysed successfully by traditional ionisation methods (EI, CI), FABMS has been used for analysing coumaric acids.[98] FABMS can be very useful for characterising coumarin glycosides.[99,100] Negative ion FABMS has been successfully employed for analysing coumarin glycosides. In addition to confirming molecular weights, this technique yields useful structural information from fragment ions corresponding to the loss of coumaric and glycosidic residues.[100]

Tandem Mass Spectrometry of Coumarins

Several MS/MS studies of furanocoumarins have been conducted.[101] Annulated furanocoumarins[102] and isomeric dimethylfuranocoumarins[84] have been characterised by different types of MS/MS. These include collision-induced dissociation, surface-induced dissociation, charge-stripping and charge inversion. The results obtained from the reported studies allowed unequivocal structural characterisation, even in cases in which the conventional EI mass spectra of isomers were identical. Distinguishing isomeric dimethyl coumarins was possible by surface-induced dissociation MS/MS only; collision-induced dissociation yielded nearly identical product ion spectra of these isomers.

Recommended Mass Spectrometric Procedures for Analysing Coumarins

LC/MS techniques, in combination with tandem MS if additional structural information is required, are probably the best mass spectrometric methods for analysing coumarins. LC/MS methods have the advantage of speed and minimal sample preparation requirements. Techniques based on APCI (or possibly electrospray for coumarin glycosides) are likely to offer the best combination of sensitivity and reproducibility.

4 Cyanogenic Compounds

Compounds that are capable of cyanogenesis, the formation of free HCN, are found in several plant families.[103] Although several cyanogenic compounds occur naturally, those most likely to be consumed by humans are the cyanogenic glycosides. The principal cyanogenic food crop is cassava, the fourth most important crop in tropical countries. This contains the cyanogenic glycoside linamarin (Figure 4.8).

Sorghum, peas, beans and grams are also cyanogenic and cyanogens are present in the seeds of many fruits, including apples, pears and peaches.[2] Cases of acute poisoning caused by consumption of seeds, sometimes in processed

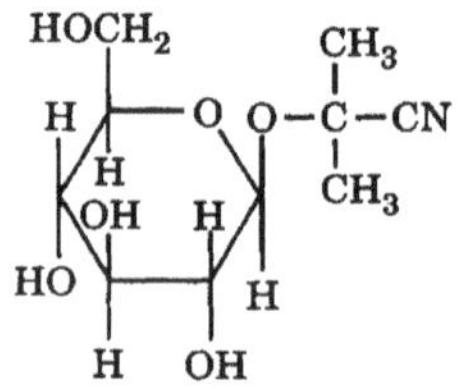

HOCH₂   CH₃

(structure diagram of Linamarin)

Linamarin

Figure 4.8   *The structure of the cyanogenic glycoside linamarin*

form, have been reported in several countries. Furthermore, cassava diets have been associated with a variety of chronic toxicity symptoms (not all of which may be related to HCN release).

Many analytical techniques for detecting cyanogens are based on the detection of cyanide above an agreed threshold level. It is important to establish this threshold because all plants produce cyanide as part of the biochemical process that generates ethylene. However, it is also important to distinguish between cyanide present in the parent cyanoglycoside and 'free' cyanide;[103] thus more sophisticated analytical techniques are required. Despite this, remarkably few reports of the use of mass spectrometry for characterising cyanogenic glycosides can be found.

Direct Probe Mass Spectrometry of Cyanogenic Glycosides

Early attempts to obtain EI mass spectra of cyanogenic glycosides have been described in a review of techniques for isolating and characterising these compounds.[104] These were largely focused on EI mass spectrometry of acetate or trimethylsilyl derivatives. These yielded spectra containing limited structural information and absent or low-intensity molecular ions. The molecular weight of trimethylsilylated amygdalin has been confirmed by ammonia chemical ionisation.[105] The earliest successes in obtaining mass spectra of free cyanogenic glycosides by direct probe methods used ammonia CI from a moving belt[106] or FD mass spectrometry.[107] Both types of spectra yielded information on the molecular weight of the intact molecules and the nature of the aglycones. CI mass spectrometry is useful for routine molecular weight determination of cyanogenic glycosides, as exemplified by studies of species of *Centaura*.[108]

GC/MS of Cyanogenic Glycosides

GC/MS of trifluoroacetyl derivatives of amygdalin yields limited information.[109] The methodology described in this study was restricted by the narrow mass range of the instrument used. In a more comprehensive study, five mandelonitrile glycosides were detected in the glycosidic fraction isolated from several Passiflora fruits by GC/EI-MS or GC/NCI-MS of trifluoroacetylated derivatives.[110]

FABMS of Cyanogenic Glycosides

Only one report of the FABMS of cyanogenic glycosides has appeared.[111] Both positive and negative ion spectra were structurally informative, yielding data on the molecular weights and aglycones of the compounds studied.

Recommended Mass Spectrometric Procedures for the Analysis of Cyanogenic Glycosides

FABMS may be the method of choice for analysing cyanogenic glycosides, especially when isolated in relatively pure form. None of the methods described above is entirely satisfactory for analysing mixtures. However, the success of FABMS suggests that LC/FAB is potentially useful in such cases. The CI data reported above indicate that APCI LC/MS may be even more appropriate (and is likely, if successful, to be more robust). ESI or possibly APCI may also be of value, although success here will be more dependent on the solvents used for HPLC.

5   Glucosinolates

The glucosinolates are a uniform class of naturally occurring anionic compounds that contain a glucose moiety attached through a β-thioglucosidic link (Figure 4.9).

Despite the common structural features of the glucosinolates, there is considerable variability in their aglycones. The variable R moiety ranges from simple and branched alkyl and alkenyl side chains, through alcohols, methylthioalkyl, methylsulfinyl, aralkyl and heterocyclic groups. A major characteristic of glucosinolates is that they are readily hydrolysed by the enzyme myrosinase, yielding a labile aglucone that rearranges spontaneously to an isothiocyanate (Figure 4.10).

Both the biological activity of the glucosinolates and the main flavour components of the plants in which they occur are a consequence of the release of these isothiocyanates. The glucosinolates and their breakdown products can possess both health benefits and (in some cases) properties that may be deleterious to health. The breakdown products are mainly responsible for the distinctive, biting taste of, for example, mustard and horseradish. Much interest has focused on the health benefits: they are known to be potent inducers of

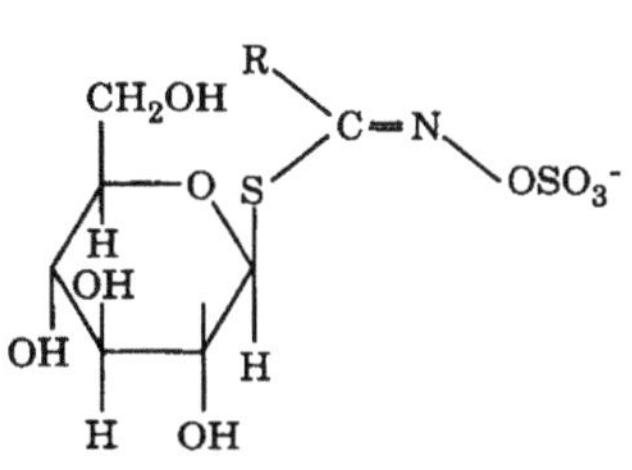

Figure 4.9   *General structure of the glucosinolates*

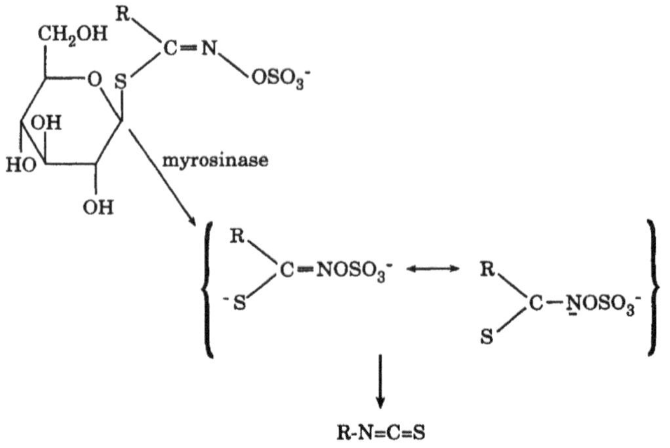

Figure 4.10 *Myrosinase catalysed deglycosylation and spontaneous rearrangement of glucosinolates to isothiocyanates*

Phase II enzymes, including quinone reductase and glutathione-*S*-transferases.[112,113] These enzymes protect against the effects of carcinogens and other potentially injurious electrophiles.

More than 100 different types of glucosinolate have been isolated from eleven families of dicotyledons.[114,115] Food and feed plants high in glucosinolates include broccoli, mustards, horseradish and rape.

Glucosinolate Breakdown Products

The first reports of mass spectrometry in glucosinolate analysis described the use of EI GC/MS to confirm the structures of volatile constituents (mainly short-chain aliphatic, arylic and allylic isothiocyanates). These were formed by breakdown of glucosinolates in horseradish roots.[116,117] The most extensive work on identifying breakdown products formed by autolysis of the aglycones of glucosinolates was reported by Cole.[118] Capillary GC/MS is now a routine method for confirming the identity of volatile glucosinolate breakdown products.[119,120] In the case of indolic breakdown products, silylation is necessary to render the compounds sufficiently volatile for GC and GC/MS.[121] More recently, GC/MS has been used to determine sulforaphane and sulforaphane nitrile in broccoli[122] and has aided the identification of bis(4-isothiocyanatobutyl) disulfide and its precursor (a glucosinolate breakdown product) in the leaves of rocket salad.[123]

GC/MS of Glucosinolate Derivatives

The first accounts of mass spectrometry of glucosinolates, rather than their breakdown products, described EI and CI GC/MS of pertrimethylsilyl derivatives of desulfoglucosinolates.[124,125] Desulfoglucosinolates are easily prepared from glucosinolates by using the enzyme sulfatase to convert the sulfate group to a hydroxyl: this step can be carried out 'on-column' as part of the sample isolation procedure.[126] Trimethylsilylation using standard procedures yields pertrimethylsilyl (TMS) derivatives that are sufficiently volatile for GC.[127,128] The EI mass spectra of desulfoglucosinolate TMS derivatives are often dominated by ions formed by C–S cleavage with charge retention on the sugar moiety. They yield little information about the nature of the side chain R, the variable part of the molecule. Only glucosinolates possessing aromatic R groups yielded ions characteristic of the side chain. However, ammonia CI GC/MS of the same derivatives was more successful as the mass spectra all contained molecular and/or fragment ions diagnostic of the side chain.[129] Capillary column GC/MS using methane as reagent gas is also a very useful qualitative analytical method for desulfoglucosinolates.[130] Ions of composition $[RC=NH]^+$ were found in the mass spectra of all the derivatives examined, aiding identification of the variable side chain R. Tandem mass spectrometry of fragment ions yielded confirmatory evidence for the structure of the side chains. An investigation of the negative ion (methane) GC/MS and GC/MS/MS of silylated desulfoglucosinolates subsequently complemented this study.[131] The negative ion method was an order of magnitude more sensitive than the positive ion technique and yielded simpler but diagnostically useful mass spectra. Neither positive nor negative ion CI yielded molecular ions of the silylated desulfoglucosinolates but sufficient information was present in both types of spectra to identify the nature of the side chain. The positive and negative ion spectra provided complementary information and both analytical techniques can be recommended for glucosinolate analysis if a GC/MS method is desirable.

Intact Glucosinolates (Direct Probe Mass Spectrometry)

Both EI and CI direct probe mass spectrometry of glucosinolates and desulfo-glucosinolates yield spectra that do not contain molecular ions. Nevertheless, they exhibit intense diagnostic ions that enable the structures of these molecules to be defined.[129,132]

In contrast, FABMS of glucosinolates and desulfoglucosinolates yields abundant molecular weight information but relatively little structurally infor-mative fragmentation.[133] The glucosinolates yielded abundant cationised and protonated molecular ions in positive ion mode and the negative ion spectra were dominated by intense $[M - H]^-$ ions. However, the molecular weight information alone is sufficient to identify the structures of many glucosinolates. The negative ion FABMS spectrum of a crude extract of *Arabidopsis thaliana* seeds exhibited $[M - H]^-$ ions of seven glucosinolates (five already known to be present). However, it was suggested one of the ions may have been an artefact generated by mass spectrometric fragmentation. Nevertheless, negative ion FABMS is a very useful preliminary screening method for determining the presence of glucosinolates in crude plant extracts. A recent review of chromato-graphic and spectroscopic methods for separating and identifying glucosino-lates includes a description of FAB and CI methods.[134]

Tandem Mass Spectrometry of Glucosinolates

Glucosinolates have been characterised by negative ion FABMS with collisional activation of the $[M - H]^-$ ion and linked scanning at constant B/E.[135] The product ion spectra contained peaks that were specific for the compound class as a whole and for the side chain. Tandem mass spectrometry has also been performed on glucosinolates under LC/MS conditions (see below).

LC/MS of Glucosinolates

The first report of LC/MS of glucosinolates used thermospray ionisation to confirm the structures of desulfoglucosinolates in seeds of *Brassica campestris*.[136] The spectra all exhibited protonated molecules, establishing molecular weights, and contained useful fragment ions diagnostic of the composition of the side chain, R. It was also possible to separate two chromatographically unresolved desulfoglucosinolate peaks (desulfogluconapin and desulfoglucoalyssin) by computer manipulation of the data, because of the specificity of the individual mass spectra. Thermospray LC/MS has subsequently been used to identify desulfoglucosinolates isolated and prepared from calabrese,[137] leaves and seeds of *Arabidopsis thaliana*,[138,139] sprout extracts,[140] rape[141,142] and broccoli.[143] Additional structural information was obtained, in one instance, by tandem mass spectrometry under LC/MS conditions.[140] Applications of GC/MS and LC/MS methods to glucosinolate analysis have been described and compared.[144]

Frit FAB LC/MS has also been used successfully to identify *intact* glucosinolates (in contrast to the desulfoglucosinolates analysed by thermospray LC/MS).[145,146] Spectra were acquired in negative ion mode and yielded $[M - K]^-$ ions and non-diagnostic fragment ions. However, high-energy collisional activation and tandem mass spectrometry of the $[M - K]^-$ ions yielded group- and compound-specific product ions that allowed structure elucidation.[146] The main disadvantage of the frit FAB LC/MS method is low liquid flow tolerance. LC FAB probes require a large flow split when using standard analytical HPLC columns (only 4 μl min^{-1} was split to the FAB probe in this example). This reduces the overall sensitivity of the technique. This effect is offset somewhat by the high concentration of glucosinolates in many plant extracts, and by the specificity of the structural information by LC/MS/MS.

Recently, more up to date LC/MS methods have been applied to the analysis of glucosinolates. For example, negative ion ESI LC/MS using volatile ion-pairing reagents has aided the identification of intact glucosinolates in mustard seed.[147,148] The negative ion electrospray spectra of sinalbin and sinapin are shown in Figure 4.11.

Desulfoglucosinolates yield informative mass spectra under APCI LC/MS conditions, with protonated molecular ions and group-specific fragment ions.[149] The APCI method was robust and reproducible and was used to identify desulfoglucosinolates in Japanese rapeseed.

(a)

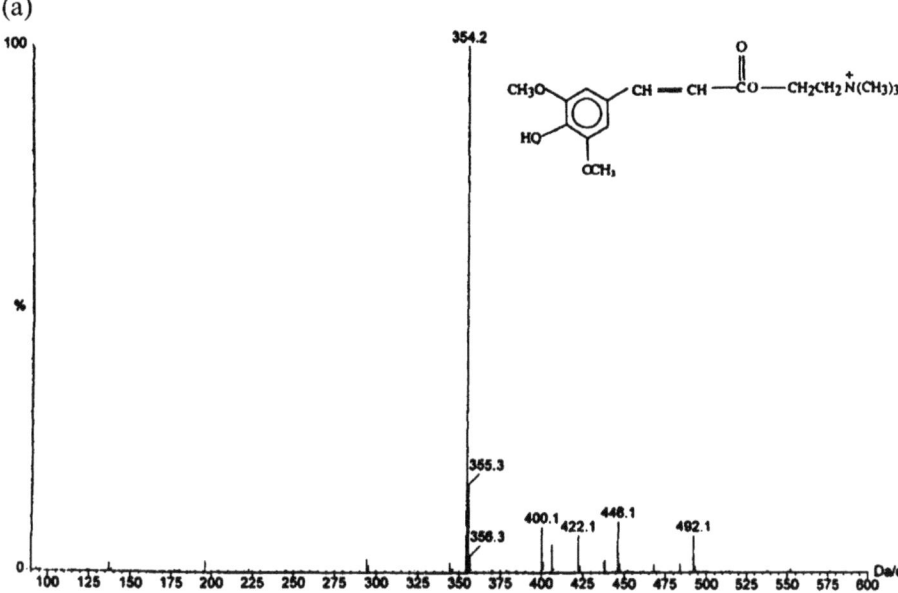

(b)

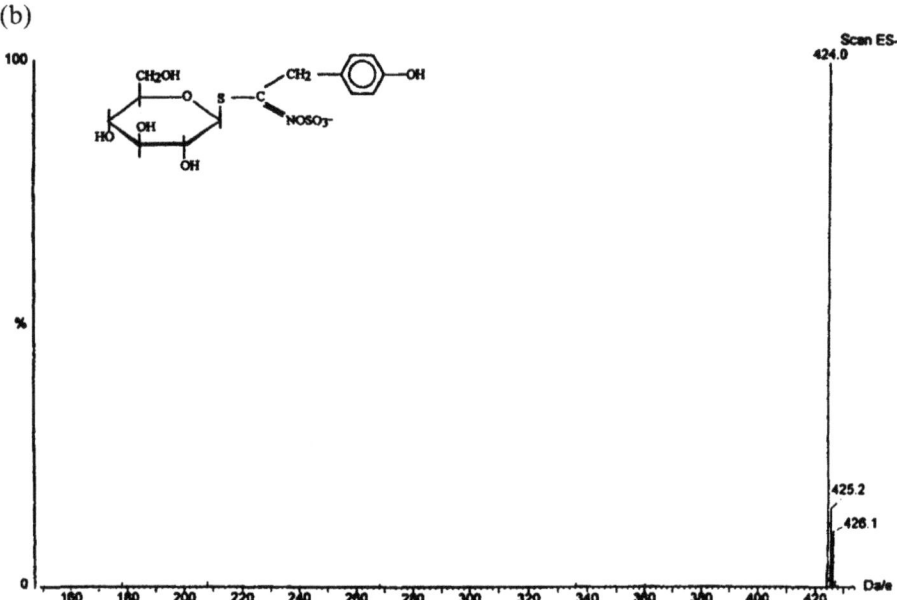

Figure 4.11 *The negative ion electrospray mass spectra of* (a) *sinalbin and* (b) *sinapin* (Reproduced from C.L. Zrybko, E.K. Fukuda and R.T. Rosen, *J. Chromatogr. A*, 1997, **767**, 43, © 1997 with permission from Elsevier Science)

Recommended Mass Spectrometric Procedures for the Analysis of Glucosinolates

GC/MS is still the main method of choice for analysing *volatile* glucosinolate breakdown products. The method is rapid, sensitive and specific and can be carried out using relatively inexpensive bench-top GC/MS systems. Negative ion FABMS is useful for rapidly screening crude plant extracts for glucosinolates. However, care is needed; analysts must be aware that spectrum suppression can occur if impurities are present in high concentration, or if a particular glucosinolate dominates the mixture. If LC/MS is unavailable, GC/MS of trimethylsilyl derivatives of desulfoglucosinolates is a useful, sensitive and specific method for identifying glucosinolates in plant extracts. However, for speed, selectivity and convenience, LC/MS is the technique of choice, especially when additional information can be obtained by tandem MS. Although the thermospray and frit FAB methods described above are adequate for obtaining good LC/MS data, more recent publications show that electrospray and APCI techniques are now the methods of choice.

6   Mutagenic Heterocyclic Amines Generated by Cooking

A number of mutagenic and carcinogenic heterocyclic amines can be formed when proteinaceous foods are heated. These compounds are typified by five heterocyclic amines commonly found in cooked food (Figure 4.12).*

They are generally formed by heating muscle meats at high temperature (by frying or broiling) and *may* be important in the aetiology of some human

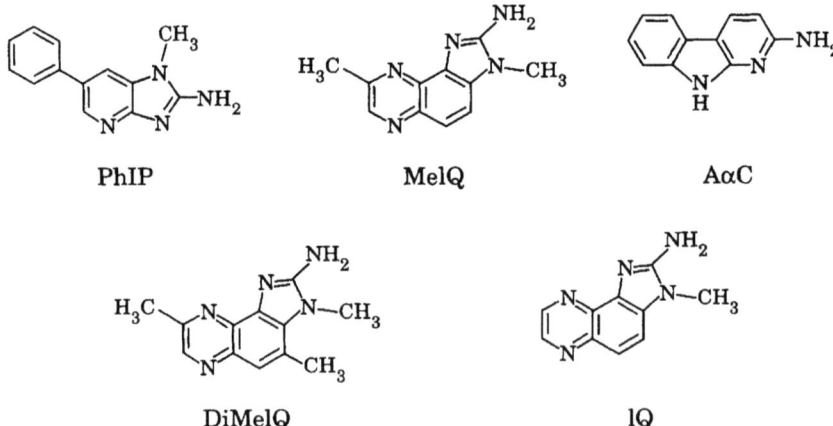

Figure 4.12   *Structures of the five main heterocyclic amines found in cooked food*

*PhIP = 2-amino-1-methyl-6-phenylimidazo[4,5-*b*]pyridine; MeIQ = 2-amino-3,8-dimethylimidazo[4,5-*f*]quinoline; AαC = 2-amino-9*H*-pyrido[2,3-*b*]indole; DiMeIQ = 2-amino-3,7,8-trimethylimidazo[4,5-*f*]quinoline; IQ = 2-amino-3-methylimidazo[4,5-*f*]quinoline.

cancers, particularly in populations consuming a western-style diet. These heterocyclic amine mutagens must undergo biotransformation in the body, for example by prostaglandin-H synthase metabolism, before they can interact with cellular macromolecules to produce nitro, nitroso and azo products.[150] Several different methods for the analysis of heterocyclic aromatic amines in foods, including mass spectrometric techniques, have recently been compared.[151] The results of the two European studies reported in this paper reveal a good correlation between the methods used. However, improvements in isolation and purification methods were required.

Mutagenic heterocyclic amines are present in cooked foods at low levels (ng g^{-1}) and require sensitive and specific techniques of analysis if they are to be assayed reproducibly.[152] Mass spectrometry, especially in combination with chromatographic techniques, is one of the most important methods used to analyse these molecules.

GC/MS of Mutagenic Heterocyclic Amines

Conventional EI mass spectrometry with solids probe sample introduction can be used to detect MeIQx* in simpler food matrices such as pork juice.[153] More importantly, it can aid the structural characterisation of new heterocyclic amine mutagens.[154] However, quantitative mass spectrometric methods require a more sophisticated approach, involving combined chromatography/MS methods. Chromatographic methods for measuring heterocyclic amines in cooked foods, including chromatographic/mass spectrometric methods, have been reviewed up to 1992.[152] An accurate and sensitive stable isotope dilution GC/negative ion electron capture method for simultaneously determining MeIQx and DiMeIQx (as their 3,5-bis(trifluoromethyl)benzyl derivatives) in fried beef was developed in 1988.[155] This was adapted to detect the same amines in human urine[156] and was further developed to determine a third amine, PhIP.[157] The method yielded detection limits of 0.05, 0.1 and 0.2 ng g^{-1} for MeIQx, DiMeIQx and PhIP respectively. The only major disadvantages of the technique were the cost of instrumentation, uncertainties in the derivatisation method and the instability of the GC capillary columns at the high temperatures required to elute the amines. Problems of instrumental cost are now less pronounced because of the development of relatively cheap, sensitive bench-top mass spectrometers with negative ion capabilities. GC/MS methods have recently been applied to the determination of mutagenic heterocyclic aromatic amines in commercial marinades,[158] the analysis of non-polar heterocyclic amines in cooked foods and meat extracts[159] and to the determination of three different mutagenic amines in the urine of volunteers who had been fed a fried meat meal.[160] The last mentioned paper also described a novel method for derivatising the mutagenic amines prior to GC/MS, involving heptafluorobuty-lation followed by methylation.

---

*MeIQx = 2-amino-3,8-dimethylimidazo[4,5-*f*]quinoxaline; DiMeIQx = 2-amino-3,7,8-trimethyl-imidazo[4,5-*f*]quinoxaline.

LC/MS of Mutagenic Heterocyclic Amines

Stable isotope dilution thermospray LC/MS methods are capable of analysing heterocyclic amines in a variety of food matrices, including salmon, sardine, beef and beef extracts, and in animal urine and faeces can attain detection limits of 0.3 ng g^{-1}.[161-163] TSP requires a complex extraction procedure but is sensitive and selective. However, a number of recent LC/MS methods use more convenient and modern electrospray or APCI techniques. One of the earliest of these reports describes a pneumatically assisted electrospray SIM LC/MS method for determining seven mutagenic heterocyclic amines.[164] Detection limits of 5.4–44 pg were obtained on standard solutions of the amines and a meat extract was analysed after solid-phase extraction purification. An automated system (including robotic sample preparation) for determining mutagenic amines in food products utilises a variety of assay techniques, including electrospray LC/MS.[165] Although HPLC/UV was the routine tool for quantifying the analytes, electrospray LC/MS proved to have some unique advantages as an alternative identification and quantification method. APCI LC/MS methods have also been developed for assaying a variety of mutagenic heterocyclic aromatic amines.[166,167] The first paper describes a comprehensive investigation of the occurrence of these mutagens in 'processed food flavours'. Some discrepancies between the Ames mutagenicity test and LC/MS analyses were found. This may be indicative of the presence of other mutagens, including isomers of the target analytes, and suggests the need to perform Ames tests as well as quantitative assays if potential health risks are to be assessed properly. The second paper utilises APCI LC/MS to determine 14 heterocyclic aromatic amines in a single analysis.[167]

LC/MS/MS methods for determining heterocyclic aromatic amines in commercially available meat products and fish have revealed the widespread distribution of PhIP and MeIQx.[168] The method, developed to assay for ten different aromatic amines, was based on selected reaction monitoring.

Novel Applications

Accelerator mass spectrometry has been used to determine [^{14}C]PhIP in animal models.[169] This metabolic study indicated that PhIP and its metabolites were present in the milk of lactating rats, and that other dietary components could affect the dosimetry of PhIP in breast-feeding offspring.

Electrospray mass spectrometry has also been used to aid the characterisation of DNA adducts of heterocyclic aromatic amines.[170,171] The major adducts were found to bind to the 8-position of guanine.

Recommended Mass Spectrometric Methods for Determining Mutagenic Heterocyclic Amines

The negative ion GC/MS methods reported here provide sensitive and specific methods for determining heterocyclic amines in foods and body fluids.

However, LC/MS and LC/MS/MS methods based on electrospray or APCI have the advantage of reducing both sample preparation and the risks of artefact formation. These are now the methods of choice wherever the specificity and selectivity of mass spectrometry are preferred over conventional assay techniques such as HPLC/UV.

7 Mycotoxins

Mycotoxins are a large and structurally diverse group of compounds, related only in that they are (a) produced by fungi and (b) cause some toxic effects in humans or animals. Although over 300 mycotoxins, produced by some 350 species of fungi, have been identified, a much smaller number has been confirmed as occurring in foods or feeds. A useful introduction and perspective has been provided[172] and current methods of analysis have been reviewed[173] as have some recent advances.[174] Two useful compilations of mass spectra of mycotoxins are available.[175,176]

Aflatoxins

The naturally occurring aflatoxins B_1, B_2, G_1 and G_2 are produced by the food spoilage fungi *Aspergillus flavus* and *A. parasiticus*, and can occur in grains, peanuts and some other commodities. Aflatoxins M_1 and M_2 are hydroxylated metabolites of B_1 and B_2 and have been found in animal tissues, especially in cow's milk from animals fed on contaminated feed. Structures of the aflatoxins are given in Figure 4.13.

Analysis for aflatoxins in foods is normally based on TLC or HPLC separation with fluorescence detection (with post-column derivatisation in the case of HPLC), and immunoaffinity columns are available for cleanup. Although these methods can provide excellent selectivity, a number of different MS approaches have been explored for confirmation.

EI spectra of the aflatoxins are readily obtained by direct probe introduction and exhibit $M^{+\bullet}$ as base peak.[177,178] The use of EI with direct probe introduction of material extracted from TLC spots for confirmatory analyses has been reported.[177,179] Low-resolution full scan spectra required extensive clean up and introduction of 10–50 ng of isolated aflatoxin.[177] At an MS resolution of 5000, and using selected ion monitoring of $M^{+\bullet}$ and some fragment ions, much less highly purified extracts could be analysed and less than 0.1 ng of aflatoxin B_1 was detectable.[179] The detection limit for aflatoxin M_1 and B_1 in complex mixtures was found to be 100-fold lower than the limit for highly purified samples under the same conditions. This was attributed to a reduction in surface bonding to the glass sample-introduction containers.

Negative ion chemical ionisation (resonance capture) of aflatoxins has been described,[180] and confirmation of residues of aflatoxin B_1 using this method was the subject of a collaborative study.[181] Aflatoxins gave relatively abundant ions corresponding to $M^{-\bullet}$, $[M - H]^-$, and $[M - CH_3]^-$. Relative intensities were temperature dependent, with higher temperatures favouring production of

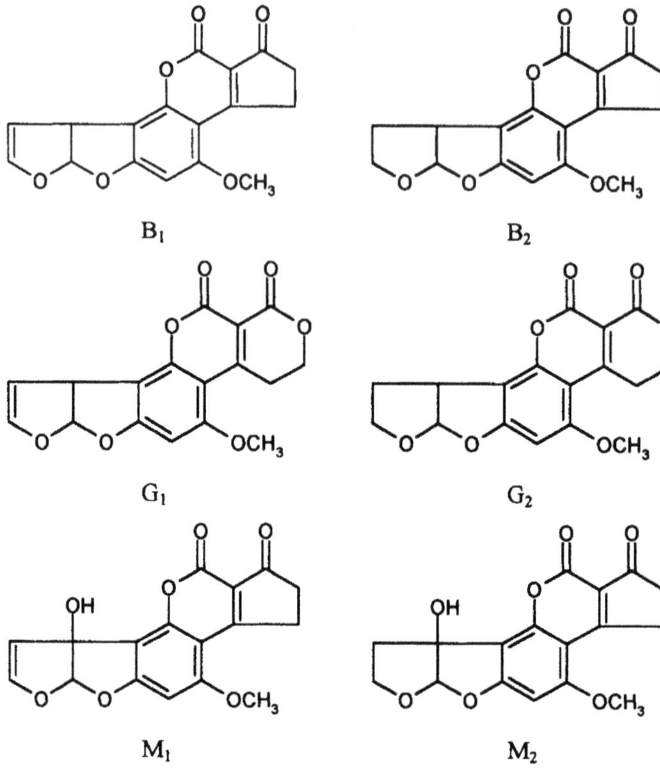

Figure 4.13   *The structures of the aflatoxins*

$[M - CH_3]^-$. The published data show a tendency for the contribution of $[M - H]^-$ to be greater in spectra of extracts than in those of pure standards, but the authors noted that the spectra produced on some instruments lacked this ion. Such differences presumably stem from differences in the concentration of Brønsted bases such as $[OH]^-$, which lead to ionisation by proton abstraction rather than by electron capture.

Subsequently, it was reported that the yield of $M^{-\cdot}$ was reported to be about 100 times greater that of $M^{+\cdot}$ in EI, or of $[M + H]^+$ in isobutane CI.[178] These authors also described the use of triple-quadrupole MS/MS to generate additional fragmentation of aflatoxins. Product ions generated by CID of $M^{+\cdot}$ formed by EI were quite similar to conventional EI spectra. Fragmentation of $M^{-\cdot}$ ions required higher pressures in the collision cell, but CID spectra could be obtained from picogram quantities of aflatoxins. Probe introduction of large amounts of crude extract was found to produce significant suppression of the aflatoxin response.

Gas chromatography of aflatoxin B_1 was reported in 1981.[182] Subsequently, a method based on the use of a short (6 m) methyl silicone coated fused silica column with on-column injection and NICI was developed for confirmation of aflatoxin B_1 in corn and peanut butter.[183] Under the conditions used the

$[M - H]^-$ peak had twice the intensity of $M^{-\cdot}$. An alternative GC/MS confirmation method employed EI and SIM at 3000 resolution.[184]

Aflatoxins can also be ionised by field desorption,[185] which yields spectra of B_1, B_2, G_1 and G_2 consisting almost entirely of $M^{+\cdot}$. FAB spectra of the B and G aflatoxins have also been obtained.[186] Using glycerol as matrix the aflatoxins produced principally $[M + H]^+$ ions, and the addition of alkali metal chlorides to the matrix resulted in the appearance of $[M + \text{metal}]^+$ ions in addition to $[M + H]^+$. FAB was used by these workers to obtain relative molecular masses of the reaction products of some aflatoxins with sodium bisulfite. An alternative matrix, glycerine/3-nitrobenzyl alcohol, also yielded good FAB spectra of aflatoxins, and MS/MS with high-energy collisions gave many daughter ions of high intensity.[187] FAB spectra have also been obtained by mixing silica gel removed from TLC plates directly with glycerol matrix; however, this method yielded spectra in which there was extensive fragmentation with $[M + H]^+$ peaks weak or absent.[188]

Several combined HPLC/MS techniques are applicable to aflatoxins. Packed capillary HPLC columns with a particle beam interface, modified for low-flow operation, yielded detection limits for pure standards in the range 1.5–3 ng under EI SIM conditions.[189]

The use of TSP LC/MS was described in 1991.[190] Spectra of aflatoxin B_1 and B_2 consisted very largely of $[M + H]^+$ while those of aflatoxin G_1 and G_2 also contained a prominent peak at $[M + H - 44]^+$. Use of 'filament-on' and discharge modes of thermospray operation provided a substantial increase in sensitivity without markedly influencing fragmentation. Full scan detection limits were between 0.5 and 2 ng and those for selected ion monitoring were between 0.04 and 0.1 ng. Thermospray has also been used to investigate the products of reaction between aflatoxins and iodine.[191]

More recently, the use of electrospray ionisation for the determination of aflatoxins in dust and urine has been reported.[192] Reverse-phase chromatography was used with the 1 ml min^{-1} solvent flow from the 4.6 mm column connected to electrospray through a 5:1 splitter. Selected ion monitoring allowed detection of 2–8 pg. MS/MS using the most intense product ion from each aflatoxin for multiple reaction monitoring resulted in slightly higher detection limits for pure compounds. However, the method yielded greater selectivity, which eliminated some additional chromatographic peaks observed in selected ion monitoring of extracts.

Ochratoxin A

Ochratoxin A (Figure 4.14) is a naturally chlorinated mycotoxin produced by some *Aspergillus* and *Penicillium* species. It is amenable to direct probe introduction. The spectrum published in one of the compilations[176] has been shown to contain errors in mass measurement of the major fragments (the incorrect spectrum persists in the NIST library). According to the later paper fragmentation of the amide bond is favoured and $M^{+\cdot}$ is of very low intensity in

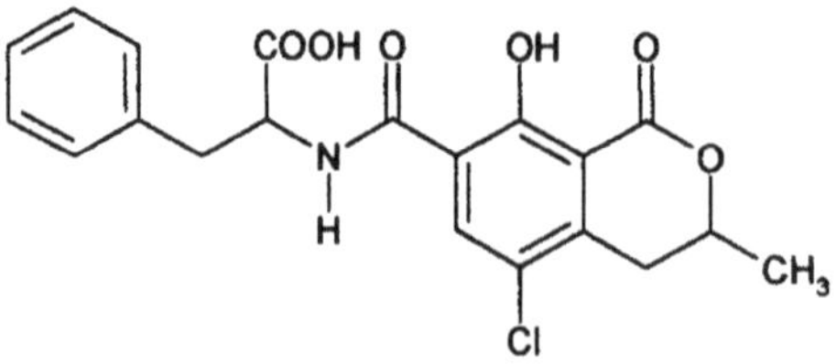

Figure 4.14 *The structure of ochratoxin A*

EI spectra. However, an EI spectrum with $M^{+\cdot}$ as base peak has been reported.[194]

A direct liquid introduction LC/MS for measurement of ochratoxin A has been described.[195] Positive ion CI spectra were dominated by $[M + H]^+$ and negative ion spectra by M^-; negative ionisation gave about 40-fold greater sensitivity.

Methylation of both the COOH and OH groups can be achieved by treatment with diazomethane to form a derivative amenable to gas chromatography. This derivatisation forms the basis of a sensitive confirmatory method using negative ion CI and selected ion monitoring.[196]

A TSP LC/MS method for the determination of ochratoxin A has been developed.[197] Under these conditions, the molecule yielded a simple spectrum containing the protonated molecule. More recently, use of electrospray LC/MS has been reported.[198] In positive ion mode, with trifluoroacetic acid in the mobile phase, the spectrum was dominated by $[M + H]^+$ with virtually no fragmentation, and a weaker $[M + Na]^+$ (about 10% of $[M + H]^+$) was also observed. Ochratoxin A was also found to be amenable to negative ion electrospray. CID of $[M + H]^+$ gave a number of product ions including an ion corresponding to loss the phenylalanine. The detection limit, using SRM in positive ion mode, was 20 pg injected on column.

Sterigmatocystin

Sterigmatocystin (Figure 4.15) can be produced by a number of *Aspergillus* and *Penicillium* species and is structurally related to aflatoxin B_1. Many reports of analytical methods based on TLC or HPLC are available, but only a small

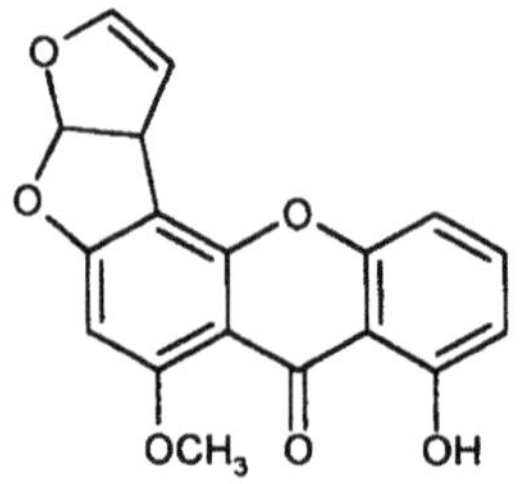

Figure 4.15 *The structure of sterigmatocystin*

number involve mass spectrometry. As with the aflatoxins, EI spectra can be obtained by direct probe introduction, giving $M^{+\bullet}$ as base peak, and molecular anions are observed under resonance electron capture conditions. A packed column GC method requires extensive deactivation of the system and elimination of metal surfaces.[199] FAB with tandem MS has been used for the identification of sterigmatocystin and related compounds.[187]

An LC/MS method for determination of sterigmatocystin in maize, bread and cheese has been described.[200] When a very weakly acidic mobile phase was used (10 mM ammonium acetate), positive ion APCI was more sensitive than negative ion APCI or electrospray. Using pure standards 0.05 ng was detectable by selected ion monitoring and the response was linear over a 100-fold concentration range. Figure 4.16 shows the positive ESI and positive and negative ion APCI spectra of sterigmatocystin under different conditions.

Simple Trichothecenes

Although trichothecin itself is produced by *Trichothecium roseum*, most of the simple non-macrocyclic trichothecenes found in foods are products of various species of the *Fusarium* genus. The trichothecenes (Figure 4.17) are all based on the tricyclic 12,13-epoxytrichothec-9-ene structure. They are classified into one of two groups, depending on the presence of a ketone function at C-8 (Group B), or of any other functional group at this position (Group A). Although a large number of such compounds have been produced in culture, only a few are of real significance as food contaminants; the latter include deoxynivalenol (DON), nivalenol (NIV) and some acetylated derivatives, and T-2 toxin. Instrumental methods for the determination of the simple trichothecenes have recently been reviewed.[201]

All of the simple trichothecenes are amenable to probe introduction and EI ionisation, and spectra of many of the underivatised compounds are available in several compilations.[175,176,202] Figure 4.18 shows the EI mass spectrum of deoxynivalenol. The relative intensity of $M^{+\bullet}$ is about 20% for DON, an 8-keto-trichothecene with no esterified OH groups, but is much less or negligible for other important trichothecenes. The spectra exhibit considerable fragmentation and although they lack obvious common fragments which would permit easy identification of the ring system, the spectra are individually highly characteristic. The prominent $[M - 48]^{+\bullet}$ rearrangement ion from DON, corresponding to loss of water and formaldehyde and apparently involving the OH moieties at C-7 and C-15, allows DON and its thermal rearrangement product, iso-DON, to be differentiated.[203]

Isobutane CI of DON gives a spectrum in which $[M + H]^+$ is base peak, but considerable fragmentation still occurs.[204] It has been reported that the $[M + H]^+$ ion accounts for 16.1, 14.7 and 10.1% of the TIC for T-2, HT-2 and DAS with isobutane as reagent gas, and 1.4, 2.8 and 1.4% with methane.[205] The same authors reported the ammonia CI spectra of several trichothecenes together with a detailed study of optimisation of the conditions. All of the main trichothecenes formed very abundant ammonium adduct ions with very

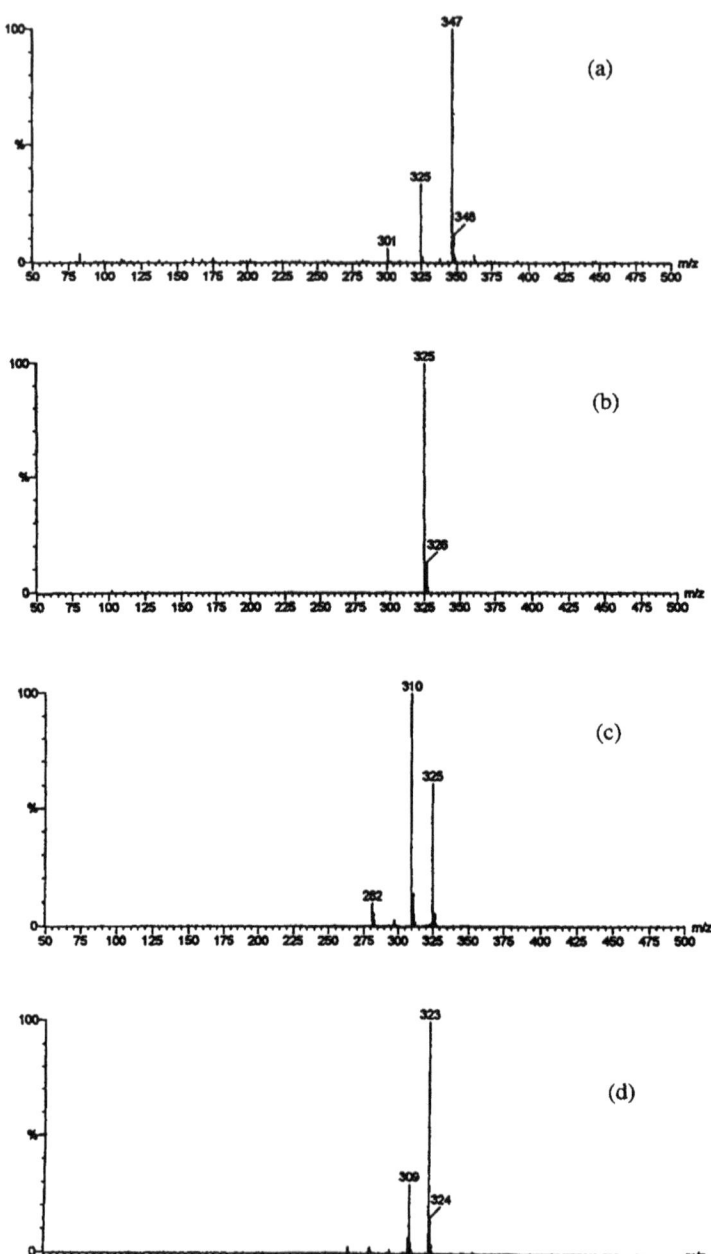

Figure 4.16 (a) *positive ion ESI spectrum of sterigmatocystin (cone voltage 45 V);*
(b) *positive ion APCI spectrum of sterigmatocystin (cone voltage 20 V);*
(c) *positive ion APCI spectrum of sterigmatocystin (cone voltage 45 V);*
(d) *negative ion APCI spectrum of sterigmatocystin (cone voltage 30 V)*
(Reproduced from K.A. Scudamore, M.T. Hetmanski, P.A. Clarke, K.A.
Barnes and J.R. Startin, *Food Addit. Contam.*, 1996, **13**, 343, with permission of the authors and publishers)

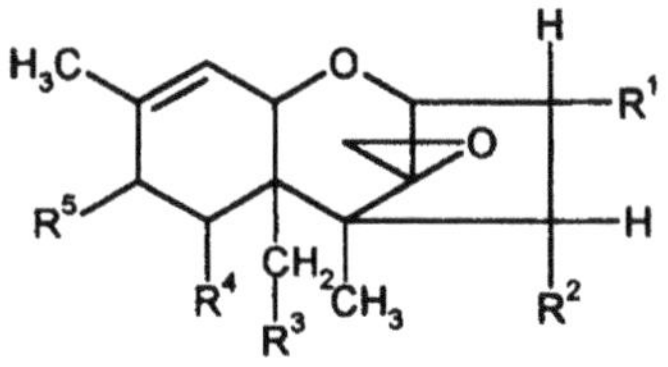

Type A

Toxin	R^1	R^2	R^3	R^4	R^5
Neosolaniol	OH	OAc	OAc	H	OH
HT-2 toxin	OH	OH	OAc	H	$OCOCH_2CH(CH_3)_2$
T-2 toxin	OH	OAc	OAc	H	$OCOCH_2CH(CH_3)_2$
Diacetoxyscirpenol	OH	OAc	OAc	H	H

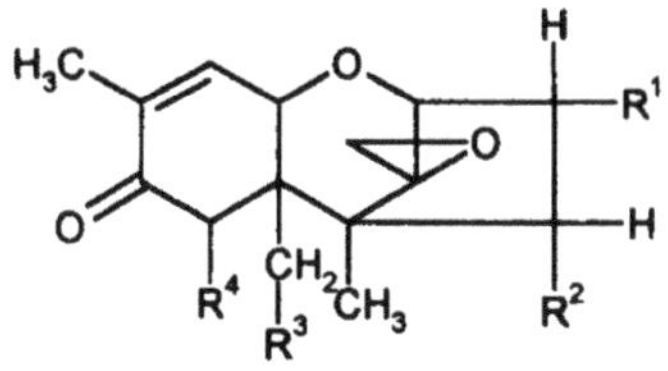

Type B

Toxin	R^1	R^2	R^3	R^4
Trichothecin	H	$O_2CCH:CHCH_3$	H	CH_3
Nivalenol	OH	OH	OH	OH
Deoxynivalenol	OH	H	OH	OH
Fusarenon-X	OH	OAc	OH	OH

Figure 4.17 *The structures of the trichothecenes*

little fragmentation. The intensity of $[M + H]^+$ was less than 10% of that of $[M + NH_4]^+$ for compounds which included at least two acyloxy groups and no more than two hydroxy groups, but over 30% with the more highly hydroxylated compounds. CI of trichothecenes with the unusual reagent gas dimethyl ether, which forms the methoxymethylene cation, has been described.[206]

Underivatised DON can also be ionised by resonance electron capture. It has been reported that $[M - 48]^-$ was the base peak and that this gave a signal around 100 times greater than the protonated molecule under positive ion CI conditions.[204] The molecular anion gave only a small signal. In the negative ion spectrum, fragments at lower mass are of considerably greater relative intensity, and some fragments appear at different m/z values.[207] Further differences in behaviour are evident in a published partial spectrum in which $[M - 1]^-$ and $[M - 2]^-$ are far more prominent.[208] These data provide a further example of the rather variable nature of resonance electron capture spectra.

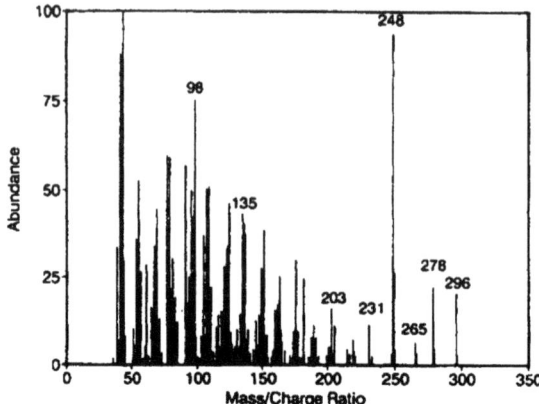

Figure 4.18 *The EI mass spectrum of deoxynivalenol*
(Reproduced from E.W. Sydenham, P.G. Thiel and R. Vleggaar, *J. AOAC Internat.*, 1996, **79**, 1365, © (1996), by AOAC International)

True chemical ionisation in negative mode has also been examined for several trichothecenes. The oxygen anion (generated in a Townsend discharge source) has been used as reagent,[209] while two publications have described ionisation by OH$^-$ (generated using CH_4 and N_2O in a conventional CI source).[210,211] The resulting proton abstraction spectra showed considerable fragmentation. It has also been shown that chloride attachment CI conditions (generated using difluorodichloromethane and methane) produced abundant $[M + Cl]^-$ from both DON and T-2 toxin, with little fragmentation.[211]

Several workers have exploited CAD and tandem MS to impart additional selectivity. The product ion spectra of the protonated molecule of DAS and of its TMS derivative have been described.[212] The fragments observed were similar to those present in the CI spectra. Product ion spectra of the ammonium ion adducts of a number of trichothecenes have also been published.[205] Comparable spectra for some of these compounds were reported in 1986 and exhibited rather fewer fragments, with smaller relative intensities.[213] Collisional association reactions between protonated molecules and ammonia have also been described.[214]

The trichothecenes are also amenable to FAB, which produces quite simple spectra with little fragmentation apart from loss of water or acetic acid from the protonated molecules,[215] as reported in a mass spectrometric method devised to screen fungal extracts. This approach is unlikely to be useful for analysis of foodstuffs however.

GC/MS has been widely adopted for determination and confirmation of trichothecenes. Fused silica capillary columns, in conjunction with on-column injection, can be used for underivatised DON[216,207,217] and a number of other trichothecenes.[209,217] However, some trichothecenes, including NIV and T-2 tetraol were found to yield poor results.[209] Consequently the trichothecenes have usually been derivatised before GC/MS, usually either by trimethyl-silylation or perfluoroacylation.

Trimethylsilyl (TMS) derivatives of Group A trichothecenes can be readily formed by treatment with volatile reagents such as *N,O*-bis-(trimethylsilyl)-acetamide (BSA) or trifluoroacetamide (BSTFA), as employed in many reported determinative methods such as that of Rosen and Rosen.[218] However, Group B trichothecenes, including the important DON, cannot be completely derivatised under such conditions and require the use of trimethyl-silylimidazole.[219, 220] Both the reagent and its by-product, imidazole, are difficult to remove by evaporation, cause rapid deterioration of capillary columns and can lead to poor reproducibility. Their presence can also lead to decomposition of the TMS ethers of other mycotoxins such as zearalenone in the GC column.[221] Despite these difficulties, GC/MS following treatment with trimethylsilylimidazole has been used successfully for surveillance.[222]

Perfluoroacyl derivatives, usually heptafluorobutyrates (HFB), although trifluoroacetates (TFA) and pentafluoropropionates (PFP) have also been used, can be prepared by treatment with either the acid anhydride or the corresponding imidazole following standard procedures.[223–226]

The EI and CI spectra of both the TMS and TFA derivatives of 14 different trichothecenes have been reported, with a detailed commentary and rationalisation of the main fragments.[227] CI spectra of the TFA derivatives of some trichothecenes have also recently been reported.[228] This publication describes a complete GC/MS analytical method with an ion-trap instrument. EI spectra of the TMS derivatives generally have weak to moderate molecular ion intensities, and considerable fragmentation occurs in both EI and CI. Although this distribution of the total ion current over many species limits sensitivity, it does lead to spectra that are highly characteristic.

The TMS ether of DON is also amenable to resonance electron capture; the molecular anion is of low relative intensity and the spectrum is dominated by a fragment ion. This is reported to be m/z 298 in one publication[227] and m/z 297 in another.[208]

The EI spectra of perfluoroacyl derivatives of some trichothecenes also exhibit considerable fragmentation, with the molecular ion sometimes weak or absent,[227] although a strong $M^{+\cdot}$ ion is given by the TFA derivative of DON.[226] The latter derivative gave a detection limit of 10 pg by selected ion monitoring with a magnetic sector mass spectrometer. The EI spectra of the HFB derivatives of DON and NIV also show intense molecular ions.[225] The perfluoroacyl derivatives have the considerable advantage of promoting sensitivity under resonant electron capture conditions.[229] Although HFB derivatives tend to provide the best sensitivity, with polyhydroxylated trichothecenes the molecular weight of the derivative exceeds the upper mass limit of some mass spectrometers so that TFA or PFP derivatives may be more appropriate. The latter are preferred.[224]

The main CID fragments of the resonance electron capture anions of heptafluorobutyrate derivatives of some trichothecenes have been described.[223] These fragments represent rather small neutral losses and provide little structural information or specificity. In contrast the daughter ions produced by CAD of a fragment ion, representing partial loss of the isovaleryl side chain, in the EI spectrum of the TFA derivative of T-2 toxin were unique and gave unequivocal proof of identity.[230]

The merits of ammonia DCI-MS/MS, isobutane DCI with reactive collisions, and GC-NCI-MS/MS of HFB derivatives have been compared.[231] It was found that the last of these approaches gave the best sensitivity, linearity and reproducibility. Unfortunately, the more widely used method of GC/NCI-MS was not included in this comparison.

LC/MS methods for the simple trichothecenes have also received attention. Direct liquid introduction and negative ion CI have been used to confirm NIV and DON in wheat kernel and wheat bran.[208] Several groups have described the use of thermospray.[197,232,233] One group used 'pure' thermospray without filament or discharge, and obtained simple spectra consisting largely of $[M + NH_4]^+$. DON also gave $[M + H]^+$ in similar intensity to the ammonium adduct, and losses of CH_2O from both $[M + NH_4]^+$ and $[M + H]^+$.[232] With selected ion, monitoring 1–10 ng could be detected. A comparison has been made of low ammonium acetate concentrations combined with filament-on mode, and pure thermospray with 0.05 M ammonium acetate.[197] The former gave more fragmentation and the latter greater sensitivity with detection limits in the range 0.1–1 ng. In another study, it was found that a number of trichothecenes gave thermospray spectra consisting almost exclusively of $[M + NH_4]^+$, with $[M + H]^+$ also evident for DON.[233] In the plasmaspray (discharge) mode ammonium adduct ions still dominated the spectra but some fragment ions were also produced.

An investigation of the use of HPLC with frit-FAB and high resolution mass spectrometry has been reported.[234] Selected ion monitoring of the protonated molecules at a resolution of 8000 allowed detection of concentrations of the order of a few hundred pg μl^{-1}, despite the introduction of only 0.5% of the HPLC flow to the mass spectrometer.

More recently, the potential of electrospray LC/MS has been demonstrated. The detection of DON, diacetoxyscirpenol (DAS) and T-2 toxin following supercritical fluid extraction of corn meal and rolled oats has been reported.[235] A mobile phase containing 50% of 3 mM ammonium acetate was used and the entire 0.2 ml min^{-1} flow was directed into the ion spray interface, without splitting. The presence of ammonium acetate resulted in the production of $[M + NH_4]^+$ ions. MS/MS with selected reaction monitoring allowed detection of the mycotoxins in extracts at concentrations corresponding to about 10 ng injected. An electrospray method using a dedicated LC/MS ion trap instrument has been developed for the concurrent determination of examples of both simple and macrocyclic trichothecenes (the latter are discussed below), although this application was directed at environmental rather than biotic sample materials.[236] Sodium acetate (0.02 mM) was added to the mobile phase in addition to ammonium acetate, leading to positive ion spectra dominated by sodiated species (which were present in the spectra even without addition of sodium salts). Under some conditions sodium-bound dimers $[2M + Na]^+$ were more abundant than $[M + Na]^+$, and $[3M + Na]^+$ ions were also observed. In-source CID was found to reduce the abundance of these ions, but not eliminate them. Detection limits for the simple trichothecenes varied considerably, from 1 pg for T-2 toxin to 1 ng for verrucarol and T2-tetraol.

The separation and identification of aflatoxins and trichothecenes in submicrogram quantity by thin-layer chromatography/fast atom bombardment (TLC/FAB) mass spectrometry has also been reported.[188]

A number of reports of applications of direct SFI/MS[237,213] and SFC/MS[238,239] to trichothecenes and some other mycotoxins have appeared. These approaches do not appear to offer any compelling advantages for the analysis of foodstuffs.

Fumonisins

The fumonisins, which have the structures shown in Figure 4.19, are a group of mycotoxins produced by *Fusarium moniliforme*, and, at least in culture, some other *Fusarium* species. The properties, occurrence, biological effects and analysis of fumonisins have been reviewed.[240] Interest has centred especially on fumonisin B_1 (FB_1), the predominant fumonisin found in contaminated corn. This causes a variety of toxic responses in animals, including equine leucoenphalomalacia and porcine pulmonary oedema, and has been associated with high incidences of human oesophageal cancer in South Africa and China. The fumonisins are both polar and thermally labile. The most widely used methods of analysis for the B fumonisins involve TLC, or HPLC with fluorescence detection after formation of a fluorescent derivative,[173,240] but considerable use has also been made of mass spectrometry. Chromatographic determination of the fumonisins has recently been reviewed.[241]

Direct probe introduction of fumonisin B_1 under either EI or isobutane CI

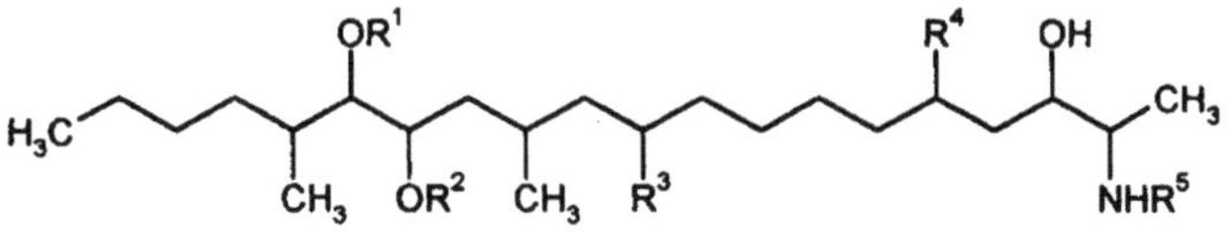

Toxin	R^3	R^4	R^5
B_1	OH	OH	H
B_2	H	OH	H
B_3	OH	H	H
B_4	H	H	H
A_1	OH	OH	CH_3CO
A_2	H	OH	CH_3CO

R^1 and R^2 are

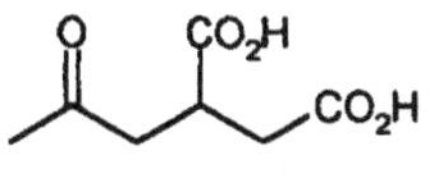

Figure 4.19 *The structures of the fumonisins*

conditions gave only decomposition or rearrangement products.[242] Following treatment with diazomethane, however, a CI spectrum with strong peaks corresponding to [M + H]$^+$ ions of the Me$_4$ to Me$_7$ derivatives was obtained. EI spectra of methylated derivatives of fumonisins B$_1$, B$_2$ and B$_3$ were also recorded using a particle beam interface, and the fragments observed were rationalised. There are no reported studies of other derivatives, or of attempts to apply GC to intact derivatised fumonisins.

Use has been made – especially for characterisation and confirmation – of methods based on hydrolysis of the fumonisins followed by derivatisation and GC/MS of the tricarballylic acid side-chains,[243] or the aminopolyol backbone.[244,245] The latter approach has also been used for quantification, where the precision of measurement has been improved by using a deuterated internal standard.[246] This method also used electron capture negative ionisation of the TFA derivative to monitor the molecular anion, since EI yielded a strong fragmentation in which the deuterated moiety was lost. The full-scan detection limit was less than 1 pg injected.

Fast atom bombardment spectra have been reported by several workers.[243,246–249] All reported that [M + H]$^+$ was the base peak. A signal-to-noise ratio of 3/1 in full-scan analysis of 50 pg was reported for this ion,[247] although other data suggest a rather larger sample requirement.[246] The same two groups explored the use of FAB with MS/MS. Daughter ion spectra show loss of the tricarballylic side chains and sequential losses of water.

Using positive ion thermospray, it was found that the [M + H]$^+$ ion, although observable, was less than 10% of the base peak, and that sensitivity was poor.[247] With negative ion thermospray, no ionisation was observed when filament, discharge and fragmenter electrodes were turned off.[250] However, a spectrum was obtained under discharge conditions, although this consisted of peaks throughout the range m/z 650–841, differing in masses by 14 and 16 Da. The authors suggest that this may be explained by formation of a double acetate adduct of the fumonisin, followed by fragmentation induced by the fragmenter electrode. Despite this unusual behaviour, by using flow injection analysis and selected ion monitoring of five of the ions produced, fumonisins were detectable in corn and muscle matrices. Detection limits, without any apparent interferences, were down to 2 ng injected and responses were linear to 200 ng.

Electrospray of fumonisins has been reported by several groups[247,251–254] as has the combination of ESI and capillary electrophoresis.[255] In positive ion mode, the spectrum is dominated by [M + H]$^+$, frequently accompanied by a sodium ion adduct and sometimes a potassium ion adduct. It was also found that in negative ion mode the deprotonated molecule was observed, but with intensity 3-fold to 10-fold lower than for positive ions.[251,252] For positive ions, the on-column detection limits reported for selected ion monitoring were 10, 2 and 5 ng for FB$_1$, FB$_2$ and FB$_3$ respectively.[251]

A detailed study of the fragmentation produced by CID from the protonated FB$_1$ molecule has been described.[254] HPLC separation, electrospray, CID and selected reaction monitoring have been combined to create a highly selective quantitative procedure that also makes use of deuterated FB$_1$ as an internal standard and that allowed quantification down to 400 pg injected.[253] Although the use of MS/MS allows greater confidence in the specificity of the determination, there was little evidence of interferences in normal SIM and the detection limits for MS and MS/MS were similar.

Zearalenone

Zearalenone, which has the structure shown in Figure 4.20, is a non-steroidal estrogenic compound produced by several species of *Fusarium*, including *F. graminearum* and *F. culmorum*, which are widely distributed throughout the world. Zearalenone has frequently been monitored in multiresidue methods together with the *Fusarium*-produced trichothecenes.

Zearalenone is quite stable to heat and is easily amenable to probe introduction and EI, giving a spectrum with a reasonably abundant molecular ion and characteristic fragmentation.[243] Chemical ionisation with isobutane gives $[M + H]^+$ with very little fragmentation, but CAD can provide a reasonably characteristic pattern of fragments.[204] In this example direct introduction, MS/MS was used for rapid screening of extracts of cereals.

GC of this dihydroxy compound usually requires derivatisation and both trimethylsilylation and perfluoroacylation have been used. The di-TMS derivative can be readily made by treatment with BSTFA (see reference 221 and references cited below). The relative intensity of $M^{+\cdot}$ in the EI spectrum is rather low, the bulk of the total ion current being carried by Me_3Si^+. However, this still yielded adequate detection limits by SIM[218,256,257] or by full-scan analysis with an ion trap instrument.[258] The spectra of TFA[226] and HFB[225] derivatives yield somewhat greater relative intensities of the $M^{+\cdot}$ ion. It has been found that the TFA derivative was insufficiently stable to allow reliable quantification,[226] but this does not seem to be the case for the HFB derivative.[225]

The thermospray spectrum of zearalenone yields $[M + H]^+$ as base peak with a less abundant $[M + NH_4]^+$ ion, and a small peak (12%) attributed to dehydration.[232] The use of thermospray for the determination of zearalenone together with some trichothecenes and other mycotoxins has also been reported.[197]

In a paper reporting the use of HPLC/FABMS using a 320 μm ID packed HPLC column, details of the FAB mass spectra of a number of zearalenone

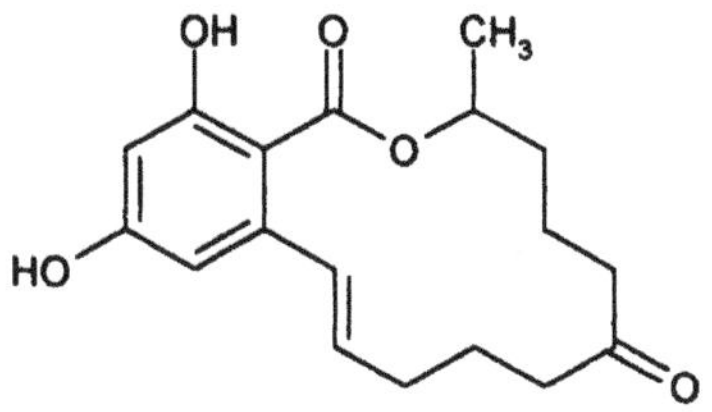

Figure 4.20 *The structure of zearalenone*

derivatives were also given.[259] Most of the fragment ions observed in EI spectra are also present in FAB spectra.

A method for the determination of zearalenone in maize using HPLC/APCI-MS has also been reported.[260] Positive and negative ionisation modes gave comparable sensitivity, but the latter was found to give more baseline noise and worse precision. Selected ion monitoring of $[M + H]^+$ ions allowed detection of injections of about 50 pg. The authors discuss in some detail optimisation of the APCI conditions.

Macrocyclic Trichothecenes

Macrocyclic trichothecenes are based on the characteristic tetracyclic 12,13-trichothec-9-ene ring structure, to which is added various ester bridges. These compounds are often both polar and labile. Examples are shown in Figure 4.21. Fungi such as *Myrothecium verrucaria* and *Stachybotrys atra* produce verrucarins, satratoxins and roridins. Plants of the genus *Baccharis* also produce certain roridins and other macrocyclic trichothecenes, including the baccharins.

EI spectra of the macrocyclic trichothecenes can be produced by direct probe introduction, but molecular ion intensities are low.[176] Some of the macrocyclics have been analysed by capillary GC/MS following trimethylsilylation of the free hydroxyl groups.[261] The high molecular weights and heat sensitivities necessitated cold on-column injection and very short columns. All of the TMS derivatives gave m/z 73 as base peak, with small molecular ion intensities.

Under ammonia CI conditions positive ion spectra mainly contain $[M + NH_4]^{+}$ [238] that are generated with similar efficiency to molecular anions by electron capture.[262] With other common reagent gases, negative ions are formed in greater abundance than positively charged protonated molecules.[262] When dimethyl ether was used as CI reagent gas, high intensities of $[M + H]^+$ and $[M + 45]^+$ were reported for verrucarin A, but fragment ions dominated for roridin A.[206]

Fragmentation induced by CID of a number of macrocyclic trichothecenes following ammonia CI[262] or negative ionisation[263,264] has been characterised, and the use of MS/MS with DCI introduction for screening and quantification of extracts described.

Analyses using thermospray ionisation have also been reported.[265] A number of roridins and baccharinoids together with 8-ketoverrucarin A were separated using a C_8 reversed-phase column with a stepwise change in the mobile phase

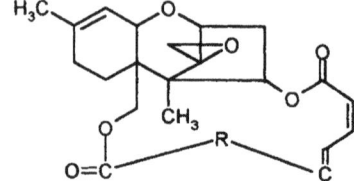

Figure 4.21 *Example structures of macrocyclic trichothecenes*

composition. The $[M + NH_4]^+$ was the most abundant ion in the spectrum for all the compounds studied. MS/MS with 'loop injections' (no HPLC column) and thermospray ionisation for detection and quantification of a variety of roridins, baccharinoids and verrucarins has been described.[266] Apart from loss of 17 Da, attributed to loss of NH_3 to form $[M + H]^+$, most of the ions produced by CID of $[M + NH_4]^+$ were formed by bond cleavage at the exocyclic ester bridges. Similar neutral losses were observed when $[M + H]^+$ ions were selected for CID.

A method for the determination of macrocyclic trichothecenes, in which they are hydrolysed to the corresponding verrucarol and analysed by NICI GC/MS after formation of the HFB derivative, has also been described (as discussed above for simple trichothecenes).[267]

Alternaria Toxins

Alternaria species, which include both plant pathogens and storage spoilage organisms, occur widely in nature and can produce a number of toxins. These include alternariol (Figure 4.22) and its methyl ether, tenuazonic acid (Figure 4.23), altenuene (Figure 4.24), and the AAL toxins (Figure 4.25), produced by *A. alternata*, which are closely related to the fumonisins.

Alternariol, a fused ring polycyclic compound, gives an EI spectrum with $M^{+\bullet}$ as base peak whereas the less aromatic altenuene gives a moderate molecular ion intensity and more fragmentation.[176] Tenuazonic acid is normally isolated as the Cu(II) salt, EI and CI spectra of which have been reported.[268] Positive and negative ion spectra have been reported and are dominated by $[M + H]^+$ and $[M - H]^-$ respectively.[269] Although this work indicated that tenuazonic acid is suitable for gas chromatography without derivatisation by using on-column injection and a short capillary column,

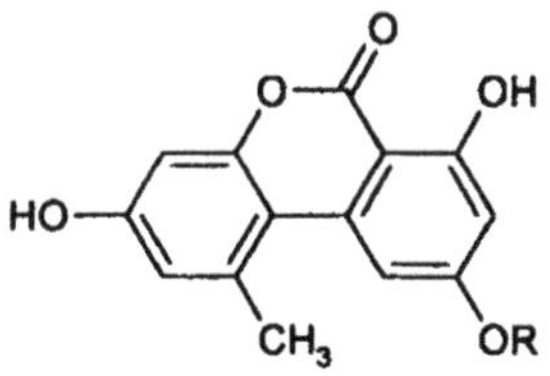

Figure 4.22 *The structure of alternariol (R = H) and its methyl ether (R = CH₃)*

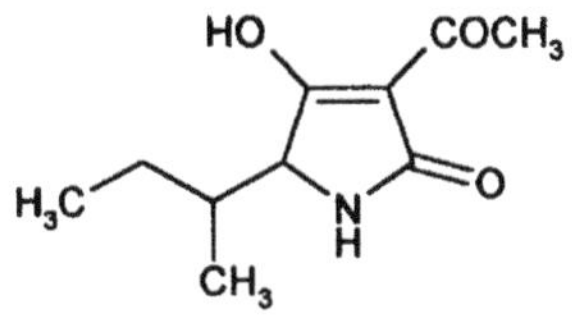

Figure 4.23 *The structure of tenuazonic acid*

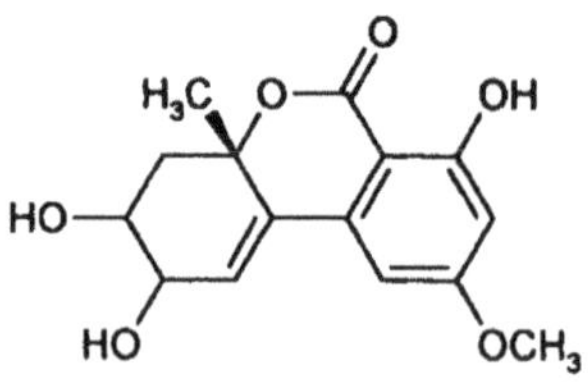

Figure 4.24   *The structure of altenuene*

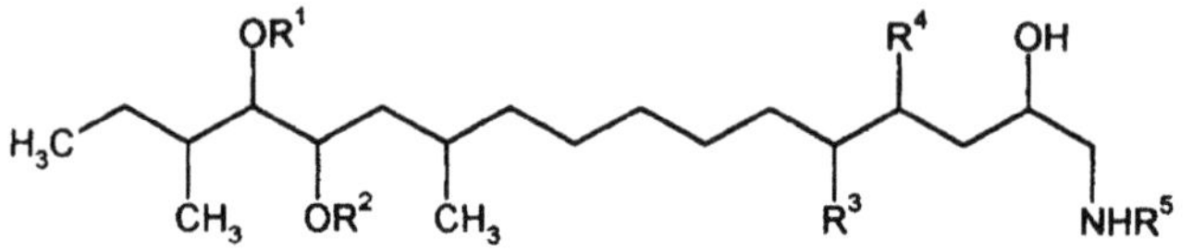

Toxin	R^1	R^2	R^3	R^4	R^5
TA	TCA	H	OH	OH	H
TB	TCA	H	H	OH	H
TC	TCA	H	H	H	H
TD	TCA	H	H	OH	COCH$_3$
TE	TCA	H	H	H	COCH$_3$

where TCA denotes tricarballylic acid

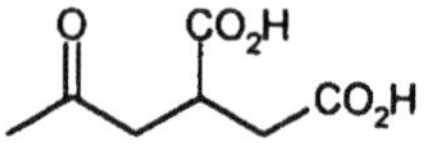

Figure 4.25   *The structures of the AAL toxins*

derivatisation is probably a better approach. This is exemplified by a GC/MS selected ion monitoring method for alternariol, alternariol monomethyl ether, altenuene, tenuazonic acid and altertoxin I using both TMS and HFB derivatives.[270]

Electrospray ionisation of AAL toxins has been reported and gave both $[M + H]^+$ and $[M + Na]^+$ in positive ion mode, and $[M - H]^-$ in negative mode.[252]

Other Mycotoxins

Patulin (Figure 4.26) is produced by several species of *Aspergillus* and *Penicillium*, and especially *P. expansum*, the principal cause of apple rot.

Determination of patulin in apple juice by GC/MS of the TMS derivative has been described and is both effective and straightforward.[271,272] Similar methodology has been described for the determination of both patulin and penicillic acid in unroasted cocoa beans (EI spectra were usefully given in full in this publication).[273] The HFB derivative is an alternative for GC/MS and both the

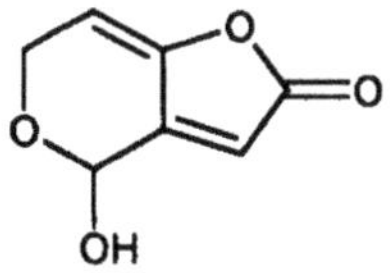

Figure 4.26 *The structure of patulin*

methodology and spectrum have been reported.[274] Detection by thermospray has also been reported.[197]

Cyclopiazonic acid (Figure 4.27) is also produced by several species of *Aspergillus* and *Penicillium*, and has been found as a contaminant in nuts, cereals and seeds, as well as in products from animals receiving contaminated feed.

With probe introduction an EI spectrum can be obtained with a reasonably intense molecular ion peak, and the use of positive ion CI and MS/MS have been described for confirmation of the presence of this toxin.[275,276] Another publication refers to GC/MS of underivatised cyclopiazonic acid.[277] There are no reports of GC/MS of derivatives, nor of HPLC/MS. Although electrospray would undoubtedly be effective, most reported HPLC analyses include zinc salts in the mobile phase,[278] and this would not be compatible with most current electrospray systems.

Moniliformin (Figure 4.28) is a *Fusarium* mycotoxin, which has been found to occur in maize. The reaction of this compound with methyl-*N*-(*tert*-butyl-dimethylsilyl)trifluoroacetamide generates an unexpected, but highly characteristic, fluorinated derivative.[279] GC/MS with selected ion monitoring allowed detection of 5 pg injected.

The EI, CI and FAB spectra of sambutoxin, produced by *Fusarium* species associated with dry rot of potatoes, have been reported.[280]

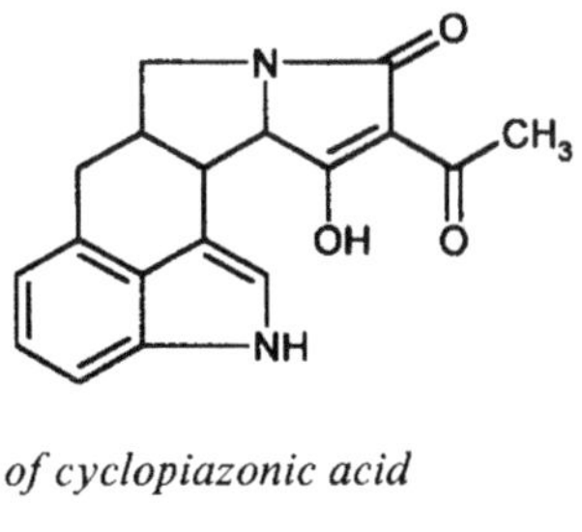

Figure 4.27 *The structure of cyclopiazonic acid*

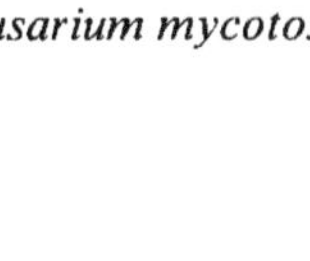

Figure 4.28 *The structure of the Fusarium mycotoxin moniliformin*

8 Phytoalexins

The phytoalexins are antimicrobial compounds with a wide range of structures (for example furanocoumarins, cinnamylphenols and phaseolin). They are produced by plants, in response to microbial infection. They exhibit toxicity across the biological spectrum, not just to microorganisms.[2] Many compounds already discussed in other subsections of this chapter can be classified as phytoalexins. Consequently, the mass spectrometric analysis of a few representative examples of this class of bioactive compounds will be discussed here only very briefly.

GC/MS is the favoured method of analysis of *cis*- and *trans*-resveratrol, (*cis*- and *trans*-3,5,4'-trihydroxystilbene), a phytoalexin produced by grapevines in response to fungal infection.[281–285] Interest in these compounds has been stimulated by claims for their protective effects against heart disease and their appearance in red and white wines. The method is rapid and sensitive (detection limits of 0.05 mg l[−1]). GC/MS has also been used as a sensitive and specific method for detecting small amounts of glyceollins (*e.g.* canescacarpin, Figure 4.29) in soybean seedlings.[286]

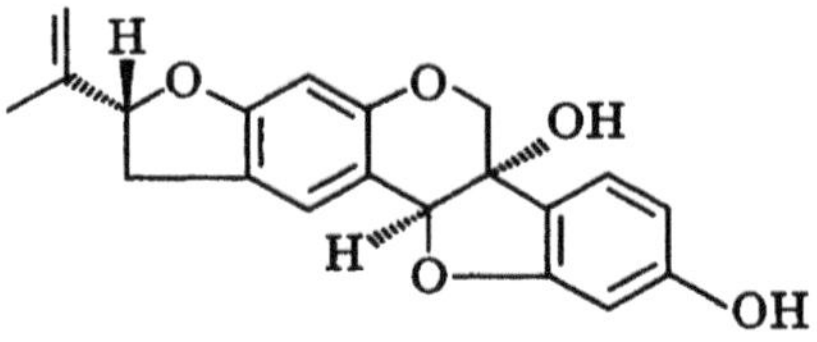

Canescacarpin

Figure 4.29 *The structure of canescacarpine*

Analysis of more complex molecules, including furanocoumarins and isoflavanoids, can be conducted using a variety of other mass spectrometric methods, including FABMS, LC/MS (particularly by electrospray) and tandem mass spectrometry. See the appropriate subsections in this chapter for a more comprehensive description.

9 Phenolic Compounds

Phenolic compounds comprise a wide range of naturally occurring substances present in the diet. Some, for example the flavonols, are believed to have beneficial properties whereas others may have both beneficial and adverse dietary effects (tannins for example), or may affect fertility or increase the risk of certain types of cancer (*e.g.* the phytoestrogens). Mass spectrometric analysis of a selection of different types of phenolics is discussed below.

The Flavonoids and Isoflavones

The flavonoids occur very widely in plants and are one of the largest groups of secondary metabolites. There was early concern regarding the mutagenicity and carcinogenicity of some flavonoids (summarised in 1987[287] and 1994[2]). However, there are now many more favourable than adverse health claims for these molecules and the focus of interest is currently on their role as protective factors.[288] This protection effect is thought to stem from their antioxidant activity,[289,290] although these molecules do have additional beneficial properties.[291] The flavonoids occur naturally as glycosides and it has been shown recently that quercetin glycosides are absorbed more efficiently than quercetin itself.[292] This emphasises the importance of accurate structural determination of food flavonoids, and mass spectrometry has shown considerable promise in this area. One of the classes of flavonoids of particular dietary importance is the flavonols. These compounds are present in foods as conjugates of kaempferol, quercetin, isorhamnetin and myricetin (Figure 4.30). The isoflavones are a class of phytooestrogens present in foods such as soya and, like the flavonoids, are found in conjugated forms of (mainly) genistein and daidzein (Figure 4.31).

The phytooestrogens, which also include the lignans, can be metabolised to biologically active forms following digestion. These compounds are of interest because of their oestrogenic activity that may have potential in the prevention of hormone related diseases.

Direct Probe Mass Spectrometry of Flavonoids

Early work, mostly EI, on the mass spectrometry of flavonoid aglycones and some glycosides has been reviewed.[293,294] These authors noted that flavonoid aglycones, in common with most aromatic compounds, yielded intense molecular ions and structurally significant fragment ions in their EI spectra. The fragment peaks can be used to establish the distribution of substituents between A and B rings. Underivatised flavonoid glycosides, on the other hand, rarely yielded molecular ions in EI; only permethylated, peracetylated or trimethylsilylated derivatives exhibited $M^{+\cdot}$ ions. These had relative abundance ranging from about 1% (flavonol derivatives) to 40–50% (3- or 7-O-glycosides). EI mass spectrometry of ten representative polymethoxyflavones has shown that the spectra are characteristic, especially for polmethoxyflavones containing the 5,7,8-trimethoxy moiety where diagnostic peaks are observed at m/z 167 and 139.[295] More recently a comprehensive account of the EI and methane CI mass spectra of a wide range of flavones, flavonols, isoflavones, flavanones and dihyroflavanols has been given.[296] The EI spectra exhibited characteristic A and B ring fragmentation that often allowed the substitution pattern and class of flavonoid to be established. The methane CI spectra were less structurally informative. Figure 4.32 shows the main A and B ring fragments in the EI spectra of flavonoids that, when used in conjunction with Table 4.1, can aid structural characterisation. Useful information about

Isorhamnetin

Kaempferol

Quercetin

Myricetin

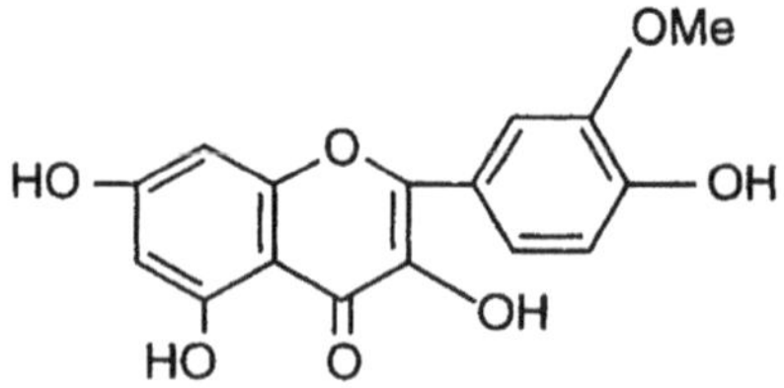

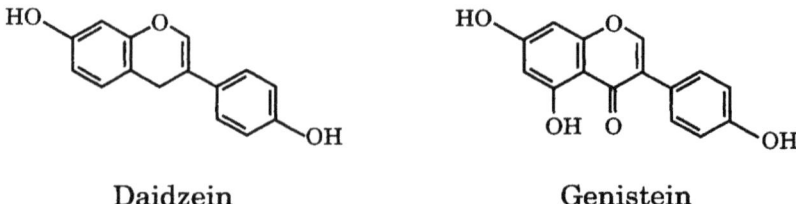

Figure 4.30 *The structures of representative food flavonoids*

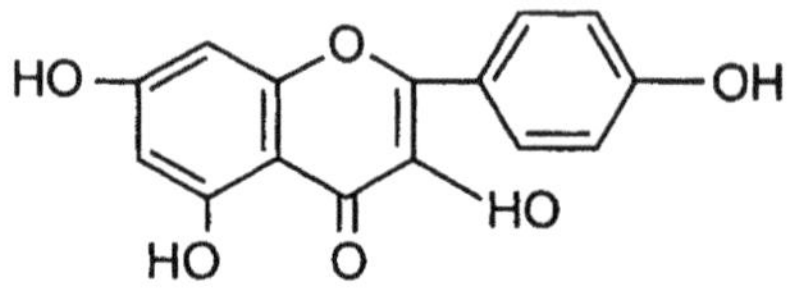

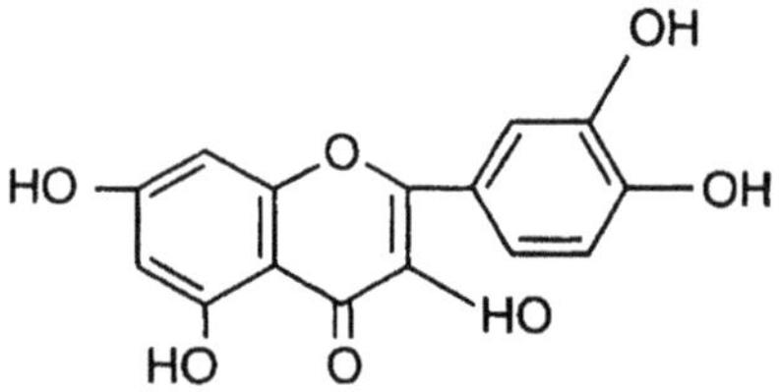

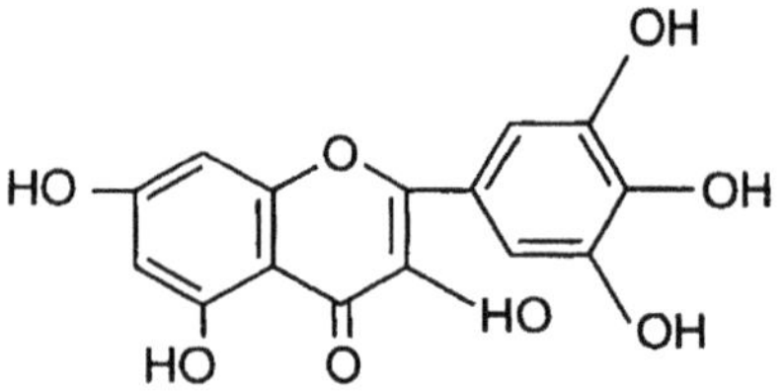

the mass spectra of flavonoids has also appeared in a number of review articles.[293–299]

Isobutane and hydrogen CI has also been used in the analysis of flavonoid aglycones.[294] Amine CI reagent gases, particularly ammonia, diethylamine and

Daidzein

Genistein

Figure 4.31 *The structures of representative food isoflavones*

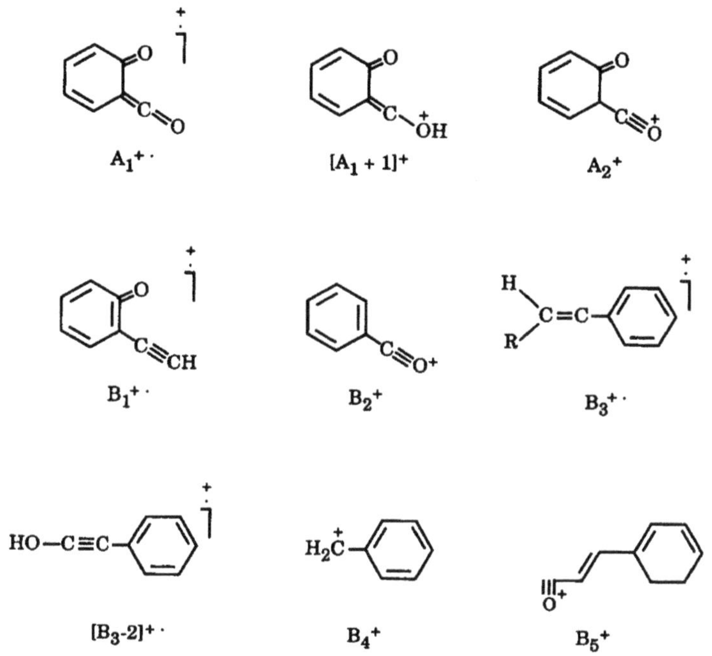

$A_1^{+\cdot}$ $[A_1 + 1]^+$ A_2^+

$B_1^{+\cdot}$ B_2^+ $B_3^{+\cdot}$

$[B_3-2]^{+\cdot}$ B_4^+ B_5^+

R = H, for flavanones and dihydrochalcones

R = OH, for dihydroflavonols

Figure 4.32 *The main A and B ring fragments in the electron ionisation mass spectra of flavonoids*
(Reproduced by permission from P.A. Hedin and V.A. Phillips, *J. Agric. Food Chem.*, 1992, **40**, 607, © 1992 American Chemical Society)

Table 4.1 *Characteristic A and B ring fragments of flavonoids*
(Reproduced from P.A. Hedin and V.A. Phillips, *J. Agric. Food Chem.*, 1992, **40**, 607, with permission of the authors and publishers (table of Hedin and Phillips modified from R.J. Grayer in *Methods in Plant Biochemistry, Vol. 1, Plant Phenolics*, ed. J.B. Harborne, Academic Press, San Diego, 1989, pp. 283–323)

Flavonoid class	$A_1/A_1 + 1$	A_2	B_1	B_2	B_3	$B_3 - 2$	B_4	B_5
Flavanones	+	−	(a)[a]	−	+	−	+	−
Dihydroflavonols	+	(+)	−	−	+	+	+	−
Dihydrochalcones	−	+	−	−	+	−	+	−
Flavones	+	−	+	(+)	−	−	−	−
Isoflavones	+	−	+	−	−	−	−	−
Flavonols	−	−	−	+	−	−	−	−
Chalcones[b]	(+)	+	(+)	−	−	−	−	+

[a] (+), sometimes present. [b] Chalcones give A_1^+ and B_1^+ on isomerisation to flavones.

triethylamine have been used to establish the molecular weights, masses of sugar residues and, occasionally, the location of the sugar residue in several naturally-occurring flavonoid glycosides.[300] Negative ion CI has also been investigated, but with limited success.[301]

Field Desorption Mass Spectrometry has provided molecular weight and elemental composition information on underivatised flavonoid glycosides.[302] However, this technique was superseded by FABMS for this class of compound. A number of quite detailed studies of flavonoid glycosides have been conducted[303-305] and reviewed.[306] The review is a general description of the mass spectrometry of underivatised, naturally occurring glycosides and includes discussion of EI, CI, DCI, FAB, FD, LDMS and LC/MS methods for analysing flavonoids and other plant glycosides.

FABMS of flavonoids has proved to be useful as an aid to molecular weight and structure determination. An early example showed the utility of high resolution accurate mass measurement in distinguishing caffeoyl residues from the isobaric glycosyl groups in the natural flavonoid pigments, anthocyanins.[307,308] A more recent FABMS study of anthocyanins describes the deduction of the chemical structures of two anthocyanins isolated from the purple sweet potato *Ipomoea batatas*.[309] A useful review of flavonoid anthocyanins summarises papers published up to the late 1980s that describe the mass spectrometric analysis of this class of compounds.[310]

GC/MS of Flavonoids

Few reports of the GC/MS analysis of flavonoid aglycones or flavonoid glycosides appeared until the 1990s, despite the widespread success of this technique in analysing complex mixtures of other natural products. Earlier work focused on GC/MS of high molecular weight, derivatised plant glycosides[311] and on TMS-ether derivatives of flavone and flavonol glycosides.[312] A suite of mass spectrometric methods, including GC/MS and particle beam LC/MS, has been used to study the mass spectrometric behaviour of methylated and silylated *Silibum mariarum* flavonoids.[313] It has now been shown that many flavones, flavonols, flavanones and chalcones can be analysed without derivatisation by GC/MS.[314-316] This method may be particularly useful for analysing complex mixtures of flavonoids in plant extracts if LC/MS methods are unavailable. Evidence for structure–retention time relationships was also presented and the combination of retention data and differences in fragment ion intensity suggested the possibility of distinguishing 6- and 8-methoxyflavones. GC/MS methods have been applied to the analysis of underivatised flavonoids in honey.[316] This method can be used to detect floral origin of honeys and may have potential in authentication studies.

Isotope dilution GC/MS techniques are now used to determine a variety of phenolic substances in urine,[317] plasma[318] and faeces.[319] These methods are designed for studying isoflavonoid (phytoestrogen) and lignan metabolism and have considerable potential in supplying firmly based scientific data on the influence of these molecules on health and disease. The metabolism of orally administered [2]H-labelled rutin, a medicinal flavonoid glycoside, has been studied in humans with the aid of GC/MS of trimethylsilylated derivatives.[320] Five urinary metabolites formed by reduction of the γ-pyrone ring and/or dehydroxylation of the B ring were identified. Metabolites of dietary isoflavones in soya are identifiable in human urine by EI GC/MS of trimethylsilyl derivatives.[321-323] Structural identification of new metabolites was enhanced by tandem mass spectrometry. Recent human studies of kaempferol and quercetin metabolism have been conducted following ingestion of *Ginkgo bioloba* tablets. These investigations have relied on negative ion GC/MS of trimethylsilyl derivatives to obtain metabolic data.[324,325]

GC/MS techniques, in general, are simple, but very powerful, methods for determining metabolic fate. The data recorded showed large individual variability in the preferred metabolic pathways of dietary isoflavones.

The metabolism of a ^{13}C-labelled synthetic flavone analogue, 5-methoxy-flavone, has been studied in animals by GC/MS/MS of plasma extracts after intravenous and peoral administration of the labelled substance.[326] The combination of GC and tandem MS methodology was sensitive (limit of quantification 1 ng ml^{-1}) and rapid (analysis time < 9 min per sample) and this technique has obvious potential in human studies of the metabolism of naturally occurring flavonoids.

LC/MS of Flavonoids

Moving belt EI and CI LC/MS techniques have been used in the analysis of flavonoid aglycones.[327] However, this early work was hampered by the lack of diagnostic fragmentation in the CI spectra and by low intensity molecular and fragment ions and high background contribution in the EI mass spectra. Belt and Direct Liquid Introduction LC/MS systems have also been used to analyse complex mixtures of flavonoid glycosides.[328,329] These techniques, which must now be regarded as obsolete, met with limited success.

More recently, TSP LC/MS has been used by several researchers to characterise both flavonoid aglycones and flavonoid glycosides. These studies have been summarised in a general review of TSP LC/MS in phytochemical analysis.[9] Flavonoid aglycones yielded intense positive ion TSP spectra, but structural information is limited by the lack of fragmentation. Many reported applications are devoted to phytochemical analysis of flavonoid glycosides. Nevertheless, the methodology is readily adaptable to food analysis. Most of the reported TSP spectra were recorded in positive ion mode, using solvent-mediated ionisation (typically 0.1 M ammonium acetate in acetonitrile/water or acetonitrile/methanol). They exhibited $[M + H]^+$ ions of low intensity and intense [aglycone + H]$^+$ ions. An example of this type of LC/MS analysis is shown in Figure 4.33.

The preference for positive ion TSP in the analysis of flavonoid glycosides is perhaps surprising. The acidic nature of the phenolic flavonoid moiety should, in theory, favour negative ion formation. Despite this, few reports of the negative ion TSP of these molecules have appeared. One exception is a

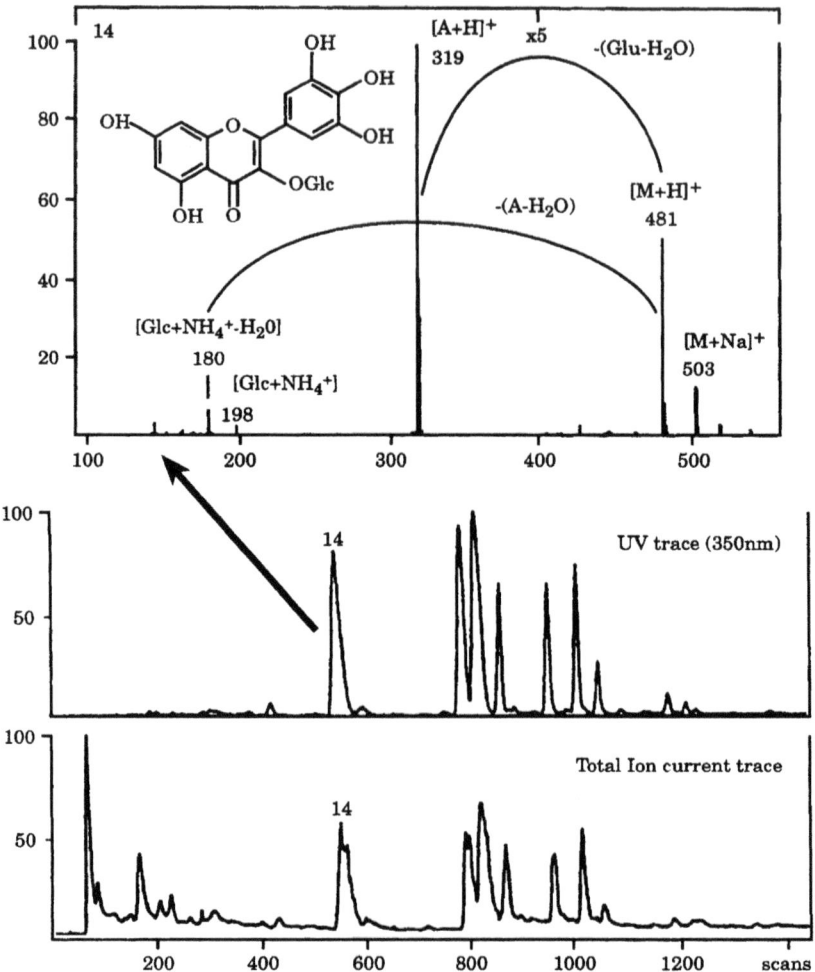

Figure 4.33 *Example thermospray LC/MS analysis of flavonoid glycosides*
(Reproduced from J.L. Wolfender, M. Maillard and K. Hostettmann,
Phytochem. Anal., 1994, **5**, 153, with permission of the authors and
publishers, © John Wiley & Sons Limited)

description of the positive and negative ion spectra of acetyl flavonoid
glucosides.[330] The negative ion TSP spectra exhibited $[M - H]^-$ and
$[aglycone - H]^-$ ions that were complementary to those seen in the positive
ion spectra. Mixtures of catechins, including catechin glycosides, extracted
from a tea plant have been analysed by positive ion TSP LC/MS/MS.[331]
Gallate esters were clearly identifiable by the loss of 168 Da from the
$[M + H]^+$ ions of catechin gallates. CID spectra yielded additional structural
information on catechins and on flavonoid glycosides, where ions due to losses
of glycosyl moieties from the $[M + H]^+$. TSP LC/MS was superior to

particle beam LC/MS because the latter did not provide molecular weight information for all the molecules studied. TSP LC/MS techniques have also been used to detect adulteration of medicinal plant extracts of flavonoid glycosides[332] and in chemotaxonomic studies of *Epilobium*, a willow herb used in folk medicine.[333]

Continuous flow secondary positive and negative ion LC/MS methods have been developed for the analysis of flavonoid glycosides in leguminous plant extracts.[334] Optimisation of instrumental parameters yielded a minimum detectable quantity of 1 ng for rutin by flow injection. The technique was used to identify flavonoid glycosides in alfalfa and chickpea and supported the controversial identification of formononetin-7-*O*-glucoside-6″-malonate as a major extract component.

More recently, the modern methods of LC/MS analysis, APCI and electro-spray have been applied to the detection and quantification of flavonoids and isoflavonoids. The first of these modern applications described the positive and negative ion APCI LC/MS analysis of isoflavones and their conjugates in soy foods.[335] Soybeans and defatted soy flour contained mostly isoflavone 6″-*O*-malonylglucoside conjugates of daidzein, genistein and glycitein. Subsequently another group of workers developed a quantitative positive ion APCI LC/MS method for quantifying daidzein and genistein in comminuted baby foods and soy flour.[336] Detection limits were 0.2 mg kg^{-1} for daidzein and 0.7 mg kg^{-1} for genistein. The method was very robust, yielding precise data over a period of ten days in which nearly 500 analyses were performed. Negative ion APCI methods have been successful in the identification and quantification of flavonols, flavones and flavanones in fruits, vegetables and beverages.[337]

A total of eleven flavonoids have subsequently been determined in red clover (a major forage crop) by positive ion electrospray LC/MS.[338] Similar techniques have been used to identify catechins in green tea extracts.[339]

LC/MS/MS methods are finding increasing favour in flavonoid and isoflavonoid analysis. Recent examples include the negative ion APCI LC/MS/MS analysis of flavonoids in extracts of *Passiflora incarnata* (a medicinal plant),[340] of rutin and other phenolic compounds in Maté tea,[341] and ESI LC/MS/MS as a general method for dereplication of flavonoid glycosides.[342] Metabolic studies have also been conducted by LC/MS/MS techniques *in vitro*[343] and *in vivo*.[344,345] The *in vitro* study used negative ion ESI LC/MS/MS to characterise plant constituents that inhibit xanthine oxidase, an enzyme that converts xanthine to uric acid, which can precipitate gout attacks. Among the compounds identified were catechin, epicatechin and quercetin 3-*O*-gluco-side. The first of the human studies successfully identified daidzein, dihydro-daidzein, *O*-desmethylangolensin and genistein in human plasma.[344] The second paper gives a useful and detailed description of negative ion ESI and APCI LC/MS/MS methods for identifying and quantifying isoflavones and their metabolites in human and animal plasma.[345] Product ion CID spectra of the isoflavones daidzein and genistein are shown in Figure 4.34. These data also indicate the benefits of high collision cell pressure: the CID spectra

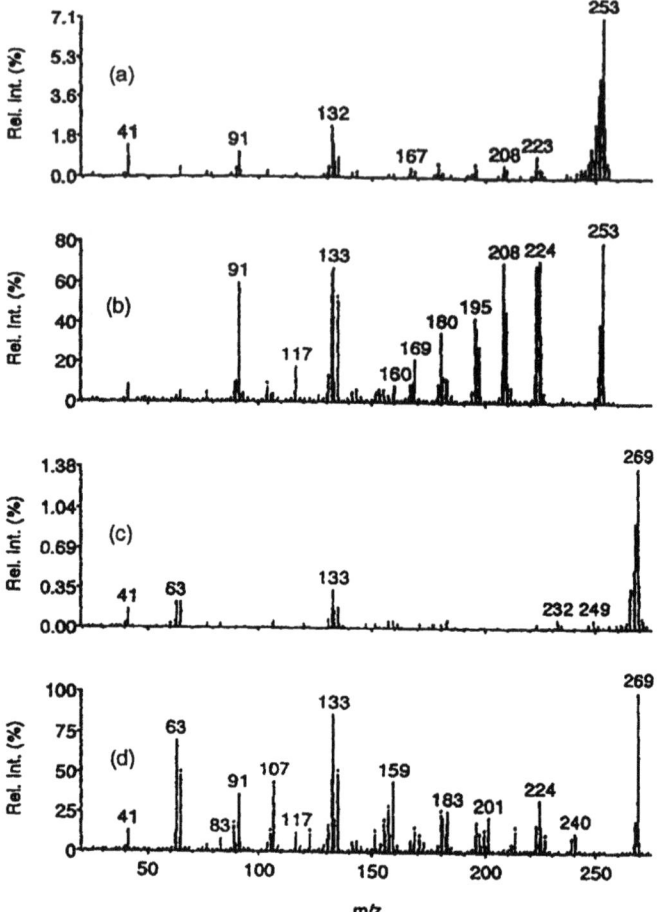

Figure 4.34   *Product ion CID spectra of the isoflavones (A and B) daidzein and (C and D) genistein. A and C were obtained using a low pressure collision cell and B and D obtained using a high pressure collision cell*
(Reproduced from S. Barnes, L. Coward, M. Kirk and J. Sfakianos, *Proc. Soc. Expt. Biol. Med.*, 1998, **217**, 254, with the permission of the authors and Blackwell Science Inc.)

obtained under these conditions yielded more diagnostic fragmentation and higher sensitivity.

Deuterium exchange and APCI LC/MS/MS has been used to determine the number of exchangeable protons in a number of isoflavones.[346]

Several studies of the anthocyanins have benefited from the application of electrospray mass spectrometry. These include anthocyanins of grapes[347] and of a variety of other plant tissues.[348,349] Additional structural information can be obtained by combining LC/MS with tandem mass spectrometry.[350]

CE/MS of Isoflavones

CE/ESI MS has recently been applied to the analysis of isoflavones and offers the prospect of yet higher sensitivity in SIM mode (sub-femtomolar detection limits).[351]

Tandem MS of Flavonoids

Besides the MS/MS methods used with GC/MS or LC/MS (discussed above), several tandem mass spectrometric studies of pure flavonoid glycosides have been conducted using FABMS as the ionisation technique. Studies have been conducted using MIKES[352] or B/E linked scanning.[353,354] However, true tandem mass spectrometry has also been employed on triple magnetic sector mass spectrometers,[355] on triple quadrupole instruments[356] and on magnetic-quadrupole hybrid low-energy instruments.[354] Low-energy MS/MS spectra can be used for determining carbohydrate sequence in O-diglycosyl flavonoids and for characterising the terminal monosaccharide in O,C-diglycosyl flavonoids. High-energy spectra, generated using multiple magnetic sector instruments or B/E linked scanning, allow differentiation of isomeric di-6,8-C-glycosides.

Continuous infusion of tea extracts into the ion source of a triple quadrupole mass spectrometer operated in negative ion APCI mode has been used to analyse catechin and related polyphenols.[357] Product ion spectra of the deprotonated molecules of epigallocatechin-3-gallate and epicatechin-3-gallate are shown in Figure 4.35, together with a rationale for the main ions seen. This work demonstrates that it is possible to obtain informative data rapidly from suitably processed plant extracts, without recourse to liquid chromatography.

Other Methods

Plasma desorption mass spectrometry (PDMS) has been used successfully in the analysis of anthocyanidins and anthocyanins.[358,359] PDMS has been superseded by laser desorption mass spectrometry (LDMS) for the analysis of small and large molecules. It is unlikely that PDMS or LDMS will be more useful than electrospray LC/MS in the analysis of anthocyanins, or flavonoids in general. However, it is possible that the solid phase desorption techniques may find a use in the analysis of crude mixtures.

Recommended Procedures for Mass Spectrometric Analysis of Flavonoids

MS/MS, LC/MS and LC/MS/MS ESI and APCI techniques have emerged as the current methods of choice for analysing isoflavones and flavonoids in food plants and in metabolic studies of food components. The methodology is sufficiently sensitive to identify and quantify a wide range of polyphenolic compounds in plants and in body fluids. CE/MS also appears to show promise,

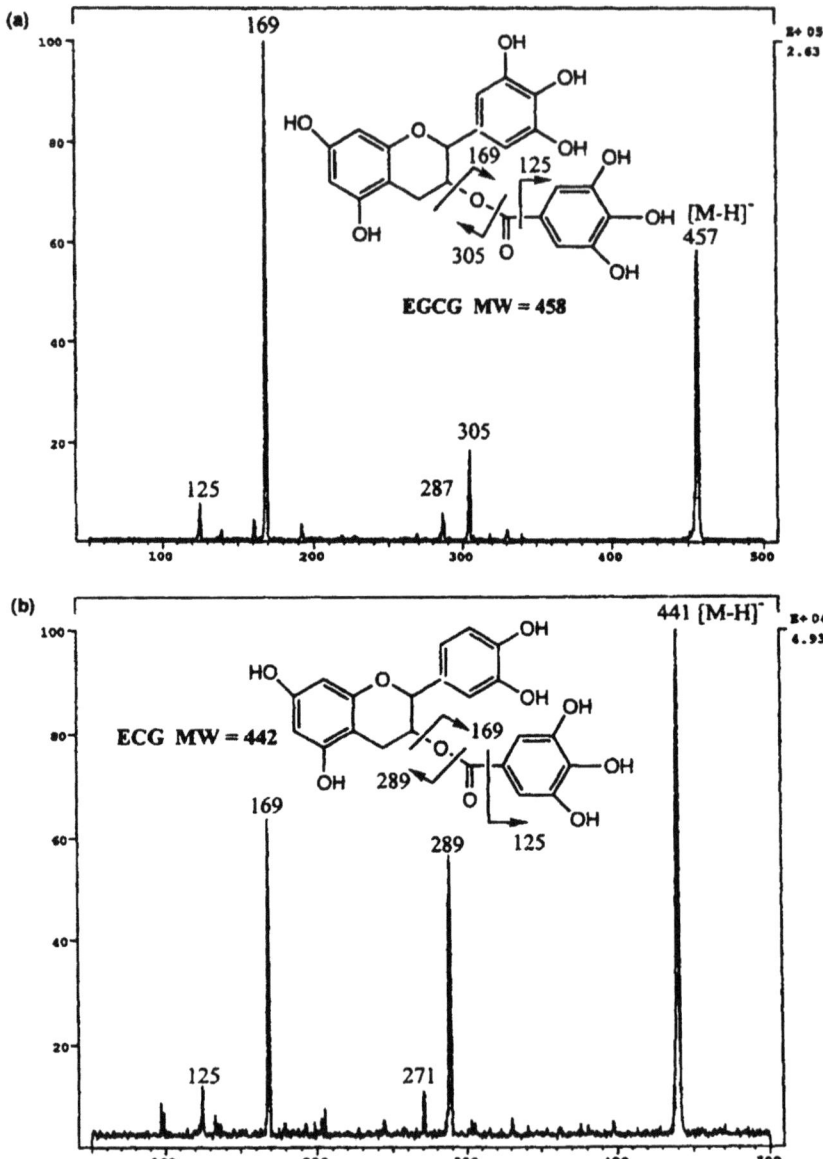

Figure 4.35   *Product ion spectra of the deprotonated molecules of (upper spectrum) epigallocatechin-3-gallate and (lower spectrum) epicatechin-3-gallate* (Reproduced from G.K. Poon, *J. Chromatogr. A*, 1998, **794**, 63, © 1998 with permission from Elsevier Science)

however this methodology is not yet as robust as the aforementioned techniques.

Phenolic Acids (Low Molecular Weight)

Simple, low molecular weight phenolic acids (for example, vanillin, ferulic acid and 4-hydroxybenzoic acid), although not necessarily bioactive in their own right, can function as useful markers of lignification. Lignin, the major non-carbohydrate component of plant cell walls, strongly influences the texture and digestibility of food plants and the breakdown products of lignans (which may be related to lignin) are sometimes biologically active (*e.g.* phytoestrogens). Simple phenolic acids can be generated by alkaline hydrolysis of food plants and used as markers of the degree and progress of lignification.

GC/MS of Phenolic Acids

Although phenolic acids may be analysed by conventional solids probe (EI or CI) mass spectrometry, this technique is not particularly useful for the complex mixtures encountered in the analysis of food plants. GC/MS of methylated or trimethylsilylated derivatives is the main mass spectrometric technique of choice: the derivatisation can be conducted simply and rapidly and the volatile derivatives can be separated easily and quickly and yield characteristic mass spectra. This type of analysis is exemplified by the determination of phenolic acids in maize grits.[360] Other examples include a GC/MS technique aimed at fingerprinting phenolic acids in wines[361] and a GC/MS method for monitoring changes in the phenolic composition of bamboo shoots during storage.[362]

Pyrolysis GC/MS of Phenolic Acids

An alternative to conventional GC/MS of derivatives of phenolic acids generated by hydrolysis of plant tissue is pyrolysis GC/MS (Py/GC/MS) performed directly on the plant tissues. Pyrolysis mass spectrometry is covered in more detail in Chapter 11 and only leading references and brief examples are given here. The technique is rapid and convenient and has been used to characterise phenolic acids generated by thermal breakdown of lignin and lignocellulose.[363-365] Free phenolic acids are generally not observed; the lignins pyrolyse to methylated and partially methylated phenols and phenolic aldehydes and ketones, such as guaiacol, 2,6-dimethoxyphenol and syringaldehyde. Among the many applications of Py/GC/MS to food plant analysis are investigations of kiwi fruit mucilage[364] and of the maturation of asparagus tissue.[365]

LC/MS of Phenolic Acids

Positive ion thermospray LC/MS, using a discharge electrode to effect solvent-mediated CI, yields $[M + H]^+$ ions and some fragmentation, principally losses of H_2O, CO and CO_2, when applied to the analysis of simple phenolic acids.[366] A more recent application is exemplified by the comparative determination of benzoic and cinnamic acid and their corresponding phenolic acids by UV, electrochemical and particle beam LC/MS.[367] However, a more promising method is ESI LC/MS, recently used to identify and quantify phenolic acids and aldehydes in dietary fibre supplements, wine and lignocellulose by-products.[368] The more robust method of APCI also shows considerable promise, exemplified by the quantitative determination of phenolic acids in negative ion SIM mode over the 0.01–1000 ng range.[369]

Recommended Methods for Mass Spectrometric Analysis of Phenolic Acids

GC/MS of volatile derivatives provides the most readily accessible method for the mass spectrometric analysis of phenolic acids. LC/MS also shows consider-able promise and the newer methods of APCI and electrospray ionisation have advantages over the published TSP methods because of their increased robust-ness and convenience. APCI is more robust than ESI and has shown promise in the analysis of a wide range of phenolic acids.[369]

Polyphenols and Condensed Polyphenols (Vegetable Tannins)

The polyphenols are complex polymeric phenols with a range of biological activity. They are dealt with separately from lower molecular weight phenolics because their complexity presents special difficulties for the analyst. They can be classified into two main groups: the hydrolysable tannins (which have a central glycosidic core esterified with gallic or hexahydrodiphenic acid) and the condensed tannins (dimers or higher oligomers of substituted flavan-3-ols). The structures of a simple hydrolysable tannin, theogallin (found in young tea leaves) and of a condensed tannin, procyanidin B-2 (isolated from cider apples) are shown in Figure 4.36.

Other subgroups of tannins occur naturally, but these are less important from the dietary viewpoint. Condensed tannins impart an astringent taste to some

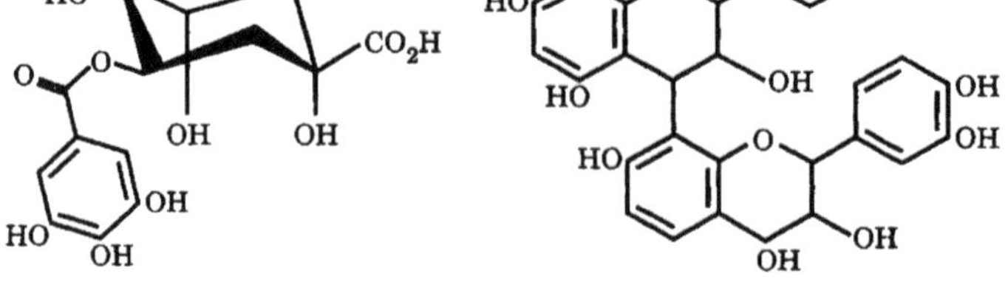

Figure 4.36   *The structures (left) of a simple hydrolysable tannin, theogallin (found in young tea leaves) and (right) of a condensed tannin, procyanidin B-2 (isolated from cider apples)*

plant foods. However, their most important dietary characteristics are their anti-nutritional properties. These are believed to stem from the ability of tannins to bind strongly to proteins and to nutrient minerals, significantly reducing their bioavailability. It is possible that tannins may also possess some beneficial properties because of their antioxidant capacity.

Direct Probe Mass Spectrometry and GC/MS of Polyphenols

The involatility of tannins makes them unsuitable for vaporisation from a heated probe or for gas chromatographic analysis unless they are derivatised. Early mass spectrometric studies were therefore based on EI or CI mass spectrometry and GC/MS of permethylated, peracetylated or pertrimethylsilyl-ated derivatives of polyphenols.[293,294,370-372] These methods met with limited success and have been superseded by condensed phase mass spectrometric techniques, particularly FABMS and LC/MS.

FABMS and MALDI of Polyphenols

Field desorption mass spectrometry provide informative spectra on oligomeric procyanidins.[373] However, the introduction of FABMS resulted in the first significant advance in qualitative mass spectrometric methods for analysing polyphenols. A landmark paper gave a comprehensive account of the negative ion FABMS of a range of polyphenols, including gallates, ellagitannins, procyanidins, flavonol gallates and novel polymeric flavonoids from sorghum.[374] The negative ion FABMS spectrum of a trimeric glyco-side $(5,7,3',4'$-tetrahydroxy-flavan-5-O-β-glucoside$(4,8)5,7,3',4'$-tetrahydroxy-flavan-5-O-β glucoside$(4,8)$-eriodictyol$)$, isolated from the food crop sorghum, is shown in Figure 4.37. Besides confirming the molecular weight of this complex molecule, the spectrum exhibits fragment ions corresponding to loss of glucose (m/z 993), a 3,4-flaven glucoside (m/z 721) and an eriodictyol unit (m/z 867). A more recent example of the application of FABMS (LSIMS) is to the determination of the composition of commercial tannin extracts.[375] Three gallotannin, three ellagitannins, a mixed hydrolysable tannin and one proantho-cyanidin powder were analysed and the main constituents of the tannin powders identified. LSIMS has also aided the characterisation of oligomeric and polymeric procyanidins from grape seeds[376] and the identification of catechin oligomers from apple.[377] The latter paper also made use of MALDI mass spectrometry. Positive and negative ion FABMS yielded evidence of the presence of a catechin undecamer and MALDI (when silver ions were added to the matrix) provided evidence for the pentadecamer. Even in the absence of silver, evidence for the presence of the dodecamer and undecamer was present in the positive and negative ion MALDI-ToF spectra. FABMS and LC/MS of polyphenols were reviewed in 1991, however, a number of significant improve-ments in LC/MS techniques have occurred since then.[378]

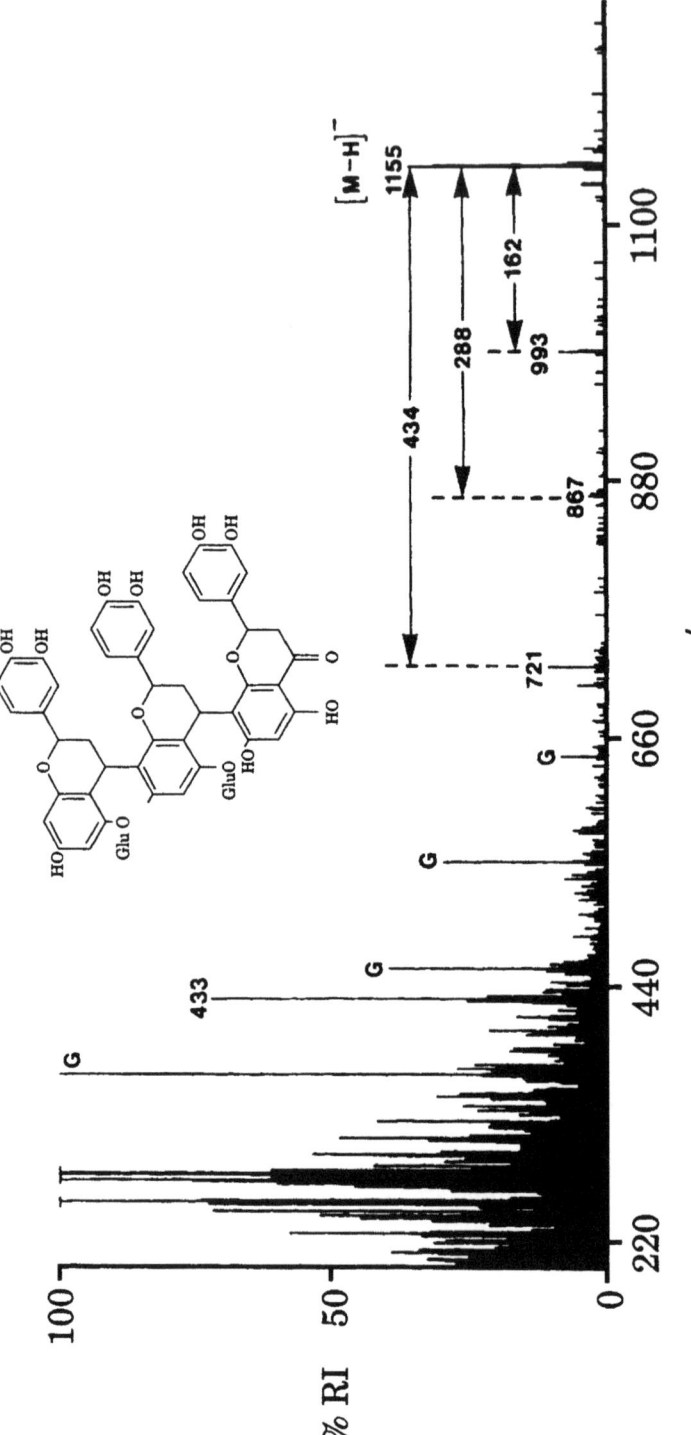

Figure 4.37 *The negative ion FABMS spectrum of a trimeric glycoside (5,7,3',4'-tetrahydroxy-flavan-5-O-β-glucoside (4,8)-5,7,3',4'-tetrahydroxy-flavan-5-O-β-glucoside (4,8)-eriodictyol)*
(Reproduced from R. Self, J. Eagles, G.C. Galletti, I. Mueller-Harvey, A.G.H. Lea, D. Magnolato, U. Richli, R. Gujer and E. Haslam, *Biomed. Env. Mass Spectrom.*, 1986, **13**, 449, with the permission of the authors and editors, © John Wiley & Sons Limited)

LC/MS of Polyphenols

An early report of the negative ion thermospray mass spectrometry of procyanidin B-2 dimer isolated from cider apples suggested the promise of this technique in the analysis of polyphenols.[136] A comparison of the negative ion FAB and TSP spectra of a procyanidin B-2 dimer isolated from cider apples is shown in Figure 4.38.

Few subsequent reports of TSP applications appeared and these focused on positive ion TSP: theory and practice would suggest that negative ionisation was more appropriate for mildly acidic compounds like polyphenols. Examples of the application of positive ion TSP LC/MS include the analysis of tea polyphenols[379] and of the polymeric thearubigin fraction of tea.[380] In the first report, only protonated aglycones were seen in the mass spectra because the labile glycosidic bonds were cleaved under TSP conditions. The second report described low mass fragments in the positive ion thermospray mass spectra of flavonol oligomers. An interesting alternative approach to the analysis of polyphenols in crude plant extracts describes the combination of LC/MS and LC/UV with post-column addition of shift reagents.[381] This combination of techniques permitted the complete structural determination of some poly-phenols.

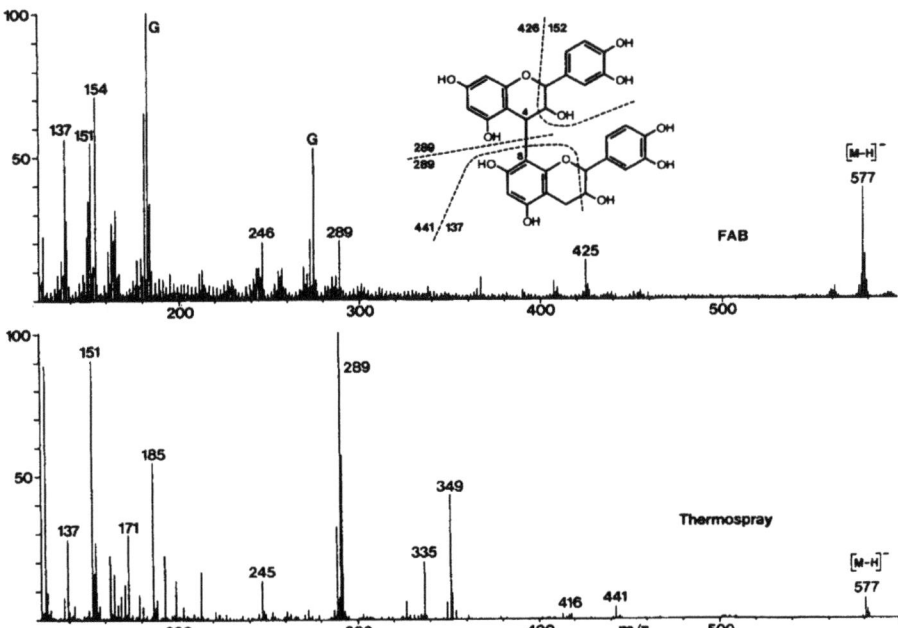

Figure 4.38 *Comparison of the negative ion FAB and TSP spectra of a procyanidin B-2 dimer*
(Reproduced from F.A. Mellon, J.R. Chapman and J.A.E. Pratt, *J. Chromatogr.*, 1987, **394**, 209, © 1987 with permission from Elsevier Science)

The greatest advance in mass spectrometric analysis of polyphenols, as in the analysis of so many other intractable compounds, has been provided by applications of ESI mass spectrometry. An early example compared ESI, APCI, TSP and continuous flow FABMS in the determination of secondary metabolites in crude plant extracts.[382] Tannins were one of the classes of secondary metabolite examined and ESI showed considerable promise in the analysis of polyphenol aglycones and glycosides, especially when additional information was obtained by MS/MS.

Since this early example of the use of ESI for analysing polyphenols and tannins, a number of instructive examples of LC/MS applications have appeared. These include the characterisation of highly polymerised procyanidins in cider apple skin and pulp.[383] The presence of singly and multiply charged ions with *m/z* ratios ranging from 1153.0 to 2305.5, together with information provided by natural abundance isotope spacing, was interpreted as evidence for the presence of a complete series of polymeric procyanidins with a degree of polymerisation up to 17. An evaluation, from the same laboratory, of ESI LC/MS in the analysis of polyphenolic oligomers concluded that the technique would lead to significant progress in the analysis of phenolic compounds, despite limitations of mass range and difficulties of interpretation.[384]

Several examples of the application of ESI and APCI to the analysis of polyphenols extracted from tea can be found in the recent literature.[385-388] Positive ion flow injection ESI aided the structural analysis of theaflavate A which was isolated from tea and was an oxidation product of epicatechin gallate,[385] and of theaflavin B, isotheaflavin-3'-*O*-gallate and neotheaflavin-3-*O*-gallate from black tea.[386] Biologically active green tea catechins, including (−)-epigallocatechin gallate, were identified in green tea and human plasma by positive ion capillary LC/MS.[387] This is a particularly significant analytical application in that it demonstrates the potential of the technique in investigating the metabolism of these compounds and in studying possible links between plasma levels, consumption and prevention of cancer. The final example, positive ion APCI LC/MS of theaflavin standards and theaflavins isolated from black tea, illustrates the potential of this method in obtaining useful structural information from theaflavins.

Monomeric and oligomeric flavan-3-ols from unripe almond fruits have been characterised with the aid of positive ion ESI LC/MS and LC/MS/MS.[389] Two monomers, (+)-catechin and (−)-epicatechin, and 15 oligomeric procyanidins were identified by gel chromatography and HPLC: mass spectrometry helped confirm identifications made by hydrolysis and was also useful in detecting minor components and in resolving peaks of co-eluting compounds.

Procyanidins in cocoa and chocolate may also be analysed qualitatively by ESI LC/MS.[390] A complex series of procyanidins containing 2–10 procyanidin units was identified. Only negative ion ESI provided useful mass spectrometric data for the higher oligomers. Negative ion ESI LC/MS is also useful for studying wine tannins.[391] Mass spectra corresponding to monocharged dimers to pentamers of trihyroxylated units were observed; higher oligomers (from pentamers upward) were observed as doubly charged ions. Heptamers corresponded to the highest masses detected. LC/MS provided new data on condensed tannins in wines, suggesting that, besides procyanidins, mixed di- and tri-hydroxylated flavonol units and pure trihydroxylated units were present. Figure 4.39 shows a 3-D map derived from mass chromatograms of wine fractions and Figure 4.40 shows the mass chromatograms of dihydroxy, trihydroxy and mixed catechin and gallocatechin units.

Pyrolysis GC/MS of Polyphenols

Py/GC/MS of phenolic acids (formed as breakdown products of lignocellulose) described above show that this methodology might also be appropriate for tannins. Several reports of the Py/GC/MS of polyphenols have appeared, exemplified by studies of wine[392] and sorghum, grape and apple tannins.[393] The Py/GC/MS analysis of highly polar phenolic breakdown products, not normally amenable to GC, can be achieved by thermally assisted hydrolysis/ methylation during pyrolysis.[394]

Recommended Mass Spectrometric Methods for Analysis of Polyphenols

ESI LC/MS is now the technique of choice for the analysis of polyphenols, particularly in negative ion mode. Additional structural information can be

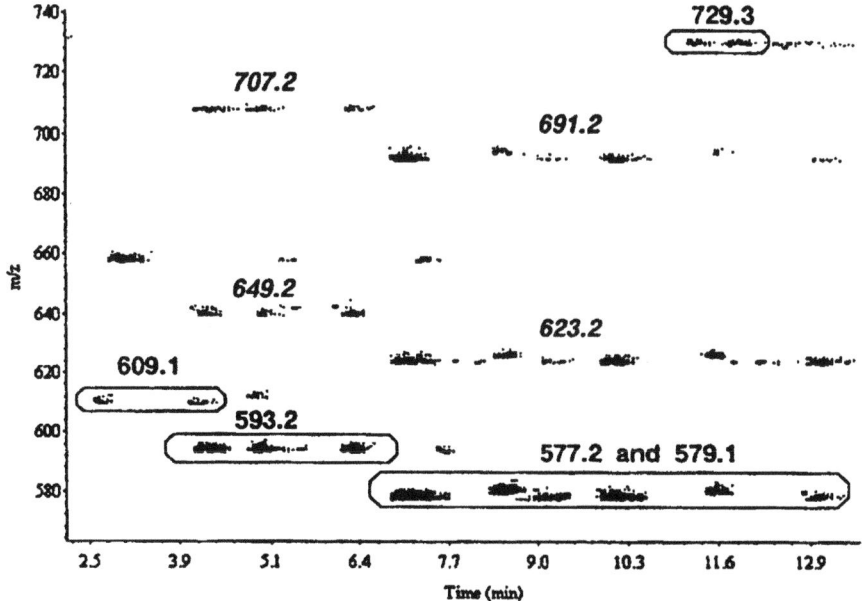

Figure 4.39 *A 3D map derived from mass chromatograms of wine fractions*
(Reproduced by permission from H. Fulcrand, S. Remy, J.-M. Souquet, V. Cheynier and M. Moutonet, *J. Agric. Food Chem.*, 1999, **47**, 1023, © 1999 American Chemical Society)

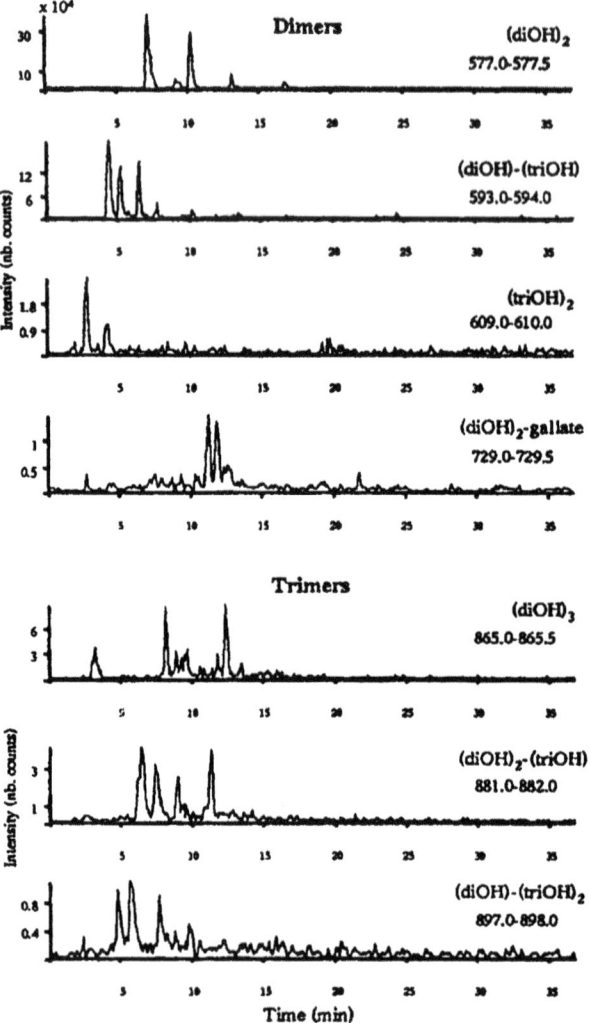

Figure 4.40 *Mass chromatograms of dihydroxy, trihydroxy and mixed catechin and gallocatechin units*
(Reproduced by permission from H. Fulcrand, S. Remy, J.-M. Souquet, V. Cheynier and M. Moutonet, *J. Agric. Food Chem.*, 1999, **47**, 1023, © 1999 American Chemical Society)

generated by LC/MS/MS. The potential of CE/MS should also be explored as the complexity of polyphenol mixtures isolated from foods and food plants makes them an obvious candidate for the application of the ultra-high resolution separation technique, CE. It is also probable that applications of MALDI to the analysis of crude mixtures of polyphenols and of higher molecular weight tannins will increase.

10   Saponins

Saponins are glycosides found in plants and some marine organisms, but only food plant saponins will be considered here. Triterpenoid saponins are the major type found in food plants and comprise an aglycone, the sapogenin, (typically based on an oleanane skeleton) to which are attached one to three oligosaccharide chains. The major saponin found in legume seeds, soyasaponin VI, is shown in Figure 4.41.[395] This structure was only identified in 1992 when very mild extraction conditions were used on the milled seed.[396] Prior to this, conventional hot solvent extraction techniques degraded this compound to soyasaponin I (a similar structure, but without the maltol group).

The saponins have soap-like properties, from which their name is derived, and are widely distributed in plants, being found in more than 500 genera. The glycoalkaloids are, strictly speaking, saponins but are dealt with above in the section on alkaloids in food plants. Examples of food plants containing saponins include legumes (*e.g.* chickpeas, haricot and kidney beans and lentils), spinach and asparagus. Some saponins have haemolytic properties, or may interact with the brush border membranes of intestinal epithelial cells, increasing their permeability.[397] This property raises the possibility that foods rich in saponins could increase absorption of substances in the diet that might be detrimental to health, such as allergenic proteins or naturally occurring toxins. Some saponins may also possess beneficial effects, including the ability to lower plasma cholesterol levels.[397] Knowledge of the effects, types and quantities of saponins found in food plants has been described as 'fragmentary'.[398] However, rapid advances in analytical technology are remedying this situation. Saponins have also been a major focus of interest of pharmaceutical and food companies because they may provide a new source of drugs, flavourings, 'functional' foods or tonics.

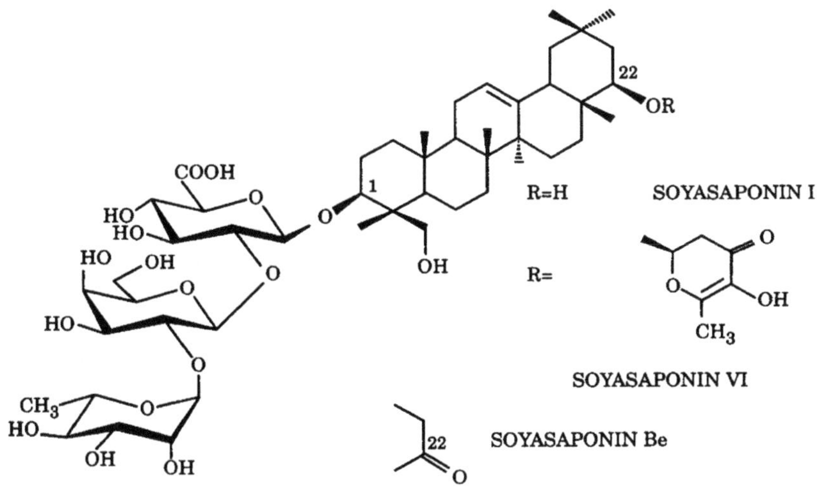

Figure 4.41   *The structure of the major saponins found in legume seeds*

A considerable increase in interest in the saponins has occurred in the past ten years because of the biological activity of these natural products. Consequently, they have also been the focus of considerable attention from analysts throughout the world and have been examined using a wide range of analytical methods, including some of the most sophisticated techniques of mass spectrometry.

Direct Probe Mass Spectrometry of Saponins

Sapogenins (saponin aglycones) yield characteristic EI mass spectra that are very useful in confirming the identity of known molecules or as an aid to the structural elucidation of unknowns.[399] Diagnostically important ions reveal important structural features in the aglycone. These include the presence or absence of a C-12–C-13 double bond and the gross substitution patterns of the A, B, D and E rings (characterised by a retro-Diels–Alder fragmentation).[400,401] Identity may also be deduced by matching unknowns against libraries of known spectra, or by comparison with published compilations of data.[402,403] The saponins are generally so involatile and thermally labile that they cannot be analysed by EI, unless they are first converted into their permethyl or peracetyl derivatives. An exception to this rule is the 'Desorption' EI (DEI) mass spectrum of hederacosides, which yield molecular ions and abundant fragments of these underivatised molecules.[404] The DEI method was, surprisingly, more successful than FABMS because detection limits were better. Molecular ions are generally not observed for derivatised saponins possessing more than four sugar residues (less for furostanol saponins).[405,406]

DCI of saponins with the commonly used reagent gases (methane, isobutane or ammonia) can yield protonated $[M + H]^+$, cationised $[M + NH_4]^+$ or, for negative ion DCI, deprotonated $[M - H]^-$ molecules. Although these are not always observed, structurally significant fragment ions that help in the characterisation and sequencing of sugar residues are usually seen.[306,407]

GC/MS of Saponins

GC/MS of silylated derivatives of saponins can yield some structural information.[311,408] A number of sapogenins have been determined by GC/MS. A recent example is provided by the identification of medicagenic acid, hederagenin, soyasapogenols B, C, D, E and F and oleanolic acid as the major components of two alfalfa cultivars.[409] SIM stable isotope dilution GC/MS methods have also been used to quantify the aglycones of glycyrrhizin and glycyrrhetinic acid in human plasma.[410] Animal studies of saponin metabolism have been conducted using similar GC/MS techniques.[411]

GC/MS is insufficiently versatile to be recommended as a method for the structural analysis of saponin derivatives. An exception may be made for complex mixtures of particularly intractable aglycones, when GC/MS of silylated derivatives can be useful. GC/MS of partially methylated alditol acetate (PMAA) derivatives of sugars is also very useful for characterising the composition and linkage of polysaccharide chains.[412,413] PMAA methodology has been successfully applied to the characterisation of the carbohydrate moieties of saponins, as summarised in 1995.[407] An alternative to the PMAA technique for analysing saponin sugars is GC/MS of methyl trimethylsilyl derivatives of hexoses and pentoses released following the hydrolysis of permethylated saponins.[414]

FABMS of Saponins

A variety of condensed phase soft ionisation techniques, including field desorption, plasma desorption and laser desorption mass spectrometry showed some promise in the analysis of saponins. However, FABMS was the first mass spectrometric technique to yield a straightforward, reliable mass spectrometric method for analysing intact saponins. FABMS is more useful in negative than in positive ion mode for the neutral and acidic saponins, in direct contrast to the case of the basic glycoalkaloids.[415] Under negative ion conditions intense $[M - H]^-$ ions, accompanied by fragments corresponding to losses of sugar residues, are usually seen. This enables the molecular weight and carbohydrate composition to be determined. Although many applications of FABMS to the analysis of plant saponins have been reported,[306,407] relatively few of these describe studies of saponins found in food plants. A notable exception is the FABMS analysis of saponins in crude extracts of 13 varieties of legume seed.[416] Soyasaponins I to IV, A_1 and A_2 were identified in the mixtures and the presence of a previously unknown saponin was deduced. The structure of this saponin was inferred from its mass spectrum and from other circumstantial evidence, an inference which subsequent NMR studies showed was correct. The method was useful in providing a fingerprint of saponin composition of bean varieties. This was an unusual instance of the success of FABMS in mixture analysis: FAB is not often successful in this type of application because of suppression and other effects. However, the 'soap-like', *i.e.* surfactant, properties of saponins make them almost ideal candidates for FABMS analysis. This is because they concentrate on the surface of the liquid matrix and are thus desorbed and ionised with high efficiency by the bombarding atom or ion beam. The matrix can markedly affect the quality of the FAB spectra; PEG 200 has advantages over glycerol in some applications.[415] Thioglycerol and 3-nitrobenzyl alcohol may also be effective.[417]

FABMS has been central in the study of potential new food crops; for example, Andean crops,[418] including bitter lupin and quinoa. Identification of potentially toxic saponins in these cultivars was achieved with the aid of FABMS. Besides providing structural information on intact food plant saponins, FABMS can be used to monitor their degradation during heat processing. Studies of the fate of avenacosides (oat saponins) during heat processing have shown that the major products are desrhamnoavenacosides A and B.[419]

LC/MS of Saponins

Because the saponins are complex, involatile thermally labile molecules, successful LC/MS could only be achieved consistently when soft ionisation techniques were combined with HPLC. Saponins with molecular weights in the approximate range 700–1100 were analysed qualitatively by LC/FAB using a frit interface.[420] The negative ion spectra were especially informative. Sensitivity was maintained by using a 1.5 mm bore HPLC column, thus reducing the post-column flow split (necessary because the frit-FABMS probe flow capacity is typically only about 5 μl min^{-1}) to 20:1. LC/FAB interfaces have the major disadvantage that they can only accommodate low solvent flows and must be used either with flow splitting from conventional columns or capillary LC columns. These practices result in reduced sensitivity and detection limits (flow splitting) or, for capillary columns, are technically difficult and reduce dynamic range. An alternative to LC/FAB, capable of accepting higher flow rates (up to 1.5–2 ml min^{-1}), was thermospray LC/MS. However, success with this technique has been very variable. Glycyrrhizin yielded only fragment ions under TSP conditions.[421] Nevertheless, other researchers did show that careful control of mass spectrometric and solvent conditions generates protonated and adduct ions of intact saponins carrying up to three sugars.[422] LC/FAB has been recommended for saponins bearing more than three sugars.[9] Both TSP and LC/FAB have been used by authors from the same laboratory to screen crude plant extracts for saponins.[423] However, a more recent publication from the same laboratory advocates the use of negative ion ESI, and MS/MS, for the analysis of triterpene glycosides.[382] Positive and negative ion electrospray LC/MS techniques have been used successfully to analyse saponins from alfalfa, clover and mungbeans[424] and ginseng extracts.[425] The convenience and versatility of electrospray as a general LC/MS technique suggests that these will be the first of many applications in the field of saponin analysis.

Tandem Mass Spectrometry of Saponins

Surprisingly few publications describe the characterisation of saponins by MS/MS techniques. Early examples include collisionally-activated dissociation of a number of non-food glycosides, including *Hedera helix* L. saponins, after FABMS[415] or DEI ionisation[404] and FAB/MS/MS of saponins isolated from jasmine and other plants.[426–428] These experiments showed that MS/MS was useful in determining sugar sequence and, with the DEI MS/MS method,[404] capable of sub ppb sensitivity. Saponins isolated from partially purified extracts of mungbean sprouts have been determined qualitatively with the aid of positive ion FAB MS/MS.[429, 430] More recently, multistage (MS^n) mass spectrometry using an ion trap has been applied to the analysis of saponins under positive ion ESI[431] and nano-ESI conditions.[432] Useful fragment ions characteristic of steroidal saponins and some aspects of the sugar linkage positions were found.

Recommended Mass Spectrometric Methods for the Analysis of Saponins

The most useful current methods for analysing intact saponins are LC/MS techniques based on electrospray. Successes with MS/MS and recent literature reports also suggest that the most fruitful approach for analysing complex mixtures will be electrospray LC/MS/MS or LC/MS^n. In cases where conventional MS/MS is not available, cone voltage induced fragmentation may be of some merit in generating structurally significant fragment ions.

11 References

1 B.N. Ames, *Science*, 1983, **221**, 1256.
2 R.C. Beier and H.N. Nigg in *Foodborne Disease Handbook, Volume 3*. ed. Y.H. Hui, J.R. Gorham, K.D. Murrell and D.O. Cliver, Marcel Dekker, Inc., New York, 1994, pp. 1–186.
3 D.L. Davis, *Environ. Res.*, 1989, **50**, 322.
4 G. Hocman, *Comp. Biochem. Physiol.*, 1989, **93**, 201.
5 N. Cotelle, J.L. Bernier, J.P. Henichart, J.P. Catteau, E. Gaydou and J.C. Wallet, *Free Radical Biol. Med.*, 1992, **13**, 211.
6 V.L. Sparnins and L.W. Wattenberg, *J. Natl. Cancer Inst.*, 1981, **66**, 769.
7 A.M. Benson, *Chem. Scripta*, 1987, **27a**, 67.
8 N. Tawfiq, R.K. Heaney, J.A. Plumb, G.R. Fenwick, S.R.R. Musk and G. Williamson, *Carcinogenesis*, 1995, **16**, 1191.
9 J.L. Wolfender, M. Maillard and K. Hostettmann, *Phytochem. Anal.*, 1994, **5**, 153.
10 S.L. Sinden and K.L. Deahl, in *Foodborne Disease Handbook, Volume 3*. ed. Y.H. Hui, J.R. Gorham, K.D. Murrell and D.O. Cliver, Marcel Dekker, Inc., New York, 1994, pp. 227–259.
11 R. Verpoorte and W.M.A. Niessen, *Phytochem. Anal.*, 1994, **5**, 217.
12 E.H. Kerns, K.J. Volk, J.L. Whitney, R.A. Rourick and M.S. Lee, *Drug Inf. J.*, 1998, **32**, 471.
13 H. Bjorndal, C.G. Hellerqvist, B. Lindberg and S. Svensson, *Angew. Chem. Int. Edn.*, 1970, **9**, 610.
14 R.R. Selvendran, in *Recent Developments in Mass Spectrometry in Biochemistry, Medicine and Environmental Research*. ed. A. Frigerio. Elsevier, Amsterdam, 1983, pp 159–203.
15 M. Friedman, N. Kozukue and L.A. Harden, *J. Agric. Food Chem.*, 1998, **46**, 2096.
16 H. Budzikiewicz, *Tetrahedron*, 1964, **20**, 2267.
17 W.M.J. van Gelder, L.G.M.T. Tunistra, J. van der Greef and J.J.C. Scheffer, *J. Chromatogr.*, 1989, **482**, 13.
18 K.R. Price, F.A. Mellon, R. Self, G.R. Fenwick and S.F. Osman, *Biomed. Mass Spectrom.*, 1985, **12**, 79.
19 R. Self, J. Eagles, F.A. Mellon and G.R. Fenwick, in *Mass Spectrometry of Large Molecules*, ed. S. Facchetti, Elsevier, Amsterdam, 1985, pp. 209–231.
20 K.R. Price, G.R. Fenwick and R. Self, *Food Add. Contam.*, 1985, **3**, 241.
21 M. Friedman, G. McDonald and W.F. Haddon, *J. Agric. Food Chem.*, 1993, **42**, 1397.
22 A. Ehmke, H.-M. Schiebe and M. McDowell, *Pharm. Weekblad Sci. Ed.*, **9**, 232.
23 R.J. Bushway, L.B. Perkins, L.R. Paradis and S. Vanderpan, *J. Agric. Food Chem.*, 1994, **42**, 2824.
24 S. Evans, R. Buchanan, A. Hoffman, F.A. Mellon, K.R. Price, S. Hall, F.C. Walls, A.L. Burlingame, S. Chen and P.J. Derrick, *Org. Mass Spectrom.*, 1993, **28**, 289.
25 S. Chen, P.J. Derrick, F.A. Mellon and K.R. Price, *Anal. Biochem.*, 1994, **218**, 157.

26 B. Domon and C.E. Costello, *Glycoconjugate J.*, 1988, **5**, 397.
27 M. Claeys, H. van den Heuvel, S. Chen, P.J. Derrick, F.A. Mellon and K.R. Price, *J. Am. Soc. Mass Spectrom.*, 1996, **7**, 173.
28 J.E. Peterson and C.C.J. Culvenor, in *Handbook of Natural Toxins*, Vol. 1, ed. I.R.F. Keeler and A.T. Tu, Marcel Dekker, New York, 1983, pp. 637–671.
29 N. Neuner-Jehle, H. Nesvadba and G. Spiteller, *Monatsh. Chem.*, 1965, **96**, 321.
30 E. Pedersen and E. Larsen, *Org. Mass Spectrom.*, 1970, **4**, 249.
31 Y.V. Rashkes, U.A. Abdulaev and S.Y. Yunusov, *J. Nat. Comp.*, 1978, **78**, 121.
32 E. Roeder, T. Bourauel and V. Neuberger, *Phytochemistry*, 1992, **31**, 4041.
33 W.F. Haddon and R.J. Molyneux, *28th Annual Conference on Mass Spectrometry and Allied Topics*, New York, 1980, paper No. RAMOA6.
34 P.A. Dreifuss, W.C. Brumley, J.A. Sphon and E.A. Caress, *Anal. Chem.*, 1983, **55**, 1036.
35 J. Karchesy, M. Deinzer, D. Griffin, and D.C. Rohrer, *Biomed. Mass Spectrom.*, 1984, **11**, 455.
36 M.L. Deinzer, B.L. Argobast, D.R. Buhler and P.R. Cheeke, *Anal. Chem.*, 1982, **54**, 1811.
37 H.J. Huising, F. de Boer, H. Hendriks, W. Balraadjsing and A.P. Bruins, *Biomed. Environ. Mass Spectrom.*, 1986, **13**, 293.
38 H. Hendriks, H.J. Huising and A.P. Bruins, *J. Chromatogr.*, 1988, **428**, 352.
39 M.M. Mossoba, H.S. Lin, D. Andrzejewski, J.A. Sphon, J.M. Betz, L.J. Miller, R.M. Eppley, M.W. Truckess and S.W. Page, *J. AOAC Int.*, 1994, **77**, 1167.
40 C.K. Winter, H.J. Segall and A.D. Jones, *Biomed. Environ. Mass Spectrom.*, 1988, **15**, 265.
41 L. Witte, P. Rubiolo, C. Bicchi and T. Hartmann, *Phytochemistry*, 1993, **32**, 187.
42 W. Sperl, H. Stuppner, I. Gassner, W. Judmaier, O. Dietze and W. Vogel, *Eur. J. Pediatr.*, **154**, 112.
43 C.E. Parker, S. Verma, K.B. Tomer, R.L. Reed and D. Buhler, *Biomed. Environ. Mass Spectrom.*, 1990, **19**, 1.
44 T.M. Chen, R.C. George, J.L. Weir and T. Leaphart, *J. Nat. Prod.*, 1990, **53**, 359.
45 C. Crews, J.R. Startin and P.A. Clarke, *Food Add. Contam.*, 1997, **14**, 419.
46 G. Lin, K.Y. Zhou, X.G. Zhao, Z.T. Wang and P.P.H. But, *Rapid Commun. Mass Spectrom.*, 1998, **12**, 1445.
47 J.A. Edgar, H.J. Lin, C.R. Kumana and M.M.T. Ng, *Am. J. Chinese Med.*, 1992, **20**, 281.
48 B. Flannigan, in *Toxic Substances in Crop Plants*, ed. J.P.F. D'Mello, C.M. Duffus and J.H. Duffus, Royal Society of Chemistry, Cambridge, 1991, pp. 226–257.
49 M. Barber, J.A. Weisbach, B. Douglas and G.O. Dudek. *Chem. Ind (London)*, 1965, 1072.
50 T. Inoue, Y. Nakahara and T. Niwaguchi, *Chem. Pharm. Bull.*, 1972, **20**, 409.
51 D. Voight, S. Johne and D. Groger, *Pharmazie*, 1974, **29**, 697.
52 J. Vokoun and Z. Rehacek, *Coll. Czech. Chem. Commun.*, 1975, **40**, 1731.
53 J. Schmidt, R. Kraft and D. Voight, *Biomed. Mass Spectrom.*, 1978, **5**, 674.
54 J.K. Porter and D. Betowski, *J. Agric. Food Chem.*, 1981, **29**, 650.
55 N. Feng, E.I. Minder, T. Grampp and D.J. Vonderschmitt, *J. Chromatogr.*, 1992, **575**, 289.
56 N. Haering, J.A. Settlage, S.W. Sanders and R. Schuberth, *Biomed. Mass Spectrom.*, 1985, **12**, 197.
57 C. Eckers, D.E. Games, D.N.B. Mallen and B.P. Swann, *Anal. Proc.*, 1982, **8**, 133.
58 D.E. Games, N. Alcock, L. Cobelli, C. Eckers, M.P.L. Games, A. Jones, M.S. Lant, M.A. McDowall, M. Rossiter, R.A. Smith, S.A. Westwood and H.Y. Wong, *Int. J. Mass Spectrom. Ion Phys.*, 1983, **46**, 181.
59 R.D. Plattner, S.G. Yates and J.K. Porter, *J. Agric. Food Chem.*, 1983, **31**, 785.
60 V. Havlíček , M. Flieger, V. Křen and M. Ryska, *Biol. Mass Spectrom.*, 1994, **23**, 57.
61 A.F. Casy, *J. Pharm. Biomed. Anal.*, 1994, **12**, 41.

62 R.A. Shelby, J. Olsovska, V. Havlíček and M. Flieger, *J. Agric Food Chem.*, 1997, **45**, 4674.
63 L. Lopez Bellido and M. Fuentes, *Adv. Agron.*, 1986, **40**, 239.
64 U. Majchrzak-Kuczyńska, M. Wiewiórowski and E. Wyrzykiewicz, *Org. Mass Spectrom.*, 1984, **19**, 600.
65 N.S. Vulfson and W.G. Zaikin, *Usp. Khim.*, 1976, **45**, 1870.
66 W.G. Zaikin and N.S. Vulfson, *Khim. Geterotsikl. Soedin.*, 1978, **11**, 1443.
67 D. Schumann, N. Neuner-Jehle and G. Spiteller, *Monatsh. Chem.*, 1968, **99**, 390.
68 M. Wink, H.M. Schiebel, L. Witte and T. Hartmann, *Planta Med.*, 1980, **38**, 238.
69 M. Wink, L. Witte, T. Hartmann, C. Theuring and V. Volz, *Planta Medica*, 1983, **48**, 253.
70 M. Wink and L. Witte, *Planta*, 1984, **161**, 519.
71 A.D. Kinghorn and M.F. Balandrin, in *Alkaloids: Chemical and Biological Perspectives*, Vol. 2, ed. W.S. Pelletier, Wiley, Chichester, 1984, pp. 105–148.
72 A. Planchuelo-Ravelo, L. Witte and M. Wink, *Z. Naturforsch.*, 1993, **48c**, 702.
73 A. Planchuelo-Ravelo and M. Wink, *Z. Naturforsch.*, 1993, **48c**, 414.
74 M. Wink, in *Methods in Plant Biochemistry*, Vol. 8, ed. P. Waterman, Academic Press, London, 1993, pp. 197–239.
75 M. Wink, C. Meissner and L. Witte, *Phytochemistry*, 1995, **38**, 139.
76 T. Aniszewski, *J. Sci. Food Agric.* 1993, **61**, 409.
77 M. Wink, H.J. Heinen, H. Vogt and H.M. Scheibel, *Plant Cell Rep.*, 1984, **3**, 230.
78 G. Rodighiero, F. Dall'Acqua, and M.A. Pathak, in *Topics in Photomedicine*, ed. K.C. Smith, Plenum, New York, 1987, p. 319.
79 F. Dall'Acqua and P. Martelli, *J. Photochem. Photobiol., B*, 1991, **9**, 235.
80 H. Budzikiewicz, C. Djerassi and D.H. Williams, in *Structure Elucidation of Natural Products by Mass Spectrometry*, Vol 2, Holden-Day, San Francisco, 1964, pp. 254–261.
81 B.T. Kiremire and P. Traldi, *Trends Heterocyc. Chem.*, 1992, **1**, 175.
82 R.T. Aplin and C.B. Page, *J. Chem. Soc. C*, 1967, 2593.
83 R.D.H. Murray, J. Mendez and S A. Brown, in *The Natural Coumarins*, Wiley, New York, 1982, pp. 45–53.
84 S.R. Horning, M.E. Bier, R.G. Cooks, G. Brusini, P. Traldi, A. Guiotto and P. Rodighiero, *Biomed. Environ. Mass Spectrom.*, 1989, **18**, 927.
85 D. Voigt, J. Schmidt, K. Schreiber, M.H.A. Elgamal and F.K. El-Bay, *J. Prakt. Chem.*, 1977, **319**, 767.
86 A.G. Harrison, *Chemical Ionization Mass Spectrometry*, CRC Press, Boca Raton, 1983.
87 R.D. Plattner and G.F. Spencer, *Org. Mass Spectrom.*, 1988, **23**, 624.
88 M. Tkaczyk, A.M. Zobel, J.R. Plomley and R.E. March, *Org. Mass Spectrom.*, 1993, **28**, 1148.
89 R.C. Beier, G.W. Ivie and E.H. Oertut, *Phytochemistry*, 1994, **36**, 869.
90 J.X. de Vries, B. Tauscher and G. Wurzel, *Biomed. Environ. Mass Spectrom.*, 1988, **15**, 413.
91 D.S. Weinberg, M.L. Manier, M.D. Richardson and F.G. Haibach, *J. Agric. Food Chem.*, 1993, **41**, 48.
92 D. Ehlers, M. Pfister, D. Gerard, K.W. Quirin, W.R. Bork and P. Toffel-Nadolny, *Int. J. Food Sci. Technol.*, 1996, **31**, 91.
93 N.J. Alcock, L. Corbelli, D.E. Games, M.S. Lant and S.A. Westwood, *Biomed Mass Spectrom.*, 1982, **9**, 499.
94 K.D.R. Setchell, M.B. Welsh and C.K. Lim, *J. Chromatogr.*, 1987, **386**, 315.
95 J.X. de Vries and K.A. Kymber, *J.Chromatogr.*, 1991, **562**, 31.
96 A. Capiello, G. Famiglini, F. Mangani and B. Tirillini, *J. Am. Soc. Mass Spectrom.*, 1995, **6**, 132.
97 M. Ganzera, S. Sturm and H. Stuppner, *Chromatographia*, 1997, **46**, 197.
98 C.G. de Koster, G.J. Niemann and W. Heerma, *Z. Naturforsch.*, 1985, **40**, 580.

99 M.H.A. Elgamal, N.M.M. Shalaby, H. Duddeck and M. Hiegmann, *Phytochemistry*, 1993, **34**, 819.
100 L.-J. Lin, L.-Z. Lin, N. Ruangrungel and G.A. Cordell, *Phytochemistry*, 1993, **34**, 825.
101 B.T. Kiremire, D. Chiarello, P. Traldi, U. Vettori, A. Guiotto and P. Rodighiero, *Rapid Commun. Mass Spectrom.*, 1990, **4**, 117.
102 B. Pelli, P. Traldi, P. Rodighiero and A. Guiotto, *Biomed. Environ. Mass Spectrom.*, 1986, **13**, 417.
103 R.H. Davis, in *Toxic Substances in Crop Plants*, ed. J.P.F. D'Mello, C.M. Duffus and J.H. Duffus, Royal Society of Chemistry, Cambridge, 1991, pp. 202–225.
104 D.S. Seigler, *Phytochemistry*,1975, **14**, 9.
105 C. Fenselau, S. Pallante, R.P. Betzinger, W.R. Benson, R.P. Barron, E.B. Sheinin and M. Maienthal, *Science*, 1978, **198**, 625.
106 T. Cairns and E.G. Siegmund, *Biomed. Mass Spectrom.*, 1982, **9**, 307.
107 P.A. Dreifuss, G.E. Wood, J.A.G. Roach, W.C. Brumley, D. Andrzejewski and J.A. Sphon, *Biomed. Mass Spectrom.*, 1980, **7**, 201.
108 L. Cardona, I. Fernandez, J.R. Pedro and R. Vidal, *Phytochemistry*, 1992, **31**, 3507.
109 T. Cairns, J.E. Froberg, S. Gonzales, W.S. Langham, J.J. Stamp, J.K. Howie and D.T. Sawyer, *Anal. Chem.*, 1978, **50**, 317.
110 D. Chassagne, J.C. Crouzet, C.L. Bayonove and R.L. Baumes, *J. Agric. Food Chem.*, 1996, **44**, 3817.
111 F.A. Mellon and G.R. Fenwick, *Biomed. Mass Spectrom.*, 1984, **11**, 375.
112 Y. Zhang, P. Talalay, C.-G. Cho and G.H.A. Posner, *Proc. Natl. Acad. Sci. USA*, 1992, **89**, 2399.
113 Y. Zhang, T. W. Kensler, C.-G. Cho, G.H. Posner and P. Talalay, *Proc. Natl. Acad. Sci. USA*, 1994, **91**, 3147.
114 E.W. Underhill, in *Secondary Plant Products*, ed. E.A. Bell and B.V. Charlwood, Springer-Verlag, Berlin, 1980, pp. 493–511.
115 G.R. Fenwick, R.K. Heaney and W.J. Mullin, *CRC Crit. Rev. Food Sci. Nutr.*, 1983, **18**, 123.
116 H.E. Nursten and J. Gilbert, *J. Sci. Food Agric.*, 1972, **23**, 527.
117 M. Kojima, M. Uchida and Y. Akahori, *Yakugaku Zasshi*, 1973, **93**, 453.
118 R. A. Cole, *Phytochemistry*, 1976, **15**, 759.
119 K. Grob and P. Matile, *Phytochemistry*, 1980, **19**, 1789.
120 M.E. Daxenbichler, G.F. Spencer, D.G. Carlson, G.B. Rose. and A.M. Brinker, *Phytochemistry*, 1991, **30**, 2623.
121 L. Latxague, C. Gardrat, J.L. Coustille, M.C. Viaud and P. Rollin, *J. Chromatogr.*, 1991, **586**, 166.
122 W.C.K. Chiang, D.J. Pusteri and E.A. Leitz, *J. Agric. Food Chem.*, 1998, **46**, 1018.
123 M.S. Cerny, E. Taube and R. Battaglia, *J. Agric. Food Chem.*, 1996, **44**, 3835.
124 K. Olsson, O. Theander and P. Aman, *Carbohydr. Res.*, 1977, **58**, 1.
125 J. Eagles, G.R. Fenwick, R. Gmelina and D. Rakow, *Biomed. Mass Spectrom.*, 1981, **8**, 265.
126 W. Thies, *Naturwissenschaften*, 1979, **66**, 364.
127 E.W. Underhill and D.F. Kirkland, *J. Chromatogr.*, 1971, **57**, 47.
128 R.K. Heaney and G.R. Fenwick, *J. Sci. Food Agric.*, 1980, **31**, 593.
129 J. Eagles, G.R. Fenwick and R.K. Heaney, *Biomed. Mass Spectrom.*, 1981, **8**, 278.
130 G.J. Shaw, D. Andrzejewski, J.A.G. Roach and J.A. Sphon, *J. Agric. Food Chem.*, 1989, **37**, 372.
131 G.J. Shaw, D. Andrzejewski, J.A.G. Roach and J.A. Sphon, *J. Agric. Food Chem.*, 1990, **38**, 616.
132 G.R. Fenwick, J. Eagles, R. Gmelin and D. Rakow, *Biomed. Mass Spectrom.*, 1980, **7**, 410.
133 G.R. Fenwick, J. Eagles and R. Self, *Org. Mass Spectrom.*, 1982, **17**, 544.

134 T. Prestera, J.W. Fahey, W.D. Holtzclaw, C. Abeygunawardana, J.L. Kachinski and P. Talalay, *Anal. Biochem.*, 1996, **239**, 168.
135 G. Bojesen and E. Larsen, *Biol. Mass Spectrom.*, 1991, **20**, 286.
136 F.A. Mellon, J.R. Chapman and J.A.E. Pratt, *J. Chromatogr.*, 1987, **394**, 209.
137 F.A. Mellon, G.R. Fenwick, J.A. Lewis and E.A. Spinks, in *Analytical Applications of Spectroscopy*, ed. C.S. Creaser and A.M.C. Davies, Royal Society of Chemistry, 1988, pp. 301–304.
138 L.R. Hogge, D.W. Red and E.W. Underhill, *J. Chromatogr. Sci.*, 1988, **26**, 348.
139 L.R. Hogge, D.W. Reed, E.W. Underhill and G.W. Haughn, *J. Chromatogr. Sci.*, 1988, **26**, 551.
140 C.E.M. Heeremans, R.A.M. van der Hoeven, W.M.A. Niessen, J. Vuik, R.H. de Vos and J. van der Greef, *J. Chromatogr.*, 1989, **472**, 219.
141 R. Lange and M. Petrzika, *Fett Wiss. Technol.*, 1991, **93**, 284.
142 R. Lange, M. Petrzika, B. Raab and F. Linow, *Nahrung*, 1991, **35**, 385.
143 B.J. Shelp, L. Liu and D. McLellan, *Can. J. Plant Sci.*, 1993, **73**, 885.
144 R. Lange and M. Petrzika, *Lebensm. Biotechnol.*, 1990, **7**, 176.
145 P. Kokkonen, J. van der Greef, W.M.A. Niessen, U.R. Tjaden, G.J. ten Hove and G. van de Werken, *Rapid Commun. Mass Spectrom.*, 1989, **3**, 102.
146 P. Kokkonen, J. van der Greef, W.M.A. Niesen, U.R. Tjaden, G.J. ten Hove and G. van de Werken, *Biol. Mass Spectrom.*, 1991, **20**, 259.
147 C.L. Zrybko and R.T. Rosen, *ACS Symp. Ser.*, 1997, **660**, 125.
148 C.L. Zrybko, E.K. Fukuda and R.T. Rosen, *J. Chromatogr. A*, 1997, **767**, 43.
149 M. Ishida, I. Chiba, Y. Okuyama, Y. Takahata and N. Kaizuma, *Jpn. Agric. Res. Quart.*, 1997, **31**, 73.
150 E. Wolz, D. Wild and G.H. Degen, *Arch. Toxicol.*, 1995, **69**, 171.
151 C. De Meester, *Z. Lebensm.-Unters. Forsch. A*, 1998, **207**, 441.
152 M.G. Knize, J.S. Felton and G.A. Gross, *J. Chromatogr.*, 1992, **624**, 253.
153 H. Lee, M.-Y. Lin and S.-C. Chan, *Mutation Res.*, 1994, **308**, 77.
154 M.G. Knize, E. Hopmans and J.A. Happe, *Mutation Res.*, 1991, **260**, 313.
155 S. Murray, N.J. Gooderham, A.R. Boobis and D.S. Davis, *Carcinogenesis*, 1988, **9**, 321.
156 S. Murray, N.J. Gooderham, A.R. Boobis and D.S. Davis, *Carcinogenesis*, 1989, **10**, 763.
157 S. Murray, A.M. Lynch, M.G. Knize and N.J. Gooderham, *J. Chrom.*, 1993, **616**, 211.
158 L.M. Tikkanen, K.J. LatvaKala and R.L. Heinio, *Food Chem. Toxicol.*, 1996, **34**, 725.
159 K. Skog, A. Solyakov, P. Arvidsson and M. Jagerstad, *J. Chromatogr.*, 1998, **803**, 227.
160 R. Reistad, O.J. Rossland, K.J. LatvaKala, T. Rasmussen, R. Vikse, G. Becher and J. Alexander, *Food Chem. Toxicol.*, 1997, **35**, 945.
161 Z. Yamaizumi, H. Kasai, S. Nishimura, C.G. Edmonds and J.A. McCloskey, *Mutation Res.*, 1986, **173**, 1.
162 R.J. Turesky, H. Bur, T. Huynh-Bah, H.U. Aeschenbacher and H. Milon, *Food Chem. Toxicol.*, 1988, **26**, 501.
163 N.J. Gooderham, D. Watson, J.C. Rice, S. Murray and D.S. Davies, *Biochem. Biophys. Res. Commmun.*, 1987, **148**, 1377.
164 M.T. Galceran, E. Moyano, L. Puignou and P. Pais, *J. Chromatogr. A*, 1996, **730**, 185.
165 L.B. Fay, S. Ali and G.A. Geross, *Mutation Res.*, 1997, **376**, 29.
166 B. Stavric, B.P.Y. Lau, T.I. Matula, R. Klassen, D. Lewis and R.H. Downie, *Food Chem. Toxicol.*, 1997, **35**, 185.
167 P. Pais, E. Moyano, L. Puignou and M.T. Galceran, *J. Chromatogr. A*, 1997, **778**, 207.
168 E. Richling, D. Haring, M. Herderich and P. Schreier, *Chromatographia*, 1998, **48**, 258.

169 R.J. Mauthe, E.G. Snyderwinde, A. Ghoshal, S.P.H.T. Freeman and K.W. Turteltaub, *Carcinogenesis*, 1998, **19**, 919.

170 W. Pfau, U. Brockstedt, C. Schulze, G. Neurath and H. Marquardt, *Carcinogenesis*, 1996, **17**, 2727.

171 W. Pfau, C. Schulze, T. Shirai, R. Hasegawa, and U. Brockstedt, *Chem. Res. Toxicol.*, 1997, **10**, 1192.

172 A.E. Pohland, *Food Addit. Contam.*, 1993, **10**, 17.

173 J.L. Richard, G.A. Bennett, P.F. Ross and P.E. Nelson, *J. Animal Sci.*, 1993, **71**, 2563.

174 J. Gilbert, *Food Addit. Contam.*, 1993, **10**, 37.

175 R.J. Cole and R.H. Cox, *Handbook of Toxic Fungal Metabolites*, Academic Press, New York, 1981.

176 L.R. Dusold, A.E. Pohland, P.A. Dreifus and J.A. Sphon, *Mycotoxins Mass Spectral Data Bank*, Association of Official Analytical Chemists, Arlington, Virginia, 1981.

177 W.F. Haddon, M. Wiley and A.C. Waiss, *Anal. Chem.*, 1971, **43**, 268.

178 R.D. Plattner, G.A. Bennett and R.D. Stubblefield, *J. Assoc. Off. Anal. Chem.*, 1984, **67**, 734.

179 W.F. Haddon, M.S. Masri, V.G. Randall, R.H. Elsken and B.J. Meneghelli, *J. Assoc. Off. Anal. Chem.*, 1977, **60**, 107.

180 W.C. Brumley, S. Nesheim, M.W. Trucksess, E.W. Trucksess, P.A. Dreifuss, J.A.G. Roach, D. Andrzejewski, R.M. Eppley, A.E. Pohland, C.W. Thorpe and J.A. Sphon, *Anal. Chem.*, 1981, **53**, 2003.

181 D.L. Park, V. Diprossimo, E. Abdel-Malek, M. Trucksess, S. Nesheim, W.C. Brumley, J.A. Sphon, T.L. Barry and G. Petzinger, *J. Assoc. Off. Anal. Chem.*, 1985, **68**, 636.

182 F. Friedli, *J. High Resol. Chromatogr. Chromatogr. Commun.*, 1981, **4**, 495.

183 M.W. Trucksess, W.C. Brumley and S. Nesheim, *J. Assoc. Off. Anal. Chem.*, 1984, **67**, 973.

184 R.T. Rosen, J.D. Rosen and V.P. DiProssimo, *J. Agric. Food Chem.*, 1984, **32**, 276.

185 J.A. Sphon, P.A. Dreifuss and H.-R. Schulten, *J. Assoc. Off. Anal. Chem.*, 1977, **60**, 73.

186 H.J. Walther, C.E. Parker, D.J. Harvan, R.D. Voyksner, O. Hernandez, W.M. Hagler, P.B. Hamilton and J.R. Hass, *J. Agric. Food Chem.*, 1983, **31**, 168.

187 D. Uyakul, M. Isobe and T. Goto, *J. Assoc. Off. Anal. Chem.*, 1989, **72**, 491.

188 D.N. Tripathi, L.R. Chauhan and A. Bhattacharya, *Anal. Sci.*, 1991, **7**, 423.

189 A. Cappiello, G. Famiglini and B. Tirillini, *Chromatographia*, 1995, **40**, 411.

190 W.J. Hurst, R.A. Martin and C.H.Vestal, *J. Liquid Chromatogr.*, 1991, **14**, 2541.

191 M. Holcomb, W.A. Korfmacher and H.C. Thompson, *J. Anal. Toxicol.*, 1991, **15**, 289.

192 A. Kussak, C.A. Nilsson, B. Andersson and J. Langridge, *Rapid Commun. Mass Spectrom.*, 1995, **9**, 1234.

193 R.T. Gallagher and H.M. Stahr, *Appl. Spectroscopy*, 1981, **35**, 131.

194 H. Xiao, R.R. Marquardt, A.A. Frohlich and Y.Z. Ling, *J. Agric. Food Chem.*, 1995, **43**, 524.

195 D. Abramson, *J. Chromatogr.*, 1987, **391**, 315.

196 Y. Jiao, W. Blaas, C. Rühl and R. Weber, *J. Chromatogr.*, 1992, **595**, 364.

197 E. Rajakylä, K. Laasasenaho and P.J.D. Sakkers, *J. Chromatogr.*, 1987, **384**, 391.

198 M. Becker, P. Degelmann, M. Herderich, P. Schreier and H.-U. Humpf, *J. Chromatogr. A*, 1998, **818**, 260.

199 A.S. Salhab, G.F. Russell, J.R. Coughlin and D.P.H. Hsieh, *J. Assoc. Off. Anal. Chem.*, 1976, **59**, 1037.

200 K.A. Scudamore, M.T. Hetmanski, P.A. Clarke, K.A. Barnes and J.R. Startin, *Food Addit. Contam.*, 1996, **13**, 343.

201 W. Langseth and T. Runderberger, *J. Chromatogr. A*, 1998, **815**, 103.

202 E.W. Sydenham, P.G. Thiel and R. Vleggaar, *J. AOAC Internat.*, 1996, **79**, 1365.
203 R. Greenhalgh, B.A. Blackwell, J.R.J. Pare, J.D. Miller, D. Levandier, R.M. Meier, A. Taylor and J.W. ApSimon, in *Mycotoxins and Phytotoxins*, ed. P.S. Steyn and R. Vleggaar, Elsevier Science Publishers, Amsterdam, 1986, pp. 137–152.
204 R.D. Plattner and G.A. Bennett, *J. Assoc. Off. Anal. Chem.*, 1983, **66**, 1470.
205 R. Kostiainen and A. Hesso, *Biomed. Environ. Mass Spectrom.*, 1988, **15**, 79.
206 E.P. Burrows, *Biol. Mass Spectrom.*, 1994, **23**, 492.
207 W.C. Brumley, M.W. Trucksess, S.H. Adler, C.K. Cohen, K.D. White and J.A. Sphon, *J. Agric. Food Chem.*, 1985, **33**, 326.
208 R. Tiebach, W. Blass, M. Kellert, S. Steinmeyer and R. Weber, *J. Chromatogr.*, 1985, **318**, 103.
209 W.F. Miles and N.P. Gurprasad, *Biol. Mass Spectrom.*, 1985, **12**, 652.
210 W.C. Brumley, D. Andrzwjewski, E.W. Trucksess, P.A. Dreifuss, J.A.G. Roach, R.M. Eppley, F.S. Thomas, C.W. Thorpe and J.A. Sphon, *Biol. Mass Spectrom.*, 1982, **9**, 451.
211 J.A.G. Roach, J.A. Sphon, J.A. Easterling and E.M. Calvey, *Biomed. Environ. Mass Spectrom.*, 1989, **18**, 64.
212 R.D. Plattner, M.N. Beremand and R.G. Powell, *Tetrahedron*, 1989, **45**, 2251.
213 H.T. Kalinoski, H.R. Udseth, B.W. Wright and R.D. Smith, *Anal. Chem.*, 1986, **58**, 2421.
214 R. Kostiainen, *Biomed. Environ. Mass Spectrom.*, 1989, **18**, 116.
215 J.R. Paré, R. Greehalgh, P. Lafontaine and J.W. ApSimon, *Anal. Chem.*, 1985, **57**, 1472.
216 K. Knauss, J. Fullemann and M.P. Turner, *J. High Resol. Chromatogr. Chromatogr. Commun.*, 1981, **4**, 641.
217 Y. Onji, Y. Aoki, N. Tani, K. Umebayashi, Y. Kitada and Y. Dohi, *J. Chromatogr. A*, 1998, **815**, 59.
218 R.T. Rosen and J.D. Rosen, *J. Chromatogr.*, 1984, **283**, 223.
219 K. Tanaka, R. Amano, K. Kawada and H. Tanabe, *J. Food Hyg. Soc. Jpn.*, 1974, **15**, 195.
220 J. Gilbert, J.R. Startin and C. Crews, *J. Chromatogr.*, 1985, **319**, 376.
221 C.E. Kientz and A. Verweij, *J. Chromatogr.*, 1986, **355**, 229.
222 J. Gilbert, M.J. Shepherd and J.R. Startin, *J. Sci. Food Agric.*, 1983, **34**, 86.
223 R. Kostiainen and A. Rizzo, *Anal. Chim. Acta*, 1988, **204**, 233.
224 T. Krishnamurthy and E.W. Sarver, *J. Chromatogr.*, 1986, **355**, 253.
225 P.M. Scott, S.R. Kanhere and D. Weber, *Food Addit. Contam.*, 1993, **10**, 381.
226 B.J. Wreford and K.J. Shaw, *Food Addit. Contam.*, 1987, **5**, 141.
227 C.J. Mirocha, S.V. Pathre, R.J. Pawlosky and D.W. Hewetson, in *Modern Methods in the Analysis and Structural Elucidation of Mycotoxins*, Academic Press, 1986, pp. 353–392.
228 M. Schollenberger, U. Lauber, H.T. Jara, S. Suchy, W. Drochner and H.-M. Müller, *J. Chromatogr. A*, 1998, **815**, 123.
229 T. Krishnamurthy, M.B. Wasserman and E.W. Sarver, *Biomed. Environ. Mass Spectrom.*, 1986, **13**, 503.
230 C.J. Mirocha, R.J. Pawlosky and H.K. Abbas, *Arch. Environ. Contam. Toxicol.*, 1989, **18**, 349.
231 R. Kostiainen, A. Rizzo and A. Hesso, *Arch. Environ. Contam. Toxicol.*, 1989, **18**, 356.
232 R.D. Voyksner, W.M. Hagler, K. Tyczkowska and C.A. Haney, *J. High Res. Chromatogr. Chromatogr. Commun.*, 1985, **8**, 119.
233 R. Kostiainen, *J. Chromatogr.-Biomed. Appl.*, 1991, **562**, 555.
234 R. Kostiainen, K. Matsuura and K. Nojima, *J. Chromatogr.*, 1991, **538**, 323.
235 R.P. Huopalahti, J. Ebel, Jr. and J.D. Henion, *J. Liq. Chrom. Rel. Technol.*, 1997, **20**, 537.
236 T. Tuomi, L. Saarinen and K. Reijula, *Analyst*, 1997, **123**, 1835.

237 R.D. Smith and H.R. Udseth, *Biol. Mass Spectrom.*, 1983, **10**, 577.
238 R.D. Smith, H.R. Udseth and B.W. Wright, *J. Chromatogr. Sci.*, 1985, **23**, 192.
239 J.C. Young and D.E. Games, *J. Chromatogr. A*, 1993, **653**, 374.
240 P.M. Scott, *Int. J. Food Microbiol.*, 1993, **18**, 257.
241 G.S. Shephard, *J. Chromatogr. A*, 1998, **815**, 31.
242 J.C. Young and P. Lafontaine, *Rapid Commun. Mass Spectrom.*, 1993, **7**, 352.
243 E.W. Sydenham, W.C.A. Gelderblom, P.G. Thiel and W.F.O. Marasas, *J. Agric. Food Chem.*, 1990, **38**, 285.
244 R.D. Plattner, W.P. Norred, C.W. Bacon, K.A. Voss, R. Peterson, D.D. Shackelford and D. Weisleder, *Mycologia*, 1990, **82**, 698.
245 P.G. Thiel, W.F.O. Marasas, E.W. Sydenham, G.S. Shephard, W.C.A. Gelderblom and J.J. Nieuwenhuis, *Appl. Environ. Microbiol.*, 1991, **57**, 1089.
246 R.D. Plattner and B.E. Branham, *J. AOAC Int.*, 1994, **77**, 525.
247 A.A. Korfmacher, M.P. Chiarelli, J.O. Lay Jr., J. Bloom, M. Holcomb and K.T. McManus, *Rapid Commun. Mass Spectrom.*, 1991, **5**, 463.
248 M. Holcomb, J.B. Sutherland, M.P. Chiarelli, W.A. Korfmacher, H.C. Thompson Jr., J.O. Lay Jr., L.J. Hankins and C.E. Cerniglia, *J. Agric. Food Chem.*, 1993, **41**, 357.
249 W.P. Xie, C.J. Mirocha and J.P. Chen, *J. Agric. Food Chem.*, 1997, **45**, 1251.
250 R.A. Thakur and J.S. Smith, *Rapid Commun. Mass Spectrom.*, 1994, **8**, 82.
251 D.R. Doerge, P.C. Howard, S. Bajic and S. Preece, *Rapid Commun. Mass Spectrom.*, 1994, **8**, 603.
252 E.D. Caldas, A.D. Jones, C.K.Winter, B. Ward and D.G. Gilchrist, *Anal. Chem.*, 1995, **67**, 196.
253 Z. Lukacs, S. Schaper, M. Herderich, P. Schreier, P. and H.-U. Humpf, *Chromatographia*, 1996, **43**, 124.
254 J.L. Josephs, *Rapid Commun. Mass Spectrom.*, 1996, **10**, 1333.
255 H.B. Hines, E.E. Brueggemann, M. Holcomb and C.L. Holder, *Rapid Commun. Mass Spectrom.*, 1995, **9**, 519.
256 J.E. Roybal, R.K. Munns, W.J. Morris, J.A. Hurlbut and W.J. Shimoda, *J. Assoc. Off. Anal. Chem.*, 1988, **71**, 263.
257 J.C. Ryu, J.S. Yang, Y.S. Song, O.S. Kwon, J. Park and I.M. Chang, *Food Addit. Contam.*, 1996, **13**, 333.
258 K. Schwadorf and H.-M. Müller, *J. Chromatogr.*, 1992, **595**, 259.
259 E.R. Filho, W. Xie, C.J. Mirocha and L.R. Hogge, *Rapid Commun. Mass Spectrom.*, 1997, **11**, 1515.
260 E. Rosenberg, R. Krska, R. Wissiack, V. Kmetov, R. Josephs, E. Razzazi and M. Grasserbauer, *J. Chromatogr. A*, 1998, **819**, 277.
261 J.D. Rosen, R.T. Rosen, and T.G. Hartman, *J. Chromatogr.*, 1986, **355**, 241.
262 T. Krishnamurthy and E.W. Sarver, *Anal. Chem.*, 1987, **59**, 1272.
263 T. Krishnamurthy and E.W. Sarver, *Biomed. Environ. Mass Spectrom.*, 1988, **15**, 13.
264 T. Krishnamurthy and E.W. Sarver, *Biomed. Environ. Mass Spectrom.*, 1988, **15**, 185.
265 T. Krishnamurthy, D.J. Beck, R.K. Isensee and B.B. Jarvis, *J. Chromatogr.*, 1989, **469**, 209.
266 T. Krishnamurthy, D.J. Beck and R.K. Isensee, *Biomed. Environ. Mass Spectrom.*, 1989, **18**, 287.
267 T. Krishnamurthy, E.W. Sarver, S.L. Greene and B.B. Jarvis, *J. Assoc. Off. Anal. Chem.*, 1987, **70**, 132.
268 A. Joshi, Z. Min, W.C. Brumley, P.A. Dreifuss, G.C. Yang and J.A. Sphon, *Biomed. Mass Spectrom.*, 1984, **11**, 101.
269 M.E. Stack, P.B. Mislivec, J.A.G. Roach and A.E. Pohland, *J. Assoc. Off. Anal. Chem.*, 1985, **68**, 640.
270 P.M. Scott, D. Weber and S.R. Kanhere, *J. Chromatogr. A*, 1997, **765**, 255.
271 J.D. Rosen and S.R. Pareles, *J. Agric. Food Chem.*, 1974, **22**, 1024.

272 M. Kellert, W. Baltes, W. Blaas and M. Wittkowski, *Fresenius' Z. Anal. Chem.*, 1983, **315**, 245.
273 J.P. Chaytor and M.J. Saxby, *J. Chromatogr.*, 1981, **214**, 135.
274 E.J. Tarter and P.M. Scott, *J. Chromatogr.*, 1991, **538**, 441.
275 T. Urano, M.W. Trucksess, J. Matusik and J.W. Dorner, *J. AOAC Int.*, 1992, **75**, 319.
276 T. Urano, M.W. Trucksess, R.W. Beaver, D.M. Wilson, J.W. Dorner and F.E. Dowell, *J. AOAC Int.*, 1992, **75**, 838.
277 T. Goto, M. Matsui and T. Kitsuwa, *Proc. Jap. Assoc. Mycotoxicol.*, 1990, **31**, 43.
278 T.J.A. Lansden, *J. Assoc. Off. Anal. Chem.*, 1984, **67**, 728.
279 J. Gilbert, J.R. Startin, I. Parker, M.J. Shepherd, J.C. Mitchell and M.J. Perkins, *J. Chromatogr.*, 1986, **369**, 408.
280 J.C. Kim and Y.W. Lee, *Appl. Environ. Microbiol.*, 1994, **60**, 4380.
281 D.M. Goldberg, J. Yan, E. Ng, E.P. Diamandis, A. Karumanchiri, G. Soleas and A.L. Waterhouse, *Anal. Chem.*, 1994, **66**, 3959.
282 D.M. Goldberg, A. Karumanchiri, E. Ng, J. Yan, E.P. Diamandis and G. Soleas, *J. Agric. Food Chem.*, 1995, **43**, 1245.
283 D.M. Goldberg, J. Yan, E. Ng, E.P. Diamandis, A. Karumanchiri, G. Soleas and A.L. Waterhouse, *Am. J. Enol. Vitic.*, 1995, **46**, 159.
284 G.J. Soleas, D.M. Goldberg, E.P. Diamandis, A. Karumanchiri, J. Tan and E. Ng, *Am. J. Enol. Vitic.*, 1995, **46**, 346.
285 P. Jeandet, R. Bessis, B.F. Maume, P. Meunier, D. Peyron and P. Trollat, *J. Agric. Food Chem.*, 1995, **43**, 316.
286 C. Kraus, G. Spiteller, A. Mithofer and J. Ebel, *Phytochemistry*, 1995, **40**, 739.
287 S. Natori and I. Ueno, in *Naturally Occurring Carcinogens of Plant Origin*, ed. I. Hirono, Elsevier, Amsterdam, 1987, pp. 53–85.
288 M.G.L. Hertog, E.J.M. Feskens, P.C.H. Hollman, M.B. Katan and D. Kromhaut, *Lancet*, 1993, **342**, 1007.
289 M. Namiki, *CRC Crit. Rev. Food Sci. Nutr.*, 1990, **29**, 273.
290 C. de Whalley, S.M. Rankin, J.R.S. Hoult, W. Jessup and D.S. Leake, *Biochem. Pharmacol.*, 1990, **39**, 1743.
291 J.V. Formica and W. Regelson, *Food Chem. Toxicol.*, 1995, **33**, 1061.
292 P.C.H. Hollman, J.H.M. de Vries, S.D. van Leeuwen, M.J.B. Mengelers and M.B. Katan, *Am. J. Clin. Nutr.*, 1995, **62**, 1276.
293 T.J. Mabry and K.R. Markham, in *The Flavonoids*, ed. J.B. Harborne, T.J. Mabry and H. Mabry, Chapman and Hall, London, 1975, pp. 78–126.
294 T.J. Mabry and A. Ululeben, in *Biochemical Applications of Mass Spectrometry*, ed. G.R. Waller and O. Dermer, Wiley, New York, 1980, pp. 1131–1158.
295 G.P. Rizzi and S.S. Boeing, *J. Agric. Food Chem.*, 1984, **32**, 551.
296 P.A. Hedin and V.A. Phillips, *J. Agric. Food Chem.*, 1992, **40**, 607.
297 K.R. Markham, in *The Flavonoids: Advances in Research*, ed. J.B. Harborne, Chapman and Hall, London, 1988, pp. 427–468.
298 K.R. Markham, in *Methods in Plant Biochemistry, Vol. 1*, ed. J.B. Harborne, Academic Press, London, 1989, pp. 197–235.
299 R.J. Grayer, in *Methods in Plant Biochemistry, Vol. 1*, ed. J.B. Harborne, Academic Press, London, 1989, pp. 283–323.
300 N.N. Mollova, V.S. Bankova and S.S. Popov, *Org. Mass Spectrom.* 1987, **22**, 334.
301 H. Itokawa, Y. Oshida, A. Ikuta and Y. Shida, *Chem. Lett.*, 1982, **1**, 49.
302 H.-R. Schulten and D.E. Games, *Biomed. Mass Spectrom.*, 1974, **1**, 120.
303 A. Sakushima, F.B. West and H. Brandenberg, *Abstr. Lyomasu Kenkyu-kai*, 1984, **9**, 217.
304 M. Stobjecki, W. Olechnowick-Stepien, H. Rzadkowska-Bodalska, W. Cisowski and E. Budko, *Biomed. Environ. Mass Spectrom.*, 1988, **15**, 589.
305 I.V. Sokolov, V.I. Gunar, M.I. Nikolaeva and G.D. Tantsyrev, *J. Anal. Chem. USSR*, 1988, **43**, 918.

306 J.-L. Wolfender, M. Maillard, A. Marston and K. Hostettmann, *Phytochem. Anal.*, 1992, **3**, 193.
307 N. Saito, C.F. Timberlake, O.G. Tucknott and I.A.S. Lewis, *Phytochemistry*, 1983, **22**, 1007.
308 P. Bridle, R.S.T. Loeffler, C.F. Timberlake and R. Self, *Phytochemistry*, 1984, **23**, 2968.
309 K. Odaka, N. Terahara, N. Saito, K. Toki and T. Honda, *Phytochemistry*, 1992, **31**, 2127.
310 D. Strack and V. Wray, in *Methods in Plant Biochemistry, Vol. 1*, ed. J.B. Harborne, Academic Press, London, 1989, pp. 325–356.
311 E.M. Martinelli, *Rec. Dev. Mass Spectrom.*, 1981, **7**, 307.
312 V. Bankova, A. Dyulgerov, S. Popov and N.Z. Marekov, *Z. Naturforsch.*, 1986, **42c**, 147.
313 E. Benfenati, R. Frassanito, N. di Toro, R. Fanelli, A. Brandt, M. di Rella and L. Cecchetelli, *Nat. Prod. Lett.*, 1994, **4**, 247.
314 T.J. Schmidt, I. Merfort and U. Matthiesen, *J. Chromatogr.*, 1993, **634**, 350.
315 T.J. Schmidt, I. Merfort and G. Willuhn, *J. Chromatogr.*, 1994, **669**, 236.
316 T. Berahia, E.M. Gaydou, C. Cerrat and J.C. Wallet, *J. Agric. Food Chem.*, 1994, **42**, 1697.
317 H. Adlercreutz, T. Fotsis, C. Bannwart, K. Wähälä, G. Brunow and T. Hase, *Clin. Chim. Acta*, 1991, **199**, 263.
318 H. Adlercreutz, T. Fotsis, J. Lampe, K. Wähälä, G. Mäkelä, G. Brunow and T. Hase, *Scand. J. Clin. Lab. Invest.*, 1993, **53** (suppl. 215), 5.
319 H. Adlercreutz, T. Fotsis, M.S. Kurzer, K. Wähälä, T. Mäkelä, and T. Hase, *Anal. Biochem.*, 1995, **225**, 101.
320 S. Baba, T. Furuta, M. Horie and H. Nakagawa, *J. Pharm. Sci.*, 1981, **70**, 780.
321 G.E. Kelly, C. Nelson, M.A. Waring, G.E. Joannou and A.Y. Reeder, *Clin. Chim. Acta*, 1993, **223**, 9.
322 G.E. Kelly, G.E. Joannou, A.Y. Reeder, C. Nelson and M.A. Waring, *Proc. Soc. Expt. Biol. Med.*, 1995, **208**, 40.
323 G.E. Joannou, G.E. Kelly, A.Y. Reeder, M. Waring and C. Nelson, *J. Steroid Biochem. Mol. Biol.*, 1995, **54**, 167.
324 D.G. Watson and A.R. Pitt, *Rapid Commun. Mass Spectrom.*, 1998, **12**, 153.
325 D.G. Watson and E.J. Oliveira, *J. Chromatogr.*, 1999, **723**, 203.
326 T.R. Baker, K.R. Wehmeyer, G.R. Kelm, L.J. Tulich, D.L. Kuhlenbeck, D.J. Dobrozsi and J.V. Penafiel, *J. Mass Spectrom.*, 1995, **30**, 438.
327 D.E. Games and F. Martinez, *J. Chromatogr.*, 1989, **474**, 372.
328 D.E. Games, M.A. McDowall, K. Levsen, K.H. Schafer, P. Dobberstein and J.L. Gower, *Biomed. Environ. Mass Spectrom.*, 1984, **11**, 87.
329 R. Schuster, *Chromatographia*, 1980, **13**, 379.
330 J. Iida and H. Murata, *Anal. Sci.*, 1990, **6**, 269.
331 Y.Y. Lin, K.J. Ng and S. Yang, *J. Chromatogr.*, 1993, **629**, 389.
332 E. Schröder and I Merfort, *Biol. Mass Spectrom.*, 1991, **20**, 11.
333 B. Ducrey, J.L. Wolfender, A Marston and K. Hostettmann, *Phytochemistry*, 1995, **38**, 129.
334 L.W. Sumner, N.L. Paiva, R.A. Dixon and P.W. Geno, *J. Mass Spectrom.*, 1996, **31**, 472.
335 S. Barnes, M. Kirk and L. Coward, *J. Agric. Food Chem.*, 1994, **42**, 2466.
336 K.A. Barnes, R.A. Smith, K. Williams, A.P. Damant and M.J. Shepherd, *Rapid Commun. Mass Spectrom.*, 1998, **12**, 130.
337 U. Justesen, P. Knuthsen and T. Leth, *J. Chromatogr.*, 1998, **799**, 101.
338 X.-G. Hue, L.-Z. Lin and L.-Z. Lian, *J. Chromatogr.*, 1996, **755**, 127.
339 P. Miketova, K.H. Schram, J.L. Whitney, E.H. Kerns, S. Valcic, B.N. Timmermann and K.J. Volk, *J. Nat. Prod.*, 1998, **61**, 461.
340 A. Rafaelli, G. Moneti, V. Mercati and E. Toja, *J. Chromatogr. A*, 1997, **777**, 223.

341 M. Carini, R. Maffei Facino, G. Aldini, M. Calloni and L. Colombo, *Rapid Commun. Mass Spectrom.*, 1998, **12**, 1813.
342 H.L. Constant, K. Slowing, J.G. Graham, J.M. Pezzuto, G.A. Cordell and C.W.W. Beecher, *Phytochem Anal.*, 1997, **8**, 176.
343 E. Gariboldi, D. Mascetti, G. Galli, P. Caballion and E. Bosiso, *Pharm. Res.*, 1998, **15**, 936.
344 L. Coward, M. Kirk, N. Albin and S. Barnes, *Clin. Chim. Acta*, 1996, **247**, 121.
345 S. Barnes, L. Coward, M. Kirk and J. Sfakianos, *Proc. Soc. Expt. Biol. Med.*, 1998, **217**, 254.
346 M.A. Armendia, I. Garcia, F. Lafont and J.M. Marinas, *J. Chromatogr. A*, 1995, **707**, 327.
347 H. Tamura, Y. Hayashi, H. Sugisawa and T. Kondo, *Phytochem. Anal.*, 1994, **5**, 190.
348 A.J. Baublis and M.D. Berber-Jimenez, *J. Agric. Food Chem.*, 1995, **43**, 640.
349 A. Baldi, A. Romani, N. Mulinacci, F.F. Vincieri and B. Casetta, *J. Agric. Food Chem.*, 1995, **43**, 2104.
350 W.E. Glaessgen, H.U. Seitz and J.W. Metzger, *Biol. Mass Spectrom.*, 1992, **21**, 271.
351 M.A. Armendia, V. Borau, I. Garcia, C. Jimenez, F. Lafont, J.M. Marinas and A. Porras, *Rapid Commun. Mass Spectrom/J. Mass Spec. Comb.*, 1995, S153.
352 M. Becchi and D. Fraisse, *Biomed. Environ. Mass Spectrom.*, 1989, **18**, 122.
353 Q.M. Li, H. van den Heuvel, L. Dillen and M. Claeys, *Biol. Mass Spectrom.*, 1992, **21**, 213.
354 Q.M. Li and M. Claeys, *Biol. Mass Spectrom.*, 1994, **23**, 406.
355 F.W. Crow, K.B. Tomer, J.H. Looker and M.L. Gross, *Anal. Biochem.*, 1986, **155**, 286.
356 M.S. Lee, D.J. Hook, E.H. Kerns, K.J. Volk and I.E. Rosenberg, *Biol. Mass Spectrom.*, 1993, **22**, 84.
357 G.K. Poon, *J. Chromatogr. A*, 1998, **794**, 63.
358 K.V. Wood, C.C. Bonham, J. Ng, J. Hipskind and R.L. Nicholson, *Rapid Commun. Mass Spectrom.*, 1993, 7, 400.
359 K.V. Wood, C. Bonham, J. Hipskind and R.L. Nicholson, *Phytochemistry*, 1994, **37**, 557.
360 R. Fenz, L. Ernst and R. Galensa, *Z. Lebensm. Unters. Forsch.*, 1992, **194**, 252.
361 G.J. Soleas, J. Dam, M. Carey and D.M. Goldberg, *J. Agric. Food Chem.*, 1997, **45**, 3871.
362 E. Kozukue, N. Kozukue, H. Tsuchida, *J. Jpn. Soc. Hort. Sci.*, 1998, **67**, 805.
363 J.J. Boon, in *Physichochemical Characterization of Plant Residues for Industrial and Feed Use*, ed. A. Chesson and E.R. Ørskov, Elsevier Applied Science, London, 1989, pp. 25–49.
364 G.C. Galletti, B. Mincione, F.A. Mellon and K.W. Waldron, *Ital. J. Food Sci.*, 1993, 389.
365 F.A. Mellon, K.W. Waldron, R.R. Selvendran and G.C. Galletti, *Org. Mass Spectrom.*, 1995, **29**, 556.
366 G.C. Galletti, J. Eagles and F.A. Mellon, *J. Sci. Food Agric.*, 1992, **59**, 401.
367 C. Bocchi, M. Careri, F. Groppi, A. Mangia, P. Manini and G. Mori, *J. Chromatogr. A*, 1996, **753**, 157.
368 A.M. Gioacchini, A. Roda, G.C. Galletti, P. Bocchini, A.C. Manetta and M. Baraldini, *J. Chromatogr. A* 1996, **730**, 31.
369 M.A. Aramendia, V. Boráu, I. Garcia, C. Jiménez, F. Lafont, J.M. Marinas and F.J. Urbano, *Rapid Commun. Mass Spectrom.*, 1996, **10**, 1585.
370 I. Horman and R. Viani, *Org. Mass Spectrom.*, 1971, **5**, 203.
371 R.J. Horvat and S.D. Senter, *J. Agric. Food Chem.*, 1980, **28**, 1292.
372 D.G.I. Kingston and H.M. Fales, *Tetrahedron*, 1973, **29**, 4083.
373 G.-I. Nonaka, S. Morimoto and I. Nishioka, *J. Chem. Soc. Perkin Trans. 1*, 1983, 2139.

374 R. Self, J. Eagles, G.C. Galletti, I. Mueller-Harvey, A.G.H. Lea, D. Magnolato, U. Richli, R. Gujer and E. Haslam, *Biomed. Environ. Mass Spectrom.*, 1986, **13**, 449.
375 N. Vivas, G. Bourgeois, C. Vitry, Y. Glories and V. de Freitas, *J. Sci. Food Agric.*, 1996, **72**, 309.
376 V.A.P. de Freitas, Y. Glories, G. Bourgeois and C. Vitry, *Phytochemistry*, 1998, **49**, 1435.
377 M. Ohnishi-Kameyama, A. Yanagida, T. Kanda and T. Nagata, *Rapid Commun. Mass Spectrom.*, 1997, **11**, 31.
378 F.A. Mellon, in *The Production and Utilization of Lignocellulosics*, ed. G.C. Galletti, Elsevier Applied Science, London, 1991, pp. 151–162.
379 A. Kiene and U.H. Engelhardt, *Z. Lebensm. Unters. Forsch.*, 1996, **202**, 48.
380 R.G. Bailey, H.E. Nursten and I. McDowell, *J. Sci. Food Agric.*, 1992, **59**, 365.
381 J.-L. Wolfender and K. Hostettmann, *J. Chromatogr.*, 1993, **647**, 191.
382 J.-L. Wolfender, S. Rodriguez, K. Hostettmann and W. Wagner-Redeker, *J. Mass Spectrom. Rapid Commun. Mass Spectrom. Comb.*, 1995, S35.
383 S. Guyot, T. Doco, J.-M. Souquet, M. Moutounet and J.-F. Drilleau, *Phytochemistry*, 1997, **44**, 351.
384 V. Cheynier, T. Doco, H. Fulcrand, S. Guyot, E. Le Roux, J.M. Souquet, J. Rigaud and M. Moutounet, *Analusis*, 1997, **25**, M32.
385 X. Wan, H.E. Nursten, Y. Cai, A.L. Davis, J.P.G. Wilkins and A.P. Davies, *J. Sci. Food Agric.*, 1997, **74**, 401.
386 J.R. Lewis, A.L. Davis, Y. Cai, A.P. Davis, J.P.G. Wilkins and M. Pennington, *Phytochemistry*, 1998, **49**, 2511.
387 J.J. Dalluge, B.C. Nelson, J.B. Thomas, M.J. Welch and L.C. Sander, *Rapid Commun. Mass Spectrom.*, 1997, **11**, 1753.
388 C.-W. Chen, Y.-W. Chang and L.S. Hwang, *J. Food Drug Anal.*, 1998, **6**, 713.
389 S. de Pascual-Teresa, Y. Gutierrez-Fernandez, J.C. Rivas-Gonzalo and C. Santos-Buelga, *Phytochem. Anal.*, 1998, **9**, 21.
390 J.F. Hammerstone, S.A. Lazarus, A.E. Mitchell, R. Rucker and H.H. Schmitz, *J. Agric. Food Chem.*, 1999, **47**, 490.
391 H. Fulcrand, S. Remy, J.-M. Souquet, V. Cheynier and M. Moutonet, *J. Agric. Food Chem.*, 1999, **47**, 1023.
392 G.C. Galletti and A. Antonelli, *Rapid Commun. Mass Spectrom.*, 1993, **7**, 656.
393 G.C. Galletti and J.B. Reeves, *Org. Mass Spectrom.*, 1992, **27**, 226.
394 G.C. Galletti and P. Bocchini, *Rapid Commun. Mass Spectrom.*, 1995, **9**, 250.
395 R.G. Ruiz, K.R. Price, A.E. Arthur, M.E. Rose, M.J.C. Rhodes and R.G. Fenwick, *J. Agric. Food Chem.*, 1996, **44**, 1526.
396 S. Kudou, M. Tonomura, C. Tsukamuto, M. Shimoyamada, T. Uchida and K. Okubo, *Biosci. Biotechnol. Biochem.*, 1992, **1**, 142.
397 D. Oakenfull and G.S. Sidhu, in *Toxicants of Plant Origin, Vol. II, Glycosides*, ed. P.R. Cheeke, CRC Press, Boca Raton, 1989, pp. 97–141.
398 G.R. Fenwick, K.R. Price, C. Tsukamoto and K. Okubo, in *Toxic Substances in Crop Plants*, ed J.P.F. D'Mello, C.M. Duffus and J.H. Duffus, Royal Society of Chemistry, Cambridge, 1991, pp. 285–327.
399 H. Budzikiewicz, C. Djerassi and D.H. Williams, in *Structure Elucidation of Natural Products by Mass Spectrometry*, Vol. 2, Holden-Day, San Francisco, 1964, pp. 110–120.
400 J. Karliner and C. Djerassi, *J. Org. Chem.*, 1966, **31**, 1945.
401 E. Heftmann, R.E. Lundin, W.F. Haddon, I. Peri, U. Mor and A. Bondi, *J. Nat. Prod.*, 1979, **42**, 410.
402 S.K. Agarwal and R.P. Rastogi, *Phytochemistry*, 1974, **13**, 2623.
403 K. Shiojima, Y. Arai, K. Masuda, Y. Takase, T. Ageta and H. Ageta, *Chem. Pharm. Bull.*, 1992, **40**, 1683.
404 R. Maffei Facino, M. Carinin, P. Traldi, B. Pelli, B. Gioia and E. Arlandini, *Biomed. Environ. Mass Spectrom.*, 1987, **14**, 187.

405 T. Komori, Y. Ida, Y. Mutou, K. Miyahara, T. Nohara and T. Kawasaki, *Biomed. Mass Spectrom.*, 1975, **2**, 65.
406 R. Higuchi, T. Komori and T. Kawasaki, *Chem. Pharm. Bull.*, 1976, **24**, 2610.
407 K. Hostettmann and A. Marston, in *Saponins*, Cambridge University Press, 1995, pp. 175–231.
408 E.M. Martinelli, *Eur. J. Mass Spectrom.*, 1980, **1**, 33.
409 A. Tava, W. Oleszek, M. Jurzysta, N. Berardo and M. Odoardi, *Phytochem. Anal.*, 1993, **4**, 269.
410 M. Itoh, N. Asakawa, Y. Hashimoto, M. Ishibashi and H. Miyazaki, *Yakugaku Zasshi*, 1985, **105**, 1150.
411 Y. Sauvaire, G. Ribes, J.C. Baccou and M.M. Loubatieresmariani, *Lipids*, 1991, **26**, 191.
412 H. Bjorndal, C.G. Hellerqvist, B. Lindberg and S. Svensson, *Angew. Chem. Int. Edn.*, 1970, **9**, 610.
413 R.R. Selvendran, in *Recent Developments in Mass Spectrometry in Biochemistry, Medicine and Environmental Research*, ed. A. Frigerio. Elsevier, Amsterdam, 1983, pp. 159–203.
414 A. Debettigniesdutz, G. Reznicek, B. Kopp and J. Jurenitsch, *J. Chromatogr.*, 1991, **547**, 299.
415 D. Fraisse, J.C. Tabet, M. Becchi and J. Raynaud, *Biomed. Environ. Mass Spectrom.*, 1986, **13**, 1.
416 K.R. Price, J. Eagles and G.R. Fenwick, *J. Sci. Food Agric.*, 1988, **42**, 183.
417 B.S. Joshi, K.M. Moore, S.W. Pelletier, M.S. Puar and B.N. Pramanik, *J. Nat. Prod.*, 1992, **55**, 1468.
418 M. Lovon, A. Osagie and K.R. Price, *J. Sci. Food Agric.*, 1995, **67**, 169.
419 G. Onning, M.A. Juillerat, L. Fay and N.G. Asp, *J. Agric Food Chem.*, 1994, **42**, 2578.
420 M. Hattori, Y. Kawata, N. Kakiuchi, K. Matsuura and T. Namba, *Shoyakugaku Zasshi*, 1988, **42**, 228.
421 Y. Arai, T. Hanai, A. Nosaka and K. Yamaguchi, *J. Liq. Chrom.*, 1990, **13**, 2449.
422 M.P. Maillard, J.L. Wolfender and K. Hostettmann, *J. Chromatogr.*, 1993, **647**, 147.
423 J.L. Wolfender, K. Hostettmann, F. Abe, T. Nagao, H. Okabe and T. Yamauchi, *J. Chromatogr.*, 1995, **712**, 155.
424 M.K. Lee, Y.C. Ling, M. Jurysta and G.R. Waller, *Abstr. Pap. ACS*, 1995, **210**, 140.
425 R.B. Van Breemen, C.R. Huang, Z.Z. Lu, A. Rimando, H.H.S. Fong and J.F. Fitzloff, *Anal. Chem.*, 1995, **67**, 3985.
426 Y.-Z. Chen, N.-Y. Chen, H.-Q. Li, F.-Z. Zhao and N. Chen, *Acta Chim. Sinica*, 1986, **44**, 1020.
427 Y.-Z. Chen, N.-Y. Chen, H.-Q. Li, F.-Z. Zhao and N. Chen, *Chem. J. Chinese Univ.*, 1986, **7**, 883.
428 Y.-Z. Chen, N.-Y. Chen, H.-Q. Li, F.-Z. Zhao and N. Chen, *Biomed. Environ. Mass Spectrom.*, 1987, **14**, 9.
429 G.R. Waller, P.R. West, C.S. Cheng, Y.C. Ling and C.H. Chou, *Bot. Bull. Acad. Sinica*, 1993, **34**, 323.
430 M.R. Lee, J.S. Lee, J.C. Wang and G.R. Waller, *Abstr. Papers ACS*, 1995, **210**, 134.
431 S. Fang, C. Hao, W. Sun, Z. Liu and S. Liu, *Rapid Commun. Mass Spectrom.*, 1998, **12**, 589.
432 D.C. van Setten, G.J. ten Hove, E.J.H.J. Wiertz, J.P. Kamerling and G. van den Werken, *Anal. Chem.*, 1998, **70**, 4401.

Amino Acids, Peptides and Proteins

1 Introduction

The mass spectrometry of amino acids, peptides and proteins is appropriately introduced by acknowledging the contribution of the late Prof. Michael (Mickey) Barber. When working at AEI Ltd on the development of mass spectrometers, he collaborated with Prof. Edgar Lederer at Gif-sur-Yvette to sequence the lipopolypeptide fortuitine by EIMS.[1] Fortuitine, as its name suggests, had a uniquely favourable structure for analysis by electron ionisation. However, it provided evidence of the potential of mass spectrometry in peptide sequencing. This initiated a period of intense activity to manipulate peptide structures chemically in order to increase their volatility and facilitate cleavage of peptide bonds. This resulted in an improvement in the quality of the amino acid sequence information derived from mass spectra. Through the introduction of FAB in 1981, Mickey Barber made an even greater contribution, enabling molecular mass information (and sometimes full or partial sequence information) to be obtained from underivatised peptides, thus eliminating the intricate chemical procedures developed for EI.[2] Complete sequence information is often unnecessary since FAB-mapping, the recording of molecular ions for mixtures of peptides in an enzyme digestion, can be used to confirm the expected peptide or protein.[3]

Tandem mass spectrometry is ideally suited to amino acid sequencing. The CID fragmentation of molecular ions in tandem MS often gives full sequence information on otherwise intractable structures. A system for classifying sequence ions of peptides was suggested in 1984,[4,5] subsequently modified[6] and is now in general use. Fragment ions formed with charge retention on the N-terminus are designated a_n, b_n, c_n and ions where charge is retained on the C-terminal fragment x_n, y_n, z_n (Figure 5.1).

Table 5.1 shows the monoisotopic and average mass for amino acid residues.

The MS/MS spectrum of [Glu'] Fibrinopeptide B, formed by collisional activation of the $[M + 2H]^{2+}$ ion at m/z 786, is shown in Figure 5.2. The 'y' type

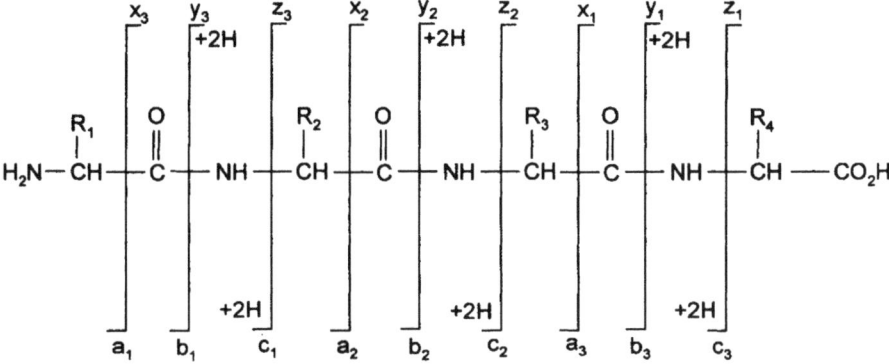

Figure 5.1 *Classification of peptide fragment ions, according to the nomenclature of Roepstorff and Fohlmann,[4,5] as modified by Biemann[6]*

Table 5.1 *Masses of common amino acid residues $NHCH(R)CO$*

Amino acid	Three letter code	One letter code	Monoisotopic mass of amino acid residue	Average mass of amino acid residue
Glycine	Gly	G	57.02146	57.0520
Alanine	Ala	A	71.03711	71.0788
Serine	Ser	S	87.03203	87.0782
Proline	Pro	P	97.05276	97.1167
Valine	Val	V	99.06841	99.1326
Threonine	Thr	T	101.04768	101.01051
Cysteine	Cys	C	103.00919	103.1448
Isoleucine	Ile	I	113.08406	113.1595
Leucine	Leu	L	113.08406	113.1595
Asparagine	Asn	N	114.04293	114.1039
Aspartic acid	Asp	D	115.02694	115.0886
Lysine	Lys	K	128.09496	128.1742
Glutamine	Gln	Q	128.05858	128.1308
Glutamic acid	Glu	E	129.04259	129.1155
Methionine	Met	M	131.04049	131.1986
Histidine	His	H	137.05891	137.1412
Phenylalanine	Phe	F	147.06841	147.1766
Arginine	Arg	R	156.10111	156.1876
Tyrosine	Tyr	Y	163.06333	163.1760
Tryptophan	Trp	W	186.07931	186.2133

ions show up clearly and can be used to confirm the partial sequence of the peptide.

The principles and practice of four-sector tandem mass spectrometry for peptide and protein structure determination and its complementarity with the Edman methodology has been discussed by Biemann.[7] Hunt, in the same volume, presented his approach to protein sequence analysis using triple quadrupole instrumentation.[8] A comprehensive review of the use of tandem

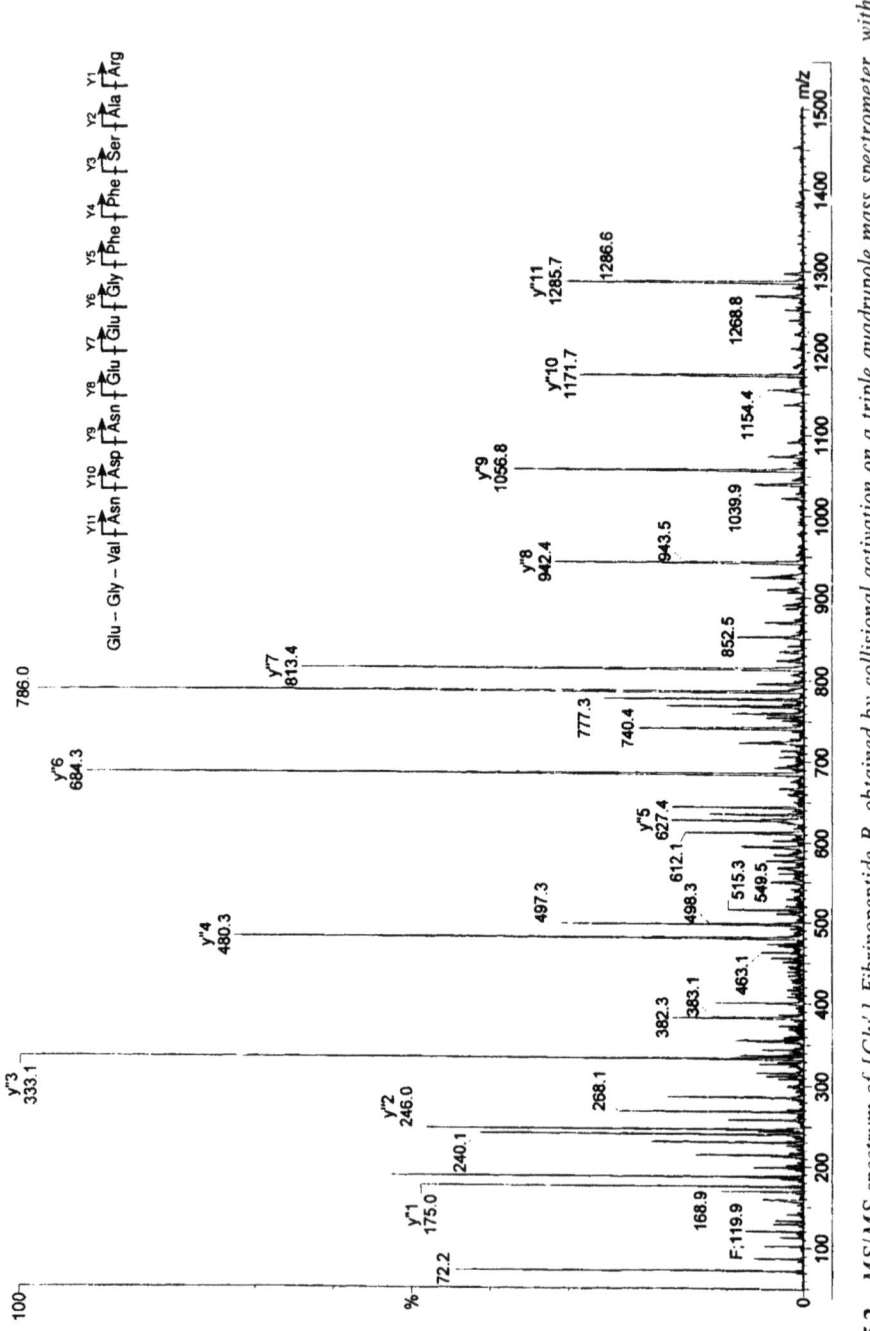

Figure 5.2 MS/MS spectrum of [Glu¹] Fibrinopeptide B, obtained by collisional activation on a triple quadrupole mass spectrometer, with the y sequence ions marked

mass spectrometry in pharmacology and toxicology shows clearly the potential for similar studies in food science.[9] An authoritative overview of hybrid tandem mass spectrometry concludes that their versatility warrants further study to confirm their potential for, *inter alia*, both low and high energy CAD experiments.[10] These reviews set the scene for the following discussion of applications relating to food science, now or in the near future.

Some amino acids (*e.g.*, Phe[11]) and peptides are sufficiently volatile to be analysed directly by EI- and CI-MS. However, the preparation of derivatives, such as the combined acetyl and permethyl derivatives, is essential for EI mass spectral analysis of peptides of up to about 2000 Da molecular weight. A practical guide to the use of EI mass spectrometry in conjunction with classical chemical degradation techniques summarised the state of the art in 1979, just prior to the introduction of FAB.[12] A recent example of the use of EI mass spectrometry for the analysis of amino acids used N(O,S)-isobutyloxycarbonylation derivatisation, followed by solid-phase extraction and *tert*-butyldimethylsilylation, to produce volatile compounds for GC/MS analysis.[13] Nineteen different food samples were screened for 17 protein and 15 non-protein amino acids. Eleven non-protein amino acids were tentatively identified. The EI mass spectra of many of the amino acid methyl esters are recorded in the commercial collections, such as the NIST database.

LSIMS was rapidly recognised by biologists as complementary to conventional chemical degradation sequencing procedures and was often used routinely, for example to check a sequence proposed from c-DNA coding, or to aid confirmation of the structures of proteins synthesised by recombinant DNA technology.

The introduction of ESI and MALDI to peptide and protein analysis has opened up new areas of application, since the practical molecular mass determining range can now be extended to beyond 300 000 Da. The gas-phase structure of electrosprayed proteins has been studied.[14] Ion–molecule reactions were carried out in an atmospheric pressure capillary inlet/reactor, based upon an ESI interface to a quadrupole mass spectrometer. Cysteine–cysteine disulfide-bound proteins were more reactive than disulfide-reduced proteins with equivalent charge, but no significant reactivity differences were found for solutions of proteins of different conformation. A detailed review of the application of ESI to protein structure elucidation was presented in the NATO ASI Series.[15]

An exciting area of renewed activity in mass spectrometry is the FTICR-MS; a single population of multiply-charged protein ions formed by ESI was trapped in an ion cell and remeasured continuously by FTICR while undergoing multiple reactions with diethylamine.[16] Electrosprayed horse heart myoglobin, with an average of 16 attached protons, formed up to three diethylamine adducts and up to nine protons were removed. FTICR has been applied to the characterisation of nisin, a natural antibiotic.[17]

The convenient determination of molecular masses up to approximately 300 000 Da by combining MALDI and ToF instrumentation with reflectron geometry has provided a fresh and relatively inexpensive approach to the analysis of peptides.[18] Considerable dissociation of MALDI-generated ions occurs after they leave the source (post source decay, PSD). This occurs in the field-free region before the reflectron and yields fragments with kinetic energies proportional to their size relative to the precursor ion mass. The reflectron treats these product ions differently from their precursors, and they appear at different locations along the time axis. Their mass measurement provides valuable precursor–product ion 'connecting' evidence, ideal for amino acid sequence determination. A novel method of sequencing peptides using PSD analysis without the necessity for scanning the reflectron has been introduced.[19] The curved-field reflectron (CFR) provides simultaneous focusing of product ions at a fixed reflectron voltage, avoiding the relatively insensitive and time-consuming need to obtain multiple mass spectra. Modifications to the CFR serve to improve performance and add automatic calibration of the product ion mass. CFR recovers the multichannel recording and implicit high sensitivity advantages of ToF mass analysis. The information content of CAD and PSD spectra of linear peptides in MALDI-TOF has been compared.[20] The authors show that CAD differentiates between Leu and Ile and reiterate that MALDI is more tolerant to sample impurities, such as buffer salts, than ESI or LSIMS.

An ion trap detector has been used for pyrolysis GC/MS studies of the amino acid-like fragments of aspartame.[21] Py-GC/MS in the presence of tetramethylammonium hydroxide produced thermally-assisted hydrolysis/methylation (THM) and this was compared with Py-GC/MS for the analysis of table-top sweeteners and a soft drink.

The entire topic of protein mass spectrometry in analytical biotechnology and biopharmaceutical research has been reviewed.[22] Topics discussed include mass determination, peptide mapping, peptide sequencing, ligand binding, determination of disulfide bonds, active site characterisation of enzymes, protein self-association and protein folding/higher-order structural characterisation. A comprehensive review of the literature on the MS of amino acids, peptides and proteins was published in 1994.[23]

Examples below are taken from the recent literature and illustrate the use of mass spectrometry in the analysis of the amino acids, peptides and proteins in foods.

2   Chemical and Enzymatic Studies

The chemistry of amino acids, peptides and proteins is of interest in several areas of food, nutrition and agricultural science. GC/MS lends itself to the analysis of the volatile products formed in amino acid/sugar reactions. A recent example is provided by antimutagenicity studies on dichloromethane extracts from eight amino acid/sugar systems to estimate Maillard reaction products.[24] This showed that pyrazines and furans were the major products in the extracts. The antimutagenicity was found to have a positive correlation with the total amounts of pyrazines and furans found in the dichloromethane extracts. FABMS was used in the study of biophenol–protein supramolecular models of hydroxytyrosol (a biophenol found in olives), and caffeine or Asp-Phe.[25] The authors observed consistently higher difference in stability constants, thus indicating a preferential molecular recognition site provided by caffeine, the biomimetic model of proline-rich mucoproteins. A detailed review of the chemistry of key intermediates in the Maillard reaction, the Amadori rearrangement products, shows how IR, NMR and MS can be used in the monitoring of their analysis, synthesis, kinetics, reactions and elucidation of their spectroscopic properties.[26] In particular, mass spectra of the free, protected, and protein-bound states under different ionisation conditions are presented. The utility of MS is put into perspective in this comprehensive review of Amadori product chemistry.

A tryptic digest of bovine β-casein was further digested with purified aminopeptidase N from *Lactococcus lactis* subsp. *cremoris* Wg2. Oligopeptides produced from the tryptic digest before and after treatment with the aminopeptidase were identified by *N*- and *C*-terminal amino acid sequences and by LC/MS.[27] The tryptic digestion of bovine β-casein has a strong bitter taste. Degradation with aminopeptidase decreased the number of hydrophobic peptides, resulting in a dramatic decrease in the bitterness of the reaction mixture.

Time-of-flight MS has been used to validate the molecular masses of components of food-grade protein hydrolysates separated and estimated by size exclusion chromatography (SEC).[28] Comparison showed that peptides of mass below 1200 Da were eluted as estimated by SEC.

A review of microbial protein cross-linking, emphasised the value of MS in the confirmation of molecular weights determined by SDS-PAGE.[29] MTGase, a transglutaminase involved in the catalysis of cross-linking of most proteins, caseins, etc. was found to have a molecular weight of approximately 38 000 Da by SDS-PAGE and 37 824 Da by MS.

Chiral phase GC/MS in the SIM mode has been used to evaluate the possible toxic effect of D-amino acids on rats fed 50 mg of the enantiomer per kg body weight daily.[30] Several D-acids were quantified in serum, liver, *etc.* An elegant technique for studying isomerisation of amino acids during cooking or food processing is based on chiral-phase GC/MS.[31] Because amino acids also isomerise during sample preparation (hydrolysis), it is necessary to distinguish between these molecules and those that racemise during cooking/processing. This was achieved by conducting the hydrolysis step in deuterated water and acid. Molecules undergoing inversion at this stage incorporate a single deuterium atom, distinguishing them from molecules inverting during cooking. The technique was subsequently used to study possible nutritional losses following food processing.[32]

3 Applications of Mass Spectrometry to Amino Acid, Peptide and Protein Analysis in Food, Nutrition and Agricultural Research

Chemical derivatisation and GC/MS continue to be used in the analysis of amino acids in foods. *N*-Trifluoroacetic-*N*-methyl ester derivatives of the amino acids of huitlacoche, an edible corn smut fungus eaten as a delicacy in Mexico, were estimated by GC/MS in the range 0.08–3.21 mg g^{-1}.[33]

In a study of seleno-amino acids by GC/MS Se-methylselenocysteine, a non-protein seleno-amino acid, was found in the plant tissue *Melilotus indica* L., a grassland legume, when grown in selenium-laden soils.[34] The authors considered that this compound might not become a toxic element in the food chain. Five Maillard products from the model reaction between selenomethionine and glucose were tentatively identified using GC and GC/MS.[35] Interest stemmed from the possible production of volatile seleno-compounds in food systems from these precursors.

ESI-MS has been compared with UV absorption as a detection method for the HPLC separation of domoic acid, a neurotoxic amino acid found in shellfish.[36] Solid-phase extraction techniques were used to clean up the sample for HPLC/UV and HPLC/ESIMS SIM comparisons down to 0.1 mg g^{-1} of domoic acid.

Urinary excretion products from children fed protein hydrolysate formulas containing modified cornstarch have been identified by GC/MS.[37] The molecular mass and fragment ions of nine compounds associated with the excretion of 2-(2′-octenyl)succinic acid (OSA) were consistent with the proposal that OSA is metabolised in human infants and children.

The glycation sites of lysozyme in a restricted water environment have been located with the aid of FAB-MS.[38] A 30 day incubation at 25 °C caused glycation at Lys-1, while a 3 day treatment at 50 °C, resulted in diglycation at Lys-1 and glycation at Lys-13 and Lys-33. Sphingomyelin was identified by FAB-MS and FAB-MS/MS as the active ingredient in Kefir, a fermented milk that enhances LFN-β secretion of a human osteosarcoma line mg-63[39] (which had been treated with a chemical inducer, poly l: poly C). Sphingosine and lysosphingomyelin also enhanced the activity while ceramide and cerebroside did not.

Phaseolin polypeptide subunits in a crystalline food protein isolate from lima beans (*Phaseolus lunatus*) were characterised by MS and partial amino acid sequences.[40] A glycosylated subunit of 26 240 Da was similar to a *C*-terminal segment of the phaseolin polypeptides of *Phaseolin vulgaris*, and a glycosylated subunit of 26 113 Da and its non-glycosylated variant of 24 249 Da were similar to an *N*-terminal segment of *P. vulgaris*.

In 1988, the structure of a food preservative, the polypeptide antibiotic Nisin, permitted in many countries in dairy products, was confirmed by FAB-MS and FAB-MS/MS[41] and by Cf-252 plasma desorption.[42,43] This work was subsequently reviewed.[44] As a non-toxic, non-odorous and colourless, heat-stable compound, nisin has ideal properties for food preservation and methods for its analysis were sought. The first study confirmed part of the amino acid sequence and the positions of the sulfur-bridged rings.[41] Later work confirmed the molecular mass and suggested the presence of two minor components.[42,43] The difference between using electrospray deposition and adsorption on nitrocellulose as sample preparation methods was also demonstrated in the last-mentioned references. Extensive mass spectrometric fragmentation occurred at inter-ring sites in samples prepared by electrospray sample deposition. Two fragment ions in the z series were formed by C-terminal cleavage of Lys-12 and Lys-22. Several a-series ions marked the C-termini of the three ring-to-chain Ala residues. Subsequent work on the effect of sample preparation on the fragmentation of nisin was reported.[44] The fragmentation patterns obtained by FAB and Cf-252 PDMS were compared, and a proposal made that fragmentation was not necessarily confined to the peptide chain.[45] However, in later work on similar Lantibiotics it was suggested that the ring locations could be reliably identified because of the lack of fragmentation of the lanthionine bridges.[46] More recently, molecular weights of two peptides purified from *Lactococcus lactis* ssp. *Lactis biovar diacetylactis* UL 719 were determined by MS as 3346.39 ± 0.4 Da and 3330.39 ± 0.27 Da. These data suggested that they might be the oxidised and native forms of nisin Z.[47] A definitive study using FTICR-MS was undertaken recently.[17] Several degradation products, variants and adducts of nisin were characterised. In particular, the [nisin + 18] component was found to be predominantly [Ser-33] nisin or 'minor nisin'. Variant nisins generated by manipulating secondary metabolism *in vivo* have also been characterised by FTICR-MS.[48,49] Part of the electrospray tandem FTICR-MS of modified nisins are shown in Figure 5.3.

The analysis of food-based peptides has been reviewed, giving LC/MS, continuous flow FABMS and tandem MS due prominence.[50]

An armoury of MS techniques was mustered to characterise protein-bound Amadori products.[51] This study of lactolated peptides in hydrolysates employed ESI-MS/MS. Decomposition of protonated molecules of these peptides yielded a product-ion spectrum that included a characteristic rearrangement and cleavage of the *O*-glycosidic bond, marked by a loss of 216 Da. Collision energy, precursor ion charge and type of peptide were studied in order to propose a strategy for detection of lactolated peptides from protein digests using LC/ESI-MS and LC/MS/MS in the neutral loss scanning mode. Picomole sensitivities were achieved.

LC/MS has been used to detect nitrated *p*-coumaric acid in a reaction mixture while studying how phenolic antioxidants prevent peroxynitrate-derived collagen modification *in vitro*.[52]

A MALDI-ToF technique has been developed as a rapid screening system to determine the presence of gliadins in food samples.[53] The method is based on the direct observation of the characteristic gliadin mass pattern in processed and unprocessed gluten-containing foods. Unfractionated protein complex mixtures gave multi-component signals between 30 and 55 kDa. This procedure is the first non-immunological alternative for the quantitative estimation of gluten gliadins in foods, with detection sensitivity equal to that of the ELISA method with which it was compared.[54] Further studies were reported in 1998,[55] including an extension to gluten avenins with a detection limit of 0.4 mg per 100 g food.[56] Figure 5.4 shows the MALDI-ToF spectrum of an unfractionated ethanol extract of wheat gliadins.

Food bacteriologists have used mass spectrometry to study the structure of immunity proteins. Characterisation of the protein conferring immunity to the

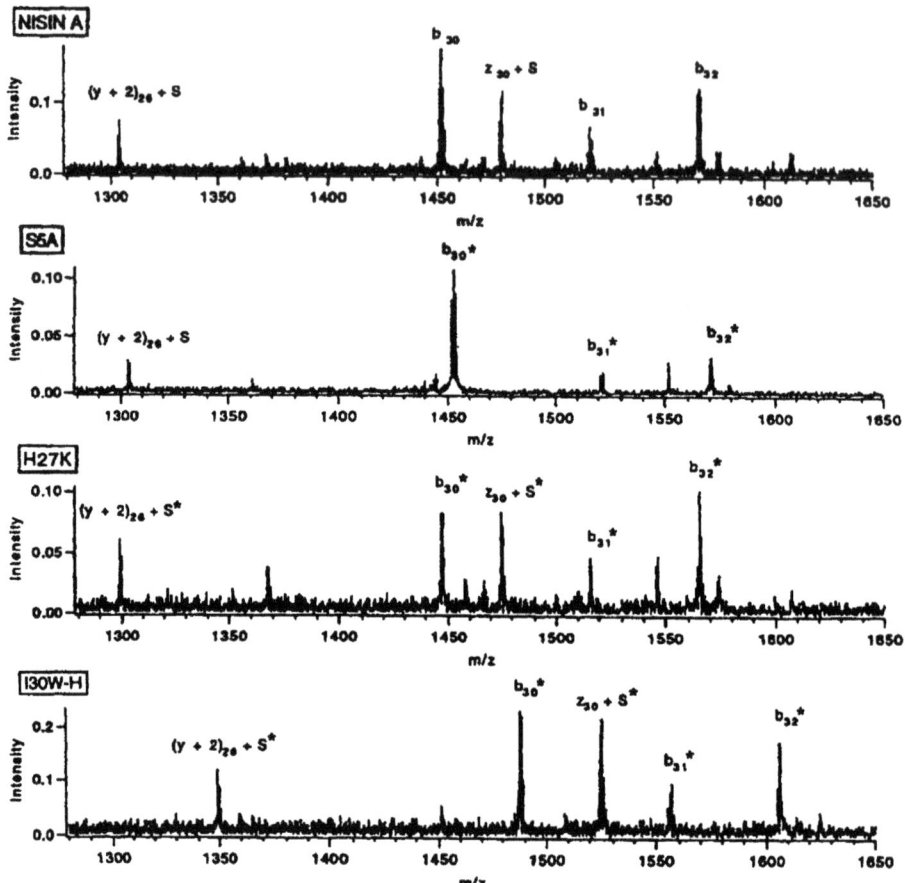

Figure 5.3   *The electrospray SORI-CAD FT-ICR mass spectra of the $[M + 3H]^+$
precursor ions of* (a) *nisin A (natural nisin) and the modified nisins S5A* (b),
H27K (c) *and I30W-H* (d) *showing the region containing doubly charged
fragment ions. The ions that are shifted in mass compared to nisin A are
marked with asterisks*
(Reproduced from H. Lavanant, A. Heck, P.J. Derrick, F.A. Mellon, A.
Parr, H.M. Dodd, C.J. Giffard, N. Horn and M.J. Gasson, *Eur. J. Mass
Spectrom.*, 1998, **4**, 405, © 1998 IM Publications)

antimicrobial peptide carnobacteriocin B2 confirmed the molecular mass as
$12\,662.3 \pm 3.4$ Da.[57] The same team used MS to determine the presence of a
disulfide bridge at Cys-22 and Cys-51 in carnobacteriocin A.[58] The molecular
mass of a 41 amino acid residue peptide, sakacin A, a bacteriocin from
lactobacillus sake Lb706 was also determined by mass spectrometry.[59] The
molecular mass of a peptide sakacin 674 from a different strain, sake Lb674, was
determined by the same group.[60]

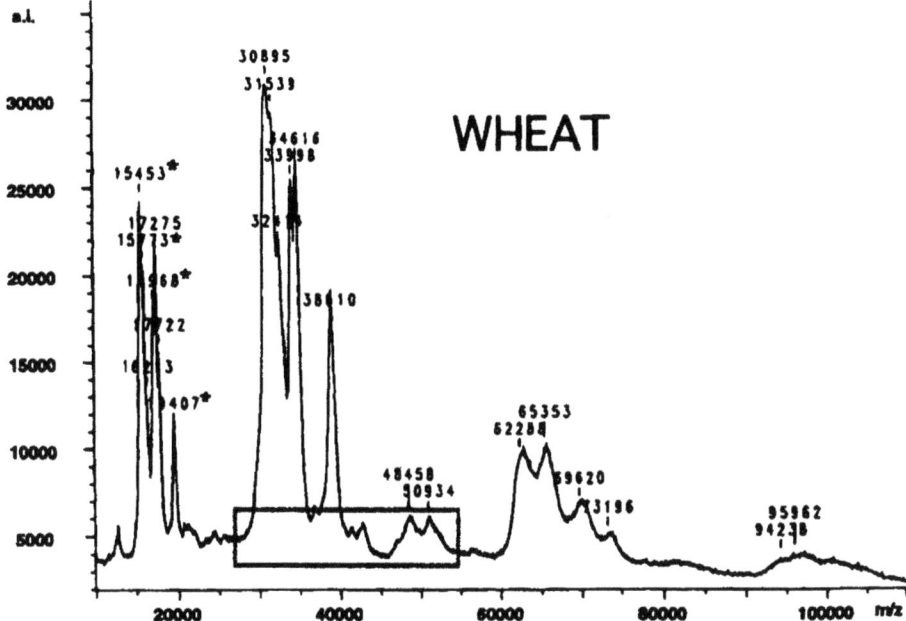

Figure 5.4 *MALDI-ToF spectrum of the unfractionated 70% ethanol extract of wheat endosperm. The characteristic gliadin region is indicated by a box. Asterisked peaks are doubly-charged ions*
Reproduced from E. Mendez, E. Camfieta, J. Sansebastian, I. Valle, J. Solis, F.J. Mayerposner, D. Suckau, C. Marfisi and F. Soriano, *J. Mass Spectrom./Rapid Commun. Mass Spectrom. (Comb. Issue)*, 1995, S123, with permission of the authors and publishers, © John Wiley & Sons Limited)

4 The Amino Acid Sequencing of Peptides and Proteins

The contribution of mass spectrometry to the sequencing of amino acids in peptides and proteins over the past 25 years has been seen as complementary to classical techniques. However, MS is unique in its capacity to provide information quickly and efficiently. For example, the sequences of peptides blocked by functional groups at the *N*-terminus are difficult to determine by classical methods but can often be read directly from the mass spectrum. The location of phosphate groups on mid-chain sites of a known peptide can be detected by the characteristic change in mass of the peptide bond fragment ion associated with the residue carrying the group. The sequence determination of unresolved mixtures is also possible. Perhaps the most useful service performed by MS in the modern protein biochemistry laboratory is to measure rapidly the molecular masses of isolated components of, for example, enzymatic hydrolysates of peptides, so that overlap sequences can be recognised for the rapid confirmation of protein structure.

In a discussion of hydrolysis and amino acid composition analysis of proteins in medical and food science research, it was claimed that '[classical] amino acid analysis is currently facing an enormous competition in the determination of the identity of proteins and amino acid homologs by the essentially faster mass spectrometry techniques'.[61]

Routinely, MS provides a reliable confirmation of molecular weight, especially useful where alternative methods yield ambiguous data. An orange cheese coryneform bacterium isolated from the surface of Gruyère produced an antimicrobial substance designated linenscin OC2. This protein inhibited the growth of Gram-positive food-borne pathogens including *Staphlycoccus aureus* and *Listeria monocytogenes*.[62] The molecular weight of linenscin OC2 was estimated by gel chromatography to be >285 000 Da. After further purification SDS-PAGE suggested a mass under 2412 Da and MS determined the definitive molecular mass of the biologically active band to be 1196.7 Da. Three antibacterial peptides have been purified from a pepsin digest of bovine lactoferrin.[63] All were cationic, *N*-terminal peptides and were characterised using MS to provide molecular masses, 3195 (residues 17–42), 2673 (residues 1–16 and 43–48, linked by a single disulfide bond) and 5851 (a disulfide-linked heterodimer from residues 1–48).

MS and MS/MS are also valuable for post-translational modification studies. A comprehensive overview of the mass spectrometry of post-translationally modified peptides contains practical details of the approach used.[64]

Although there are relatively few reports of the sequencing of food proteins and peptides by MS, the experiences from other biological areas are discussed in order to stimulate future work on food matrices. The major research groups have access to several mass spectrometers, covering the range of ionisation techniques and mass analysers, in conjunction with GC, HPLC and CE separation methods. Discussions below are partitioned according to the main techniques being used.

Laser Desorption Ionisation and Time-of-flight Mass Spectrometry

A high-resolution MALDI mass spectrometer with reflectron geometry has been used to produce laser-desorbed ions which underwent metastable decay in the first field-free region (1FFR).[18] Fragment ions from metastable decay were then mass analysed by adjusting the potentials of the reflectron according to the kinetic energies of the ions. In a later paper, the team went on to study the differences between nitrogen and helium dominated residual gas pressures in relation to ion stability and total fragment yield in ToF-MS.[65] In an interesting experiment with reflecting ToF correlation methods, a 2000 Da peptide with an acetyl group blocking the *N*-terminus and a primary alcohol at the *C*-terminus, was bombarded with 25 keV I ions while deposited on the surface of a solid target.[66] Precursor molecular ions and 'prompt' fragment ions (ions formed at or very near to the surface of the target) were accelerated and their metastable products studied to reveal the sequence of the original peptide. These data were recorded off-line on magnetic tape, providing a complete ion decay record for further study.

Sequences of biologically active peptides isolated from support-bound combinatorial peptide libraries have been obtained by MALDI.[67] 'A termination synthesis approach has been developed to encode each resin bead in support-bound combinatorial peptide libraries with the information needed to establish the sequence of the full-length products also contained on the beads'. MALDI was then used to rapidly sequence the appropriate peptides. An anti-gp120 monoclonal antibody was screened against a hexapeptide library and eight active peptides were isolated. Six of the eight peptides were shown to possess the exact recognition sequence for the antibody. A tryptic digestion of β-lactoglobulin was fractionated by IEC and PAGE. One fraction with an approximate molecular weight of 6.7 kDa was analysed by laser desorption MS. After dithiothreitol reduction, the exact masses of two peptides were established allowing the *C*-terminal lysine cleavage sites to be confirmed as Lys-100 and Lys-101 (residues 41–100 and 48–101).[68] Further MS and *C*-terminal analysis established that, prior to reduction, β-lactoglobulin *C*-terminal residues 149–162 were connected to core domain fragments. The inhibition of hydroxyl radical-induced crosslinking of peptides by food antioxidants has been studied by MALDI/ToF-MS.[69]

Exploitation of MALDI in conjunction with partial acid hydrolysis has yielded high femtomole to low picomole sensitivity and rapid molecular weight confirmation (and often partial sequence information).[70] This has earned the technique a place in their armoury of instrumentation vigorously applied to solving advanced biological problems.

State-of-the-art 7 Tesla FTMS with MALDI has been investigated to determine its efficacy in sequencing peptides.[71] Molecular ions of two nine-residue peptides were detected at 8000–17 900 resolution, with mass measurement accuracy of 8–81 ppm for all C-12 isotope ions. SID of the molecular ion species produced structurally useful information. Sample consumption was at the rate of 3 pmol per 100 laser shot spectrum. 10 μl of 1 nmol per μl solutions of peptide in water were added to 200 μl of 50 mM 2,5-dihydroxybenzoic acid in 0.1% trifluoroacetic acid in methanol solution and sprayed onto the probe tip.

CE separation and off-line MALDI-ToF detection in conjunction with dot-blot separation and immunological detection techniques were used to identify four allergens from hen egg white.[72]

Particle Desorption Ionisation

The problems associated with FABMS analysis of mixtures of peptides of different hydrophobicity has been studied.[73] In practice, the enzymic digestion of proteins and peptides gives rise to mixtures of oligopeptides, some of which do not yield FAB mass spectra. This was thought to result from their lower hydrophobicity, preventing them from competing successfully for a place in the surface of the matrix. It was found that, although hydrophobicity is a factor, it could not explain various observations made with enzyme digests of 18 proteins ranging from 10 to 67 kDa in molecular weight. The presence of specific amino acid side chains, capable of hydrogen bonding with the matrix, at strategic points in the peptide chain was instead considered to have a major bearing on whether or not the mass spectrum of a mixture component was apparent.

LSIMS has been successfully used to identify polar oligopeptides from the water-soluble fraction of goat cheese.[74] Some components in extracts purified by GFC, DEAE, and RP-HPLC were tentatively identified. Linked-scanning MS aided elucidation of the sequence of a hexapeptide and the sequence was confirmed by comparison with caprine α (S2) casein.

Phosphorylation of the *N*-terminus of oligopeptides produces intense molecular ions in FABMS.[75] Fragment ions from the *N*-phosphoryl terminal were intense whilst the *C*-terminal ion series was suppressed. Both the molecular ion and the *N*-phosphoryl fragment ions were accompanied by losses of alkene, said to reaffirm the molecular mass of the sample and the validity of the fragments formed from it.

Applications of LSIMS applied to the study of *O*-sulfation of single Tyr residues, an important post-translational modification reaction, have been reviewed.[76] Studies of LSIMS of multiple tyrosine-*O*-sulfates found in nature were described and a fragmentation mechanism for the proton catalysed desulphation process was proposed.

Electrospray Ionisation (and Ion-spray Techniques)

Two protein isoforms of 265 and 280 kDa mass have been studied by microbore reverse-phase HPLC/ESI-MS after separation by PAGE.[77] The sensitivity of the MS allowed 90% of the sub-microgram sample to be diverted by a flow splitter and used for further analyses. Sub-picomole quantities of protein entering the MS provided verification of peptide mass. The authors advocated the use of the MS method for error-free data interpretation in chemical, high sensitivity peptide sequence analysis. CE/ESI-MS is a perfect combination for high sensitivity separation and detection of natural products. Negative ionisation was chosen for the analysis of food dyes.[78]

The average mass of mundticin, a novel broad-spectrum antimicrobial peptide produced by vegetable-associated *Enterococcus mundtii*, was determined by ESI mass spectrometry to be 4287.21 ± 0.59 Da.[79] Its good solubility in water and its stability over a wide pH and temperature range indicated its potential as a natural preservation agent for foods.

Glycopeptide antibiotics, *e.g.* vancomycin, bind specifically to the *C*-terminal sequence X-D-Ala-D-Ala of cell-wall peptide ligands, providing a form of molecular recognition. Non-covalent binding of antibiotics with peptide ligands in solution 'a key molecular recognition phenomenon in antibacterial chemotherapy' can be recognised in the gas phase by ion-spray mass spectrometry.[80] Binding constants in reasonable agreement with literature values could be determined from relative ion abundances correlated with ligand concentrations in solution. ESI-MS has been proposed for the detection of exogenous bovine somatotropine (BST), a cow milk-enhancing hormone.[81] This approach is feasible because the recombinant form, authorised for administration in several countries, differs in the composition of the *N*-terminal amino acids from pituitary BST. This difference cannot be detected in plasma BST assays.

High-resolution ESI magnetic sector mass spectrometry has been used to obtain accurate mass measurements on 37 synthetic peptides.[82] Data were acquired using external calibration over a wide mass range during magnetic scanning. Molecular weights were measured in the 5–60 ppm range at $R = 2500–9000$ (10% valley) and isotopic clusters for charge states up to $10+$ were resolved at high resolution. CID in the ESI interface generated product ions from which sequence information was obtained for most unknown peptides tested with masses below 2000 Da. High-resolution mass spectrometry differentiated between Lys and Glu, a mass difference of 0.0364 Da, and CID showed differences between peptides with identical monoisotopic masses.

Tandem Mass Spectrometry

The increasing uses of desorption and desolvation ionisation techniques to obtain molecular weight information is an obvious development in the verification of the molecular structure of peptides and proteins. Since there is a paucity of fragmentation in desorption and desolvation mass spectra it is necessary artificially to decompose the molecular ions, typically protonated molecules, to produce fragmentation from which amino acid sequences can be derived. Tandem mass spectrometry is an ideal method and was quickly developed for this purpose. An overview summarises the approach adopted in one particular laboratory.[83] This publication also clearly presents the principle and practice of MS/MS to peptide sequencing. The methodology has been adapted to characterise PAH diolepoxide adducted peptides.[84] High-energy CID tandem mass spectra are dominated by peptide–adduct bond cleavage ions with charge retention on the adducting moiety, and neutral loss scans can be used to identify adducted peptides in a mixture. The same authors compared the use of various chemical derivatives in the literature, to impart a fixed charge to the chosen terminus of a peptide to simplify the CID fragmentation pattern.[85] They found that the trimethylammoniumacetyl derivatised peptide was the easiest to synthesise and purify and therefore to be recommended for general practice, while the dimethyloctylammoniumacetyl derivative was better for hydrophilic peptides. In view of the importance of methionine residues in the stabilisation of the 3-D structure of proteins through hydrophobic interactions, the discovery of sequence ions specific to methionine in the CID spectra of peptides recorded at high collision energies is an important recent development.[86]

Various amino acid substitutions were made in a *C*-terminal heptapeptide to establish substrate specificity and the reaction products, and the new sequences, were monitored by neutral loss tandem MS.[87]

A general strategy for peptide analysis, combining the use of preparative two-dimensional PAGE and various mass spectrometric techniques, has been described.[88] Subpicomole samples of tryptic peptides yielded mass and sequence information from MALDI and tandem MS. In a detailed account of several biochemical experiments, the authors illustrate the power of tandem MS especially where co-migrating and covalently modified proteins are isolated.

The Combined Classical and Mass Spectrometric Approach

A novel approach to chemical degradation peptide sequencing, using mass spectrometry to analyse nested sets of fragments from ragged-end polypeptide chains, has been published.[89] The characterisation of a new hypothalamic satiety peptide, cocaine and amphetamine regulated transcript (CART) produced in yeast, employed N-terminal amino acid sequencing and mass spectrometry.[90]

Thirteen low molecular weight phosphopeptides likely to influence the flavour profile of Compte cheese have been studied with the aid of mass spectrometry.[91] The peptides were purified from water-soluble fractions using GPC and reversed-phase HPLC and then sequenced using the Edman/mass spectrometry approach. In particular, the purified peptides corresponded to fragments of the Val-13–Lys-38 of β-casein and the Glu-5–Lys-21 of α (s2)-casein sequences. The authors concluded that proteolysis by plasmin, followed by further endopeptidase, aminopeptidase and possibly carboxypeptidase digestions, were probably involved.

The view that the classical Edman degradation and mass spectrometry are complementary techniques[92] has been put on a firmer footing by the designing of a computer algorithm that utilises both Edman and tandem mass spectral data.[93] Gas-phase sequencing is usually followed by a chromatographic step in order to determine the actual residue lost at each stage in the chemical degradation. However, under favourable circumstances tandem MS can analyse amino acid sequences of peptides, even if they are present in mixtures. On the other hand, tandem MS often does not procure the full set of sequence determining fragment ions and requires assistance from Edman data to bridge the gap. The algorithm sets up a list of possible sequences based on the Edman data and then assigns each possibility a figure of merit according to how well it fits the fragment ion data. The technique has been applied to the analysis of purified bovine caseins labelled with a radioactive site-specific probe to determine transglutaminase acyl donor sites.[94]

Plasma desorption MS and chemical sequencing have been employed to sequence the naturally acetylated N-terminus of a cysteine-containing peptide.[95] Six cysteine-containing peptides were sequenced and some were found to differ from the c-DNA structure proposed. The number of cysteine residues was determined by MS before and after alkylation with 4-vinylpyridine. High-resolution PDMS forms the basis of a technique developed to sequence peptides subjected to partial acid hydrolysis, which produces consecutive sequence-specific segments but some sequence ions may be missing.[96] The method was shown to work with peptides containing an internal S–S bond and with those containing a non-amino acid constituent. PDMS and ESI were used to establish the cyclic nature of a 70-residue peptide antibiotic after the complete primary structure had been obtained by chemical degradation analysis.[97]

Carboxymethylation of bovine skimmed milk and purification of the α s2-casein dimer has shown that all four cysteine residues in the protein were engaged in disulfide linkages.[98] A combination of MS and sequence analysis of the cysteine-containing tryptic peptides showed the presence of two interchain disulfide bridges.

One of the fastest developing areas of mass spectrometry applied to peptide analysis is the use MALDI to obtain rapid confirmation of the molecular weight of a sample peptide. MALDI has been used alongside conventional peptide chemistry in a study of caudodorsal cell hormone-1 (CDCH-1) in which conventional methods revealed five additional peptides, and MALDI was able to confirm these five plus two more peptides from pro-CDCH-1.[99] Less than 5 picomole quantities of peptides were sequenced following purification on 1.0 mm i.d. RP-HPLC columns by combined Edman–MALDI (MS) techniques.[100] Although great care was advocated in the loading of such small samples onto the HPLC column, the overall sensitivity of the system was thought to surpass that of chemical microsequencing, and to be more convenient to use.

A novel example of combined MS and chemical degradation is in the use of HPLC/ESI-MS to detect the phenlythiohydantoins released in the stepwise chemical degradation of amino acids from the peptide.[101] Baseline separation of the isobaric leucine and isoleucine derivatives was demonstrated and additional selectivity from mass analysis enhanced the certainty of interpretation.

Other Developments, Including Proteomics

A peptide-mass fingerprinting system has been set up for the rapid identification of proteins using inexpensive ToF mass spectrometers.[102] Using the Daresbury molecular weight search peptide-mass database, which at that time contained over 50 000 proteins, peptide-mass fingerprints were as discriminating as linear peptide sequences, while using smaller amounts of protein in assays taking only fractions of the time.

A suite of data processing software to help in the interpretation of ESI and FAB mass spectra was presented as early as 1991.[103] Deconvolution of spectra containing multiply-charged ions from pure compounds or mixtures and programmes for LC/MS, peptide sequence confirmation from multiple-stage MS/MS data from tandem quadrupole MS and sequencing unknown peptides from MS/MS experiments are described.

A further aid to reducing the number of candidate structures is to determine the number of exchangeable hydrogens by fully-deuterating the peptide and re-running in the ESI tandem mode.[104] If this information is used in published algorithms for interpreting mass spectra, along with fragment ion data, the reliability of sequence determination improves. A similar strategy for use with MALDI utilises post-source decay.[105] The deuteration can be performed reversibly on the same sample after normal mass analysis. The so-called exchangeable-hydrogen spectroscopy was found to be 98.5% efficient. It has also been used to investigate fundamental fragmentation mechanisms.

The systematic mapping of the total protein complement of a genome, the proteome, has, as one reviewer has pointed out, been theoretically feasible since the development of 2-D gel electrophoresis over 20 years ago.[106] However, it is only since the sequencing of complete genomes, and the parallel development of ultra-sensitive mass spectrometric techniques for protein identification, that the full potential of proteomics has begun to be realised.[106,107] The gene sequence does not yield a complete description of an organism and rapid characterisation of proteins is therefore important for a number of reasons. Among these are (i) post- (or co-) translational modifications can occur, (ii) 40–60% of new bacterial open reading frames cannot be assigned a function based on homology searches,[106] (iii) detection of quantitative changes in protein expression when an organism is put under different stresses requires accurate techniques for identifying individual proteins in a complex array. Two mass spectrometric techniques have been developed for proteomic studies. The first involves excision and specific enzymatic or chemical digestion of individual spots on a 2-D gel. The resulting peptide mixture is examined by MALDI-ToF mass spectrometry. This 'mass fingerprinting' approach yields the molecular weights of individual peptides in a protein digest that are unique to the protein of interest and can be searched against protein databases.[108–114] The second

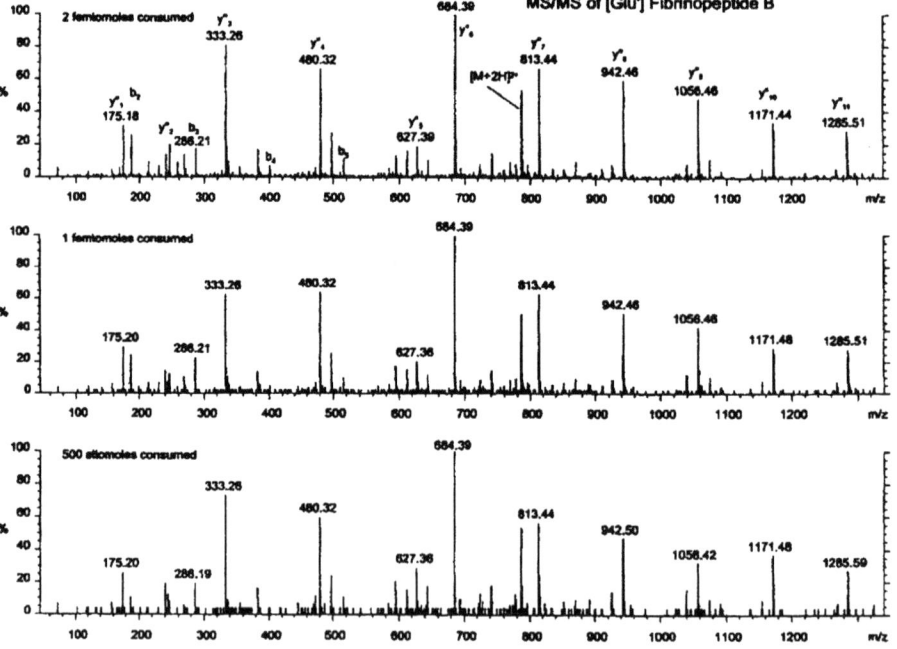

Figure 5.5   *MS/MS spectra obtained on a commercially available quadrupole/time-of-flight mass spectrometer (Q-ToF^{TM}) at three different dilutions of [Glu'] Fibrinopeptide B*
(Reproduced from material supplied by Micromass UK Ltd., with permission)

approach involves partial sequencing by MS/MS of one or preferably more peptides generated by specific digestion of a protein spot, the 'peptide sequence tag'.[115] This technique was subsequently automated by correlation of uninterpreted peptide tandem mass spectra with sequence databases.[116,117] Some subsequent developments have included development of an integrated workstation for high throughput protein characterisation[118] and nanoelectrospray following rapid proteolysis.[119] Although early efforts utilised triple quadrupole mass spectrometry, higher sensitivity has been attained by using quadrupole/time-of-flight instruments.[120] It is also possible that modern FTICR instruments will attain similar levels of sensitivity, with the added bonus of very accurate mass determination of the molecular weights of individual peptides. As an example of the degree of sensitivity that can be obtained, the quadrupole/time-of-flight MS/MS spectrum of an example peptide is shown in Figure 5.5 at several dilutions. Even the spectrum obtained on the lowest amount of sample, 500 attomoles, yields usable sequence ions.

The relevance of these techniques to food science lies in the area of food safety. Proteomics has considerable potential in investigating the response of food poisoning organisms to different environmental stresses, and has potential applications for investigating possible mechanisms by which modified DNA might be transmitted from one type of GM plant to another.

5   References

1 M. Barber, P. Jolles, E. Vilkas and E. Lederer, *Biochem. Biophys. Res. Commun.*, 1965, **18**, 469.
2 M. Barber, R.S. Bordoli, R.D. Sedgwick and A.N. Tyler, *J. Chem. Soc. Chem. Commun.*, 1981, 325.
3 H.R. Morris, M. Panico and G.W. Taylor, *Biochem. Biophys. Res. Commun.*, 1983, **117**, 299.
4 P. Roepstorff and J. Fohlman, *Biomed. Mass Spectrom.*, 1984, **11**, 601.
5 P. Roepstorff and J. Fohlman, *Biomed. Mass Spectrom.*, 1985, **12**, 631
6 K. Biemann, *Biomed. Environ. Mass Spectrom.*, 1988, **16**, 99.
7 K. Biemann, in *Mass Spectrometry of Biological Materials*, Practical Spectroscopy Series, Vol. 8, ed. C.N. McEwen and B.S. Larsen, Marcel Dekker, Inc., New York, 1990, p. 1.
8 D.F. Hunt, J. Shabanowitz, J.R. Yates, P.R. Griffin and N.Z. Zhu, in *Mass Spectrometry of Biological Materials*, Practical Spectroscopy Series, Vol. 8, ed. C.N. McEwen and B.S. Larsen, Marcel Dekker, Inc., New York, 1990, p. 169.
9 C. Fenselau, *Ann. Rev. Pharm. Toxicol.*, 1992, **32**, 555.
10 S.J. Gaskell, in *Mass Spectrometry in Biomolecular Sciences*, ed. R.M. Caprioli, A. Malorni and G. Sindona, NATO ASI Series, Kluwer Academic Publications, Dordrecht, 1996, p. 299.
11 R.A.W. Johnstone and M.E. Rose, in *Mass Spectrometry for Chemists and Biochemists*, Cambridge University Press, 1996, p. 81.
12 R.E. Lovins, in '*Mass Spectrometry – Part A* Practical Spectroscopy Series, Vol. 3, ed. C Merritt and C.N. McEwen, Marcel Dekker, Inc., New York, 1979, p. 20.
13 C.H. Oh, J.H. Kim, K.R. Kim and T.J. Mabry, *J. Chromatogr. A*, 1995, **708**, 131.
14 R.R.O. Loo and R.D. Smith, *J. Am. Soc. Mass Spectrom.*, 1994, **5**, 207.

15 T.B. Farmer and R.M. Caprioli, in *Mass Spectrometry in Biomolecular Sciences*, ed. R.M. Caprioli, A. Malorni and G. Sindona, NATO ASI Series, Kluwer Academic Publications, Dordrecht, 1996, p. 61.

16 Z.Q. Guan, J.J. Draper, V.L. Campbell and D.A. Laude, *Anal. Chem.*, 1995, **67**, 1453.

17 H. Lavanant, P.J. Derrick, A.J.R. Heck and F.A. Mellon, *Anal. Biochem.*, 1998, **255**, 74.

18 B. Spengler, D. Kirsch, R. Kaufmann and E. Jaeger, *Rapid Commun. Mass Spectrom.*, 1992, **6**, 105.

19 T.J. Cornish and R.J. Cotter, *Rapid Commun. Mass Spectrom.*, 1994, **8**, 781.

20 T. Kosada, T. Ishikawa and T. Kinoshita, *Rapid Commun. Mass Spectrom.*, 1995, **9**, 1342.

21 G.C. Galletti, G. Chiavari and P. Bocchini, *J. Anal. Appl. Pyrolysis*, 1995, **32**, 137.

22 D.N. Nguyen, G.W. Becker and R.M. Riggin, *J. Chromatogr. A*, 1995, **705**, 21.

23 A.L. Burlingame, R.K. Loyd and S.J. Gaskell, *Anal. Chem.*, 1994, **66**, 634R.

24 S.N. Jenq, S.J. Tsai and H. Lee, *Mutagenesis*, 1994, **9**, 483.

25 K. Vekey, A. Malorni, G. Pocsfalvi, A. Piperno, C. Romeo and N. Uccella, *J. Agric. Food Chem.*, 1997, **45**, 2447.

26 V.A. Yaylatan and A. Huyghuesdespointes, *Crit. Rev. Food Sci. Nutr.*, 1994, **34**, 321.

27 P.S. Tan, T.A. Kessel, F.L. van der Veerdonk, P.F. Zuurendonk, A.P. Bruins and W.N. Konings, *Appl. Environ. Microbiol.*, 1993, **59**, 1430.

28 E.M. Fujinari and J.D. Manes, *J. Chromatogr. A*, 1997, **763**, 323.

29 K. Seguro, N. Nio and M. Motoki, *ACS Symp. Ser.*, 1996, **650**, 271.

30 A. Schieber, H. Bruckner, M. Rupclassen and W. Specht, *J. Chromatogr. B*, 1997, **691**, 1.

31 R. Liardon, S. Lederman and U. Ott, *J. Chromatogr.*, 1981, **203**, 385.

32 R. Liardon and R. F. Hurrell, *J. Agric. Food Chem.*, 1983, **31**, 432.

33 R. Lizarragaguerra and M.G. Lopez, *J. Agric. Food Chem.*, 1996, **44**, 2556.

34 X. Guo and L. Wu, *Ecotoxicol. Environ. Safety*, 1998, **39**, 207.

35 J.H. Tsai, R.D. Hiserodt, C.T. Ho, T.G. Hartman and R.T. Rosen, *J. Agric. Food Chem.*, 1998, **46**, 2541.

36 J.F. Lawrence, B.P. Lau, C. Cleroux and D. Lewis, *J. Chromatogr A.*, 1994, **659**, 119.

37 R.I. Kelley, *Pediatr. Res.*, 1991, **30**, 564.

38 H. Wu, T.G. Hartman, S. Govindarajan, P.C. Kahn, C.T. Ho and J.D. Rosen, *Proc. Natl. Sci. Counc. Repub. China B*, 1991, **15**, 140.

39 K. Osada, K. Nagira, K. Teruya, H. Tachibana, S. Shirahata and H. Murakami, *Biotherapy*, 1994, **7**, 115.

40 I. Alli, B.F. Gibbs, M.K. Okoniewska, Y. Konishi and F. Dumas, *J. Agric. Food Chem.*, 1994, **42**, 2679.

41 M. Barber, G.J. Elliot, R.S. Bordoli, B.N. Green and B.W. Bycroft, *Experientia*, 1988, **44**, 266.

42 P. Roepstorff, P.F. Nielsen, I. Kamensky, A.G. Craig and R. Self, *Biomed. Environ. Mass Spectrom.*, 1988, **15**, 305.

43 P.F. Nielsen and P. Roepstorff, *Biomed. Env. Mass Spectrom.*, 1988, **17**, 137.

44 R. Self and F.A. Mellon, in *Analytical Applications of Spectroscopy*, ed. C.S. Creaser and A.M.C. Davies, Royal Society of Chemistry, London, 1988, p. 294.

45 A.G. Craig, *Biolog. Mass Spectrom.*, 1991, **20**, 195.

46 M. Liptak, K. Vekey, W.D. van Dongen and W. Heerma, *Biol. Mass Spectrom.*, 1994, **23**, 701.

47 J. Megrous, C. Lacroix, M. Bouksaim, G. LaPointe and R.E. Simard, *J. Appl. Microbiol.*, 1997, **83**, 133.

48 A.J.R. Heck, P.J. Derrick, H. Lavanant and F.A. Mellon, *Adv. Mass Spectrom.*, 1998, paper MoOr10, p. 9.

49 H. Lavanant, A. Heck, P.J. Derrick, F.A. Mellon, A. Parr, H.M. Dodd, C.J. Giffard, N. Horn and M.J. Gasson, *Eur. J. Mass Spectrom.*, 1998, **4**, 405.

50 T. Herraiz, *Analyt. Chim. Acta*, 1997, **352**, 119.

51 D. Molle, F. Morgan, S. Bouhallab and J. Leonil, *Anal. Biochem.*, 1998, **259**, 152.

52 Y. Kato, Y. Ogino, T. Aoki, K. Uchida, S. Kawakishi and T. Osawa, *J. Agric. Food Chem.*, 1997, **45**, 3004.

53 E. Mendez, E. Camfieta, J. Sansebastian, I. Valle, J. Solis, F.J. Mayerposner, D. Suckau, C. Marfisi and F. Soriano, *J. Mass Spectrom./Rapid Commun. Mass Spectrom. (Comb. Issue)*, 1995, S123.

54 E. Camafeita, P. Alfonso, T. Mothes and E. Mendez, *J. Mass Spectrom.*, 1997, **32**, 940.

55 E. Camafeita, J. Solis, P. Alfonso, J.A. Lopez, L. Sorell and E. Mendez, *J. Chromatogr. A*, 1998, **823**, 299.

56 E. Camafeita and E. Mendez, *J. Mass Spectrom.*, 1998, **33**, 1023.

57 L.E. Quadri, M. Sailer, M.R. Terebiznik, K.L. Roy, J.C. Vederas and M.E. Stiles, *J. Bacteriol.*, 1995, *177*, 1144.

58 R.W. Worobo, T. Henkel, M. Sailer, K.L.Roy, J.C. Vederas and M.E. Stiles, *Microbiology*, 1994, **140**, 517.

59 A.L. Holck, L. Axelsson, S.E. Birkeland, T. Aukrust and H. Blom, *J. Gen. Microbiol.*, 1992, **138**, 2715.

60 A.L. Holck, L. Axelsson, K. Huhne and L. Krockel, *FEMS Microbiol. Lett.*, 1994, **115**, 143.

61 M. Fountoulakis and H.W. Lahm, *J. Chromatogr. A*, 1998, **826**, 109.

62 S. Maisnierpatin and J. Richard, *Appl. Env. Microbiol.*, 1995, **61**, 1847.

63 D.A. Dionysius and J.M. Milne, *J. Dairy Sci.*, 1997, **80**, 667.

64 S.A. Carr, G.D. Roberts and M.E. Hemling, in *Mass Spectrometry of Biological Materials*, Practical Spectroscopy Series, Vol. 8, ed. C.N. McEwen and B.S. Larsen, Marcel Dekker, Inc., New York, 1990, p. 87.

65 B. Spengler, D. Kirsch and R. Kaufmann, *J. Phys. Chem.*, 1992, **96**, 9678.

66 N. Poppeschriemer, W. Ens, J.D. O'Niel, V. Spicer, K.G. Standing, J.B. Westmore and A.A. Yee, *Int. J. Mass Spectrom. and Ion Processes.*, 1995, **143**, 65.

67 R.S.Younquist, G.R. Fuentes, M.P. Lacey and T. Keough, *Rapid Commun. Mass Spectrom.*, 1994, **8**, 77.

68 S.X. Chen, C.C. Hardin and H.E. Swaisgood, *J. Protein Chem.*, 1993, **12**, 613.

69 H.J. Kim and S.J. Park, *Abst. Am. Chem. Soc.*, 1998, **216**, 9.

70 O. Vorm and P. Roepstorff, *Biol. Mass Spectrom.*, 1994, **23**, 734.

71 J.A. Castoro, C.L. Wilkins, A.S. Woods and R.J. Cotter, *J. Mass Spectrom.*, 1995, **30**, 94.

72 M. Besler, H. Steinhart and A. Paschke, *Food Agric. Immunol.*, 1998, **10**, 157.

73 P. Pucci, C. Sepe and G. Marino, *Biolog. Mass Spectrom.*, 1992, **21**, 22.

74 U. Sommerer, C. Septier, C. Salles and J.L. LeQuere, *Sci. Aliments*, 1998, **18**, 537.

75 H.J. Yang, M.Y. He, Y.H. Ye and Y.F. Zhao, *Org. Mass Spectrom.*, 1992, **27**, 746.

76 T. Yagami, K. Kitagawa and S. Futaki, *Rapid Commun. Mass Spectrom.*, 1995, **9**, 1335.

77 D. Hess, T.C. Covey, R. Winz, R.W. Brownsey and R. Aebersold, *Protein Science*, 1993, **2**, 1342.

78 T.E. Wheat, K.A. Lilley and J.F. Banks, *J. Chromatogr. A*, 1997, **781**, 99.

79 M.H.J. Bennik, B. Vanloo, R. Brasseur, L.G.M. Gorris and E.J. Smid, *Biochim. Biophys. Acta – Biomembranes*, 1998, **1373**, 47.

80 H.K. Lim, Y.L. Hsieh, B. Ganem and J. Henion, *J. Mass Spectrom.*, 1995, **30**, 708.

81 M.L. Scippo, G. Degand, A. Duyckaerts and G. MaghuinRogister, *Ann. Med. Vet.*, 1997, **141**, 381.

82 P.A. Dagostino, J.R. Hancock, L.R.Provost, P.D.Semchuk and R.S.Hodges, *Rapid Commun. Mass Spectrom.*, 1995, **9**, 597.

83 K. Biemann, in *Mass Spectrometry of Biological Materials*, Practical Spectroscopy Series, Vol. 8, ed. C.N. McEwen and B.S. Larsen, Marcel Dekker, Inc., New York, 1990, p. 1.

84 J. Zaia and K. Biemann, *J. Am. Soc. Mass Spectrom.*, 1994, **5**, 649.

85 J. Zaia and K. Biemann, *J. Am. Soc. Mass Spectrom.*, 1995, **6**, 428.

86 K.M. Downward and K. Biemann, *J. Mass Spectrom.*, 1995, **30**, 25.

87 T. Kupke, C. Kempter, G. Jung and F. Gotz, *J. Biol. Chem.*, 1995, **270**, 11 282.

88 K.R. Clauser, S.C. Hall, D.M. Smith, J.W. Webb, L.E. Andrews, H.M. Tran, L.B. Epstein and A.L. Burlingame, *Proc. Natl. Acad. Sci. USA*, 1995, **92**, 5072.

89 M. Bartlet-Jones, W.A. Jeffrey, H.F. Hansen and D.J. Pappin, *Rapid Commun. Mass Spectrom.*, 1994, **8**, 737.

90 L. Thim, P.F. Nielsen, M.E. Judge, A.S. Andersen, I. Diers, M. EgelMitani and S. Hastrup, *FEBS Lett.*, 1998, **428**, 263.

91 F. Roudotalgaron, D. Lebars, L. Kerloas, J. Einhorn and J.C. Gripon, *J. Food Sci.*, 1994, **59**, 544.

92 M.W. Rees, M.N. Short, R. Self and J. Eagles, *Biomed. Mass Spectrom.*, 1974, **1**, 237.

93 R.S. Johnson and K.A. Walsh, *Protein Sci.*, 1992, **1**, 1083.

94 B.M. Christensen, E. Sorensen, P. Hojrup, T.E. Petersen and L.K. Rasmussen, *J. Agric. Food Chem.*, 1996, **44**, 1943.

95 T.A. Egarov, V.K. Kazakov, A.K. Musolyamov, V.N. Pustobaev and G.K. Kovaleva, *Bioorganicheskaya Khim.*, 1993, **19**, 1158.

96 R.A. Zubarev, V.D. Chivanov, P. Hakansson and B.U.R. Sundqvist, *Rapid Commun. Mass Spectrom.*, 1994, **8**, 906.

97 B. Samyn, M. Martinez Bueno, B. Devreese, M. Maqueda, A. Galvez, E. Valdivia, J. Coyette and J. Van Beeumen, *FEBS Lett.*, 1994, **352**, 87.

98 L.K. Rasmussen, P. Hojrup and T.E. Petersen, *Eur. J. Biochem.*, 1992, **203**, 381.

99 K.W. Li, C.R. Jimenez, P.A. Vanveelen and W.P.M. Geraerts, *Endocrinology*, 1994, **134**, 1812.

100 C. Elicone, M. Lui, S. Germanos, H. Erdjumentbromage and P. Tempst, *J. Chromatogr. A*, 1994, **676**, 121.

101 D. Hess, H. Nika, D.T. Chow, E.J. Bures, H.D. Morrison and R. Aebersold, *Anal. Biochem.*, 1995, **224**, 373.

102 D.J.C. Pappin, P. Hojrup and A.J. Bleasby, *Current Biol.*, 1993, **3**, 327.

103 P.J.F. Watkins, I. Jardine and J.X.G. Zhou, *Biochem. Soc. Trans.*, 1991, **19**, 957.

104 N.F. Sepetov, O.L. Issakova, M. Lebl, K. Swiderek, D.C. Stahl and T.D. Lee, *Rapid Commun. Mass Spectrom.*, 1993, **7**, 58.

105 B. Spengler, B. Lutzenkirchen and R. Kaufmann, *Org. Mass Spectrom.*, 1993, **28**, 1482.

106 P. James, *Biochem. Biophys. Res. Commun.*, 1997, **231**, 1.

107 J.R. Yates, *Nippon Iyo Masu Supekutoru Gakkai Koensm.*, 1998, **23**, 11.

108 W. Henzel, T. Billeci, J. Stults, S. Wong, C. Grimley and C. Watanabe, *Proc. Natl. Acad. Sci. USA*, 1993, **90**, 5011.

109 D. Pappin, P. Hojrup and A. Bleasby, *Current Biol.*, 1993, **3**, 327.

110 M. Mann, P. Hojrup and P. Roepstorff, *Biol. Mass Spectrom.*, 1993, **22**, 338.

111 J.R. Yates, S. Speicher, P.R. Griffin and T. Hunkapillar, *Anal. Biochem.*, 1993, **214**, 397.

112 P. James, M. Quadroni, E. Carafoli and G. Gonnet, *Biochem. Biophys. Res. Comm.*, 1993, **195**, 58.

113 U. Hellmann, C. Wernstedt, J. G'Onez and C. H. Heldin, *Anal. Biochem.*, 1995, **224**, 451.

114 A. Shevchenko, M. Wilm, O. Vorm and M. Mann, *Anal. Chem.*, 1996, **68**, 850.

115 M. Mann and M. Wilm, *Anal. Chem.*, 1994, **66**, 4390.

116 J. Eng, A.L. McCormack and J.R. Yates, *J. Am. Soc. Mass Spectrom.*, 1994, **5**, 976.

117 J.R. Yates, J.K. Eng, A.L. McCormack and D. Schieltz, *Anal. Chem.*, 1995, **67**, 1426.

118 A. Ducret, I. Van Oostveen, J.K. Eng, J.R. Yates and R. Aebersold, *Protein Sci.*, 1998, **7**, 706.
119 R.K. Blackburn and R.J. Anderegg, *J. Am. Soc. Mass Spectrom.*, 1997, **8**, 483.
120 H.R. Morris, T. Paxton, A. Dell, J. Langhorne, M. Berg, R.S. Bordoli, J. Hoyes and R.H. Bateman, *Rapid Commun. Mass Spectrom.*, 1996, **10**, 889.

CHAPTER 6

Lipids

1 Introduction

Lipids are significant bulk constituents of foods (along with proteins, carbohydrates and water) and major sources of energy in the diet. They are also important contributors to the texture and 'mouth feel' of foods. Oxidation of lipids during food storage is a major contributor to the development of off-flavours. There is greatly increased interest in the relationship between the consumption of fats and the development of chronic disease, for example coronary heart disease. The growing interest in all these areas is placing greater demands on analytical techniques at both the qualitative and quantitative level. It has been pointed out that the complex mixtures of lipids in a typical organic extract, with different molecular masses, degrees of unsaturation and variable functional groups, cannot be characterised completely by any single analytical technique.[1] Nevertheless, a variety of mass spectrometric methods is central to the structural confirmation of known lipids and the qualitative analysis of unknown lipids.

Lipids are chemically diverse and can be divided into three groups: simple lipids, compound lipids and derived lipids.[2] Simple lipids are fatty acid esters that produce two classes of compound on hydrolysis, compound lipids produce three or more classes of compound on hydrolysis and derived lipids cannot be hydrolysed to fatty acids. Triglycerides, also commonly known as triacylglycerols, are the main simple lipids; their general structure is shown in Figure 6.1.

The most important group of compound lipids in foodstuffs is the phospholipids (Figure 6.2).

Derived lipids include the fatty acids themselves, fat-soluble vitamins and pro-vitamins (covered in Chapter 9), terpenoids, ethers, sterols and other

$$CH_2OCOR_1$$
$$CHOCOR_2$$
$$CH_2OCOR_3$$

Figure 6.1 *General structure of triglycerides*

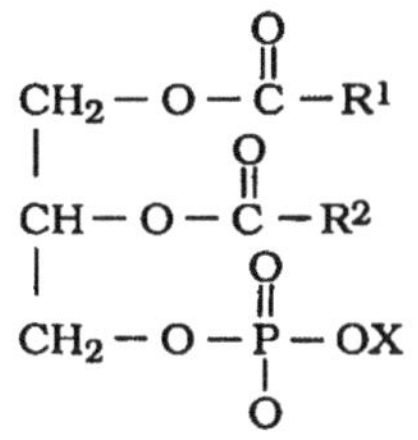

$$CH_2 - O - \overset{\overset{O}{\|}}{C} - R^1$$

$$CH - O - \overset{\overset{O}{\|}}{C} - R^2$$

$$CH_2 - O - \overset{\overset{O}{\|}}{\underset{\underset{O}{|}}{P}} - OX$$

Figure 6.2 *General structure of phospholipids*

alcohols. In addition to the above categories, lipid oxidation products are also important from both a health and a commercial perspective. Mass spectrometric analysis of these compounds is also discussed below.

Mass spectrometry possesses high sensitivity and dynamic range and is therefore an excellent technique for analysing lipids. Because lipids are commonly isolated in very complex mixtures, mass spectrometry, in combination with high-resolution separation techniques such as, particularly, capillary GC and HPLC, is often used to aid identification. Tandem mass spectrometry is also very useful for enhancing the quality of structural information, particularly for the more complex lipids. This chapter describes the qualitative and quantitative analysis of lipids and lipid metabolites. Stable isotope studies of lipid metabolism are covered in Chapter 10.

Some useful and informative reviews of the analysis of lipids have appeared recently.[1,3,4] A review of the LC/MS of lipids, although accurately reflecting the state of the art at the time of publication, is focused mainly on thermospray and other methods.[5] Electrospray and APCI LC/MS have largely superseded these methods. An up to date overview of LC/MS applications in food analysis contains an informative section on lipids.[6]

2   Derived Lipids

Fatty Acids and Fatty Acid Esters

Although free fatty acids may be analysed by EI MS or GC/MS, it is more common to determine them as their methyl esters. This is because the esters yield more informative mass spectra (usually containing a molecular ion) and have superior chromatographic properties. Silyl derivatives of fatty acids are also occasionally prepared, although these derivatives are less popular. Nevertheless, *tert*-butyldimethylsilyl derivatives have some advantages in the analysis of complex hydroxy fatty acids, and other derivatives can be used to improve sensitivity.[3] The use of stable isotope dilution GC/MS for quantification of lipids (mainly fatty acids) and their oxidation products has been reviewed.[7] The author notes the need to develop stable isotope methods for determining multiple analytes (most IDMS methods for lipids focus on single compound assays, or on a small group of compounds).

The methyl esters of linear, saturated fatty acids usually yield a molecular ion, $M^{+\cdot}$, a peak at $[M - 31]^+$ due to loss of a methoxyl radical and low mass ions of general formula $(CH_2)_nCOOCH_3^+$. Other characteristic peaks, for example a McLafferty rearrangement ion at m/z 74 that is often the base peak, are also commonly observed.[8] The EI mass spectra of branched chain fatty acids and their esters exhibit characteristic ions that can be used to locate the branching points. However, food fatty acids and their esters are not often found in branched forms ('soft' fats are a notable exception): a more useful feature of their EI spectra is that the presence of an identifiable $M^{+\cdot}$ ion yields the molecular weight, and thence directly the degree of unsaturation.

Although GC/MS using EI or other ionisation techniques are generally the method of choice for characterising complex fatty acid or fatty acid ester mixtures, LC/MS is also useful. This type of application is particularly appropriate when a combination of mass spectrometric and photodiode array UV information is used to detect conjugated dienes. These molecules have very characteristic UV spectra and the simultaneous acquisition of UV and mass spectra yields a highly specific detection technique. This is exemplified by the analysis of conjugated diene fatty acids in partially hydrogenated fat, where APCI LC/MS was the mass spectrometric technique used.[9] The data reported in this study are important because detection of conjugated dienes in tissues or body fluids of humans or experimental animals is very often adduced as evidence of lipid peroxidation. However, the demonstration of the presence of these dienes in partially hydrogenated oils that are widely consumed by human populations suggests that this conclusion is unwarranted.

Double bond location in unsaturated fatty acids and their esters is not possible by inspection of EI mass spectra. This is because extensive bond migration occurs as a result of the high internal energies absorbed during ionisation. Methods for double bond location involving chemical modification and mass spectrometry have been developed and these have been reviewed comprehensively.[10] A common approach has been to convert the double bonds to polyols, derivatise by trimethylsilylation, and perform GC/MS. The polyol-TMS ethers yield characteristic fragment ions that enable the original position of the double bond to be derived. Other chemical approaches are possible;[10] however an alternative is to form chemical derivatives that localise charge and thus prevent charge-induced double bond migration. Pyrrolidide derivatives are a notable example of this approach.[11-13] These derivatives yield clusters of ions 14 mass units apart, reducing to 12 mass units when a double bond occurs. The main disadvantages of pyrrolidides and similar heteroatom derivatives are twofold. Firstly, they are insufficiently volatile for GC and cannot be applied to GC/MS of complex mixtures; secondly, fragment ions are often of such low abundance that definitive structures cannot be assigned.[14,15] It has recently been shown that conventional mass spectrometry with resonance electron capture in the 'high energy' region (*i.e.* an electron energy of *ca.* 7 eV) of free acids and their methyl esters and pyrrolidides yields abundant, charge-remote fragment ions that allow double bond location.[16] Pyrrolidide derivatives performed particularly well under these conditions, even with fatty acids containing multiple sites of unsaturation.

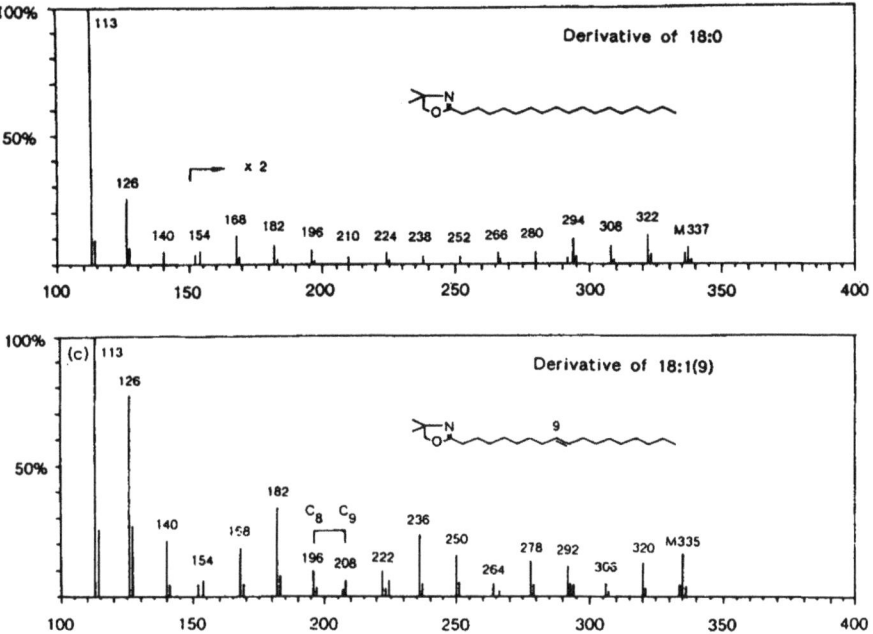

Figure 6.3 *Mass spectra of 2-substituted-4, 4-dimethyloxazolines derived from 18:0 (top) and 18:1 (bottom) fatty acids*
(Reproduced from J.Y. Zhang, Q.T. Yu, B.N. Liu and Z.H. Huang, *Biomed. Environ. Mass Spectrom.*, 1988, **15**, 33, with permission of the authors and publisher, © John Wiley & Sons Limited)

A derivative that yields ions characteristic of double bond location, but is still sufficiently volatile for GC/MS, is the 2-substituted 4,4-dimethyloxazoline formed by condensation of the fatty acid with 2-amino-2-methylpropanol.[17,18] The EI spectra of two of these derivatives are shown in Figure 6.3. The double bond position in the unsaturated fatty acid is clearly defined by the 12 Da (instead of 14 Da) gap in the lower spectrum, between m/z 196 and 208.

An alternative technique for locating double bonds is 'charge-remote' fragmentation of molecules ionised by CI. High-energy CID of free fatty acids in negative ion mode yields ion sequences that enable double bond position to be determined.[19] This is because the charge resides on the acid moiety and cannot participate in double bond rearrangement processes. The CID spectra of aliphatic acid carboxylate anions exhibit characteristic ions spaced at 14 mass unit intervals: these are formed by the process outlined in Scheme 6.1.[19]

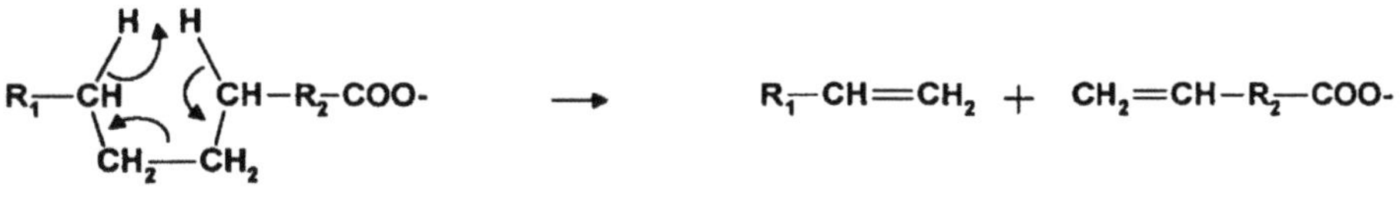

$$R_1-CH \overset{H \; H}{\underset{CH_2-CH_2}{CH-R_2-COO^-}} \longrightarrow R_1-CH{=}CH_2 + CH_2{=}CH-R_2-COO^-$$

Scheme 6.1 *Characteristic mass spectrometric decompositions in the CID spectra of aliphatic acid carboxylate anions*

Double bonds interrupt the regular sequence of ions formed by this process and allow their position to be determined. This works well for monounsaturated fatty acids but a more complex technique, involving CID and MS/MS after deuterium reduction of the double bonds[20] or attachment of cations,[21] has been adopted. High energy MS/MS instruments are not widely available and are expensive and demanding to operate. Unfortunately, the same information is not apparent in the low-energy CID spectra obtained on cheaper, less complex instruments (triple quadrupole mass spectrometers for example); consequently the technique has not been widely adopted. It is possible that a new generation of instruments will encourage wider adoption of this methodology.

Although FAB has be used as an alternative ionisation technique for introducing fatty acids into the gas phase, prior to high-energy CID MS/MS, this approach failed with mixtures because of the presence of isomeric ions.[22] This has led to the development of GC/MS/MS methods, combining the separation efficiency of GC with the selectivity of tandem MS, to determine structures of pentafluorobenzyl derivatives of fatty acids in complex mixtures.[23,24] These derivatives not only confer desirable chromatographic properties; they also generate carboxylate anions by the process shown in Scheme 6.2. The techniques were applied to the analysis of mycobacterial fatty acids, but could also be used in food studies and food analysis.

$$e^- + C_nH_{n-m}COO\text{-}CH_2\text{-}C_6F_5 \rightarrow C_nH_{n-m}COO^- + C_6F_5\text{-}CH_2^\cdot$$

Scheme 6.2 *Generation of carboxylate anions from pentafluorobenzyl derivatives of fatty acids under negative ion electron capture conditions*

Recommended Mass Spectrometric Methods for Analysis of Fatty Acids and Their Esters

Despite the development of alternative mass spectrometric techniques, EI GC/MS of methyl esters is still the principal method for analysing fatty acids. Fatty acid methyl esters are easy to chromatograph on high-resolution capillary columns and comprehensive EI mass spectral data bases are available. The combination of mass spectrometric and retention index data yields a powerful method for characterising fatty acids. In unusual cases, where fatty acids of novel structure are presented for analysis, additional information is obtainable by tandem mass spectrometry of suitable derivatives.

3   Simple Lipids

Acylglycerols

Direct Probe Mass Spectrometry

Although the EI spectra of acylglycerols usually exhibit a molecular ion accompanied by an $[M - H_2O]^+$ fragment, both these may be of low intensity, especially for triacylglycerols.[4] The nature and degree of unsaturation of acyl groups can be determined from the masses of characteristic fragment ions corresponding to $[M - RCO_2]^+$ and $[RCO]^+$. However, the position of the groups and the location of any double bonds in the fatty acid chains cannot be determined from EI spectra. CI spectra yield more abundant ions that can be used to determine molecular weight, for example ammonia CI generates intense $[M + NH_4]^+$ adduct ions and diagnostic fragment ions corresponding to $[MH - RCOOH]^+$.[25-27] FABMS is also useful for determining the molecular weights of triacylglycerols if sodium salts are added to the matrix to generate $[M + Na]^+$ adduct ions.[28,29] Complementary information is revealed by $[M - RCOO]^+$ and $[RCOO]^-$ ions in the positive and negative ion spectra, respectively. FABMS and principal component analysis can be used to classify edible oils according to their food source (olives, sunflower oil, soybean, corn and peanut).[30]

GC/MS of Acylglycerols

Mono and diacylglycerols are generally analysed by GC/MS of their trimethyl-silyl or *tert*-butyl-dimethylsilyl derivatives.[1,31] These derivatives not only possess desirable chromatographic properties, they also (especially TBDMS esters) yield characteristic fragment ions that establish the molecular weight of the mixture components and of the acyl groups.

Triacylglycerols represent a more challenging proposition. It has been noted[4] that acylglycerols, especially triacylglycerols, place heavy demands on GC/MS. The triacylglycerols are usually isolated in complex mixtures that require highly efficient chromatographic separation. The development of high-temperature apolar and polar capillary columns has permitted the analysis of a variety of acylglycerols, including triacylglycerols, more easily than in the past.[4,32] High-temperature (HT) GC/MS of intact acylglycerols has been described and reviewed in useful detail by Evershed,[4] who has emphasised the advantages of using high-temperature polar phases. The lower molecular triacylglycerol components of a volatile butter oil fraction of bovine milk fat have been characterised by EI HT GC/MS.[33] Figure 6.4 shows the mass chromatograms of $[M - RCOOH]^+$ fragment ions from mixed, saturated triacylglycerols of milk fat.

Positive ion ammonia CI has already been mentioned as a useful adjunct to EI spectra and has been employed in GC/MS analyses.[26] Ammonia NICI has also shown promise as a method for determining bovine milk fat triglycerides.[34] The NICI spectra contained intense $[RCOO]^-$, $[RCOO - 18]^-$ and $[RCOO - 19]^-$ ions. The 'Biller–Biemann' technique[35] was used to enhance mass resolved chromatograms and increase chromatographic resolution. The improvement in chromatographic resolution that can be attained is illustrated by comparing the lower and upper traces in Figure 6.5.

The negative ion spectra of two closely eluting triacylglycerols are shown in Figure 6.6. The ions at m/z 255, 281 and 283 in the upper spectrum are due to $[RCO_2]^-$ ions, here derived from $C_{16:0}$, $C_{18:1}$ and $C_{18:0}$ respectively, whereas the lower spectrum contains ions at 255 and 281, indicating the presence of two $C_{18:1}$ and one $C_{16:0}$ moieties. Additional structural evidence is provided by the fragment ions of general formulae $[RCO_2 - H_2O]^-$ and $[RCO_2 - H_2O - H]^-$.

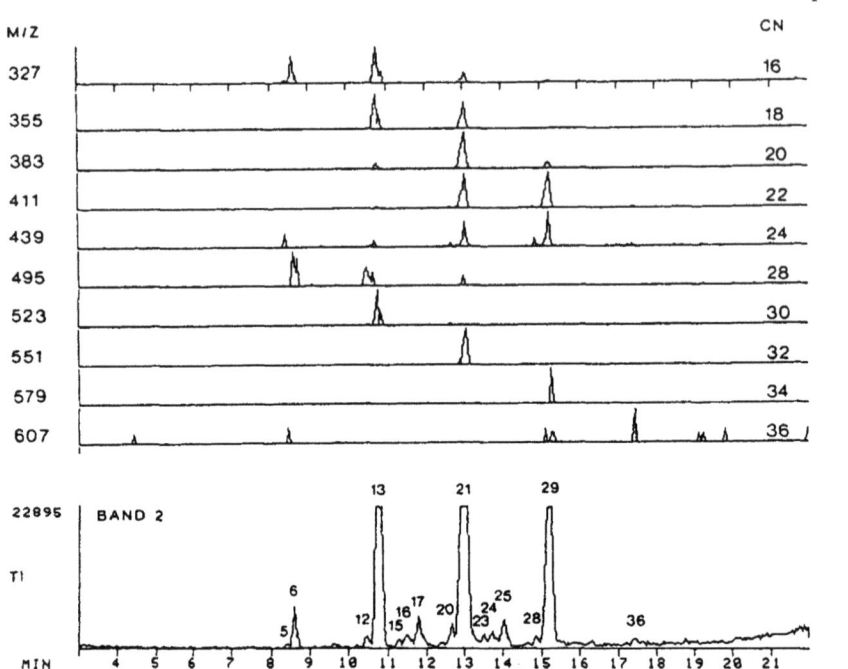

Figure 6.4 *Mass chromatograms of mixed, saturated triacylglycerols. Peak identification by carbon number (CN) and $[M - RCOOH]^+$ fragment ions*
(Reproduced from J.J. Myher, A. Kuksis, L. Marai and P. Sandra, *J. Chromatogr.*, 1988, **452**, 93, © 1988 with permission from Elsevier Science)

The weaknesses of the HT GC/MS methods have also been highlighted.[4,34] Loss of components can occur because of the poor thermal stability of some analytes, or technical limitations (particularly cold spots in transfer lines) of the equipment used. The importance of optimising HT GC/MS methods using authentic standards has been emphasised and the use of catalytic reduction to identify possible losses of unsaturated species has been advocated.[4]

Determination of double bond position in acylglycerols is as difficult as in fatty acids or fatty acid esters. Recent studies have shown the potential of nicotinyl derivatives of mono- and di-acylglycerols for this purpose.[36,37] These derivatives are sufficiently volatile for GC/MS. However, it has not yet been demonstrated that these derivatives are suitable for the unambiguous determination of mixed diacylglycerols. Furthermore, they cannot be used to determine triacylglycerols, where tandem mass spectrometry is the technique that displays most promise (see below).

LC/MS of Acylglycerols

An overview of LC/MS methods in the analysis of naturally occurring substances in foods includes comprehensive coverage of lipids, especially acylglycerols.[6] Early applications of LC/MS to the analysis of acylglycerols

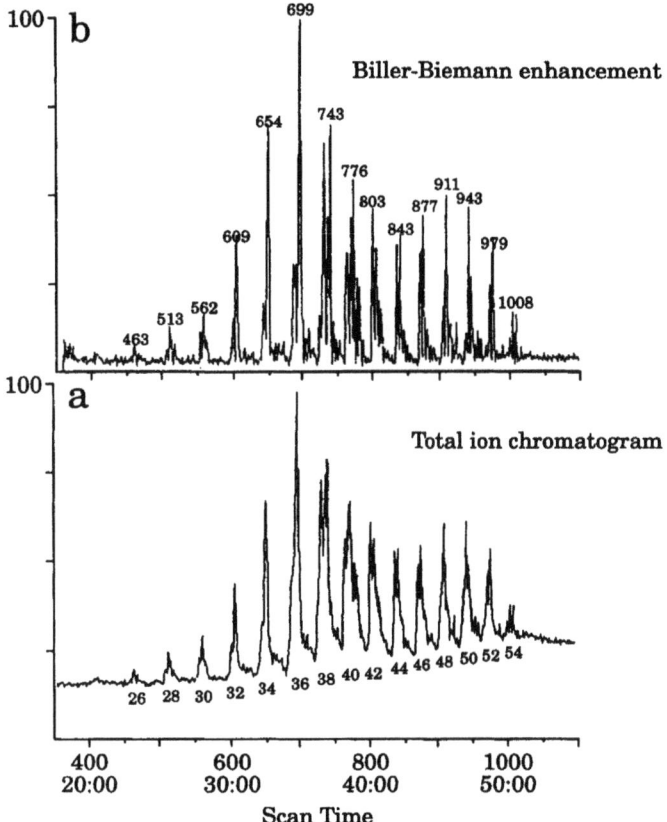

Figure 6.5 *Total ion current chromatogram* (a) *and Biller–Biemann mass-resolved chromatogram* (b) *for negative ion HT GC/MS analysis of butter fat triacylglycerols*
(Reproduced by permission of Elsevier Science from R.P. Evershed, *J. Am. Soc. Mass Spectrom.*, 1996, **7**, 350, © 1996 by American Society for Mass Spectrometry)

relied on techniques that have now been superseded by methods that are more modern and only these more recent developments will be considered here. An early study of the ESI of triglycerides showed the potential of this technique.[38] The ESI mass spectra contained protonated molecules, with no fragment ions (and therefore no structural information other than the degree of unsaturation, deducible from the molecular weight). However, MS/MS could probably be used to generate diagnostic fragment ions (see below). Although conventional ESI has limitations as a technique for the qualitative analysis of triacylglycerol structures, the intense protonated molecules formed suggest that ESI has potential in metabolic studies of known, isotopically-labelled compounds. In such cases, the appearance of most of the ion abundance in the molecular ion cluster is a positive advantage. More recently, silver-phase HPLC combined with ESI has been used successfully to quantify triacylglycerols in vegetable

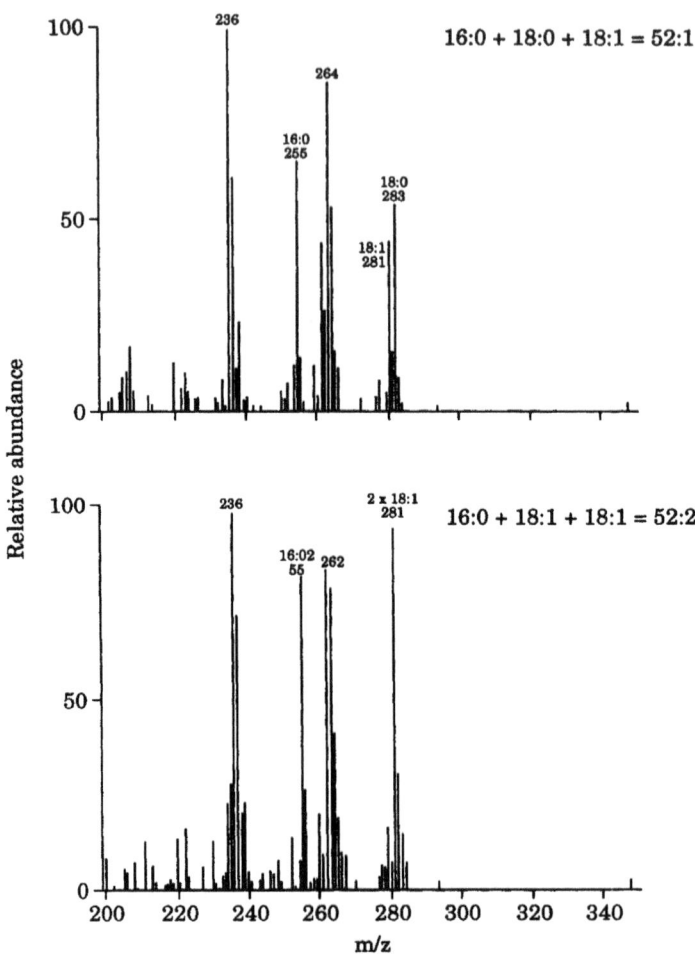

Figure 6.6 *Ammonia negative ion CI mass spectra for the triacylglycerol peaks maximising at scan numbers 975* (a) *and 979* (b) *in Figure 6.5*
(Reproduced from R.P. Evershed, *J. Am. Soc. Mass Spectrom.*, 1996, 7, 350, with permission of the author and publisher)

oils.[39] The lipids were characterised according to their carbon number, degree of unsaturation and position of the most unsaturated fatty acid (the latter information is related to the chromatographic performance of silver-phase columns, not to the mass spectrometry).

More recent studies have conclusively demonstrated the potential of APCI LC/MS in the structural and quantitative analysis of triglycerides.[40–42] APCI LC/MS of a mixture of homogeneous triglyceride standards, containing from zero to three double bonds, yielded intense $[M + H]^+$ ions for those molecules containing unsaturated fatty acids: those with fully saturated side chains only yielded 'diglyceride' ions.[40] Abundant fragment ions of composition

[M − RCOO]$^+$ were also formed. When the fatty acids contained two or three double bonds the [M + H]$^+$ ion was the base peak and the [M − RCOO]$^+$ fragments appeared with abundances varying from 13% to 25%. In contrast, the [M − RCOO]$^+$ fragment was the most abundant peak in the spectrum if the side chains contained one double bond (the [M + H]$^+$ ion was then 20–28% of base peak). An interesting recent application of APCI LC/MS was in the analysis of soybean triacylglycerols from genetically modified soybean lines.[43] The modified soybean lines, when compared with non-GM material, were shown to contain a higher proportion of more oxidatively stable triacylglycerols.

EI SFC/MS has also shown promise in triacylglycerol analysis.[44] Separation on capillary SFC columns was achieved by using an apolar siloxane stationary phase that grouped triacylglycerols according to carbon number. The separation was successful because of the low degree of unsaturation of triacylglycerols in the butter fat sample examined. Some differentiation between fatty acids at the nutritionally important *sn*-2 position and those at the *sn*-1/3 positions was possible by monitoring the fragment ion [M − RCO$_2$CH$_2$]$^+$ in reference compounds. Ions of this general formula are only formed by fragmentation of *sn*-1/3 positions. More recently, APCI SFC/MS (with ammonia as reagent gas) has been used successfully to characterise milk fat triacylglycerols.[45] A report from the same laboratory has shown that silver ion high performance APCI LC/MS techniques can be used to characterise seed oils containing γ-linolenic acid, and to separate α- and γ-linolenic acid triacylglycerols.[46]

Tandem Mass Spectrometry of Triacylglycerols

Tandem mass spectrometric methods for analysing triacylglycerols in a variety of foodstuffs have been described in detail.[47-55] Triacylglycerols were ionised by ammonia negative ion CI and selected molecular ions were subjected to collisional activation in a triple quadrupole mass spectrometer. A detailed account of the method has been presented and only the salient features will be described here.[48] The technique requires some form of pre-fractionation, for example by silver ion HPLC, to improve the accuracy of the analysis because triacylglycerol mixtures can be extremely complex. The MS/MS method also yields only the molecular weight of the constituent fatty acids and GC/MS analysis of total fatty acids is necessary to define detailed composition. MS/MS yields the molecular weights of the triacylglycerols and their proportions; the fatty acids, their distribution, combinations and proportions in the individual triacylglycerols, and the nature of the acyl group of position *sn*-2. As an example, the product ion spectra of the [M − H]$^-$ ions of three triacylglycerols containing at least one unsaturated fatty acid residue are shown in Figure 6.7.[47] Four distinct groups of ions, corresponding to low intensity ions of general formula [M − H − RCO$_2$H]$^-$, [M − H − RCO$_2$H − 74]$^-$, [M − H − RCO$_2$H − 100]$^-$ and RCO$_2^-$, are formed. Thus, the spectra are greatly simplified (although some additional fragment ions were formed in the spectrum of the highly unsaturated triarachionoylglycerol). The first three types

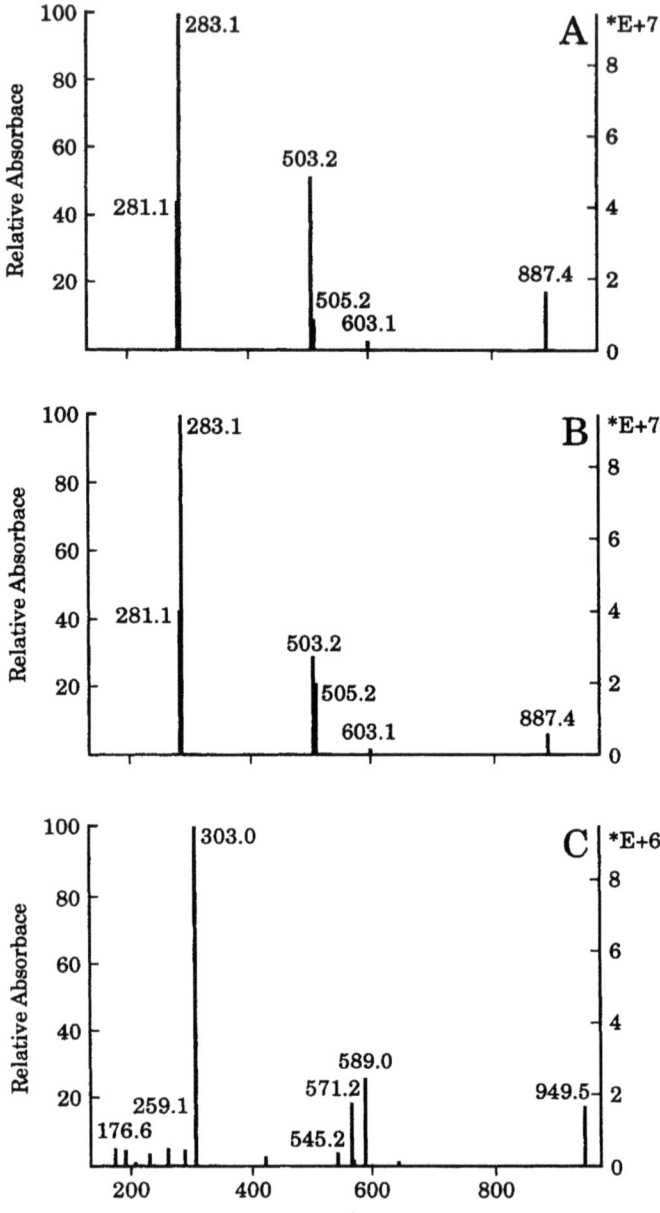

Figure 6.7 *Collisionally activated product ion spectra of the [M − H]⁻ ions of A, 1,3-distearoyl-2-oleoylglycerol (18:0-18:1-18:0); B, 1,2-distearoyl-3-oleoyl-rac-glycerol (18:0-18:0-18:1); and C, triarachidonoylglycerol (20:4-20:4-20:4)* (Reproduced from H. Kallio and G. Currie, *Lipids*, 1993, **28**, 207, with permission of the authors and publisher)

of ion are formed, in very low or zero abundance, from *sn*-2 groups, allowing determination of the substituents at this position.

Recommended Mass Spectrometric Methods for the Analysis of Acylglycerols

GC/MS of underivatised or TBDMS derivatives is an accessible and useful technique for analysing mono- and di-acylglycerols. Free triacylglycerols may also be analysed successfully by EI and positive and negative ion CI GC/MS using high-temperature stationary phases, provided the limitations of the chromatographic approach are recognised and precautions are taken to identify possible losses. However, the most promising technique for determining triacylglycerol mixtures is APCI LC/MS and further development of this methodology, with enhancement of information by MS/MS, is desirable. ESI LC/MS has a potential role in metabolic studies using stable isotopes.

4   Compound Lipids

From the perspective of food composition and nutritional value, phospholipids are the most important group of compound lipids. They are present in the cell membranes of foods, in edible oils (although processing of the oils reduces their concentration significantly) and are widely used by the food industry as emulsifiers. The last mentioned property has resulted in an increased requirement for accurate and specific methods for analysing phospholipids in foods. This is because of the need to characterise food additives and endogenous compounds in raw materials that might affect processing properties. There is also interest from a dietary perspective, in that lipid intake can affect the composition of the phospholipid bilayer in cell membranes. This may, in turn, have beneficial or deleterious consequences for the health of individuals.

A shift from the popular approach to phospholipid analysis, *i.e.* class separation by TLC followed by chemical degradation prior to GC/MS analysis of fatty acid esters, to FABMS and LC/MS was noted in 1994.[4] A good example of the conventional GC/MS approach to phospholipid analysis is provided by a study of the effect of diets supplemented with $n - 3$ fatty acids on the arachidonic acid content of major phospho- and other lipids isolated from salivary glands.[56]

FABMS and LDMS of Phospholipids

The FABMS of phospholipids has been reviewed.[57] Positive[58–63] and negative ion[60,62,64,65] FABMS of phospholipids are all capable of yielding molecular weight structural information. The spectra generally contain $[M + H]^+$ or $[M + cation]^+$ peaks in the positive ion spectra and $[M - H]^-$ ions in negative ion mode. An exception is the negative ion spectra of phosphatidylcholines. These contain the $PO_3^- - CH_2CH_2N(CH_3)_3^+$ zwitterionic group, and the highest mass ion seen in the negative FABMS spectra of phosphatidylcholines is formed by elimination of methyl from this head group. Any ambiguity in interpreting the negative ion spectra of phosphatidylcholines (there is potential for mis-identifying them as phosphatidylethanolamines) is readily overcome by record-ing the positive ion spectrum. It is important to use a 'surface precipitation technique' to ensure successful recording of intense, reproducible negative ion FABMS spectra of phospholipids[64,66] The procedure entails sample dissolution in a good phospholipid solvent and addition of this solution to the glycerol matrix used in FABMS analysis.[64] Although an 86:14:1 mixture of chloroform, methanol and water was originally used, chloroform alone may also be used successfully.[65] The major fragmentation processes yielding structural informa-tion in positive and negative FABMS are shown in Figures 6.8 and 6.9 respectively.

Abundant low mass fragments, characteris   a polar head group, are also observed.[58,59,64] The main drawback of FAE   ias been its unsuitability for the analysis of complex mixtures and for accurately quantifying phospholipids.

LDMS has shown some promise in the analysis of phospholipids.[67] However, the complex nature of the spectra indicated that this technique would be unsuitable for quantitative analysis (as is also the case with FABMS).

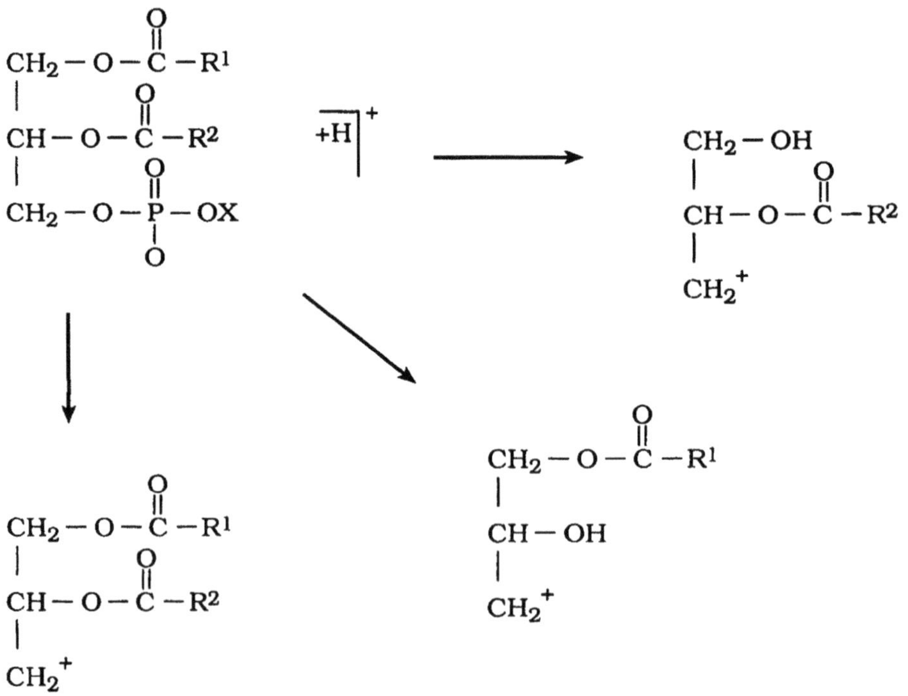

Figure 6.8   *Major, structurally informative ions in the positive ion FABMS spectra of phospholipids*

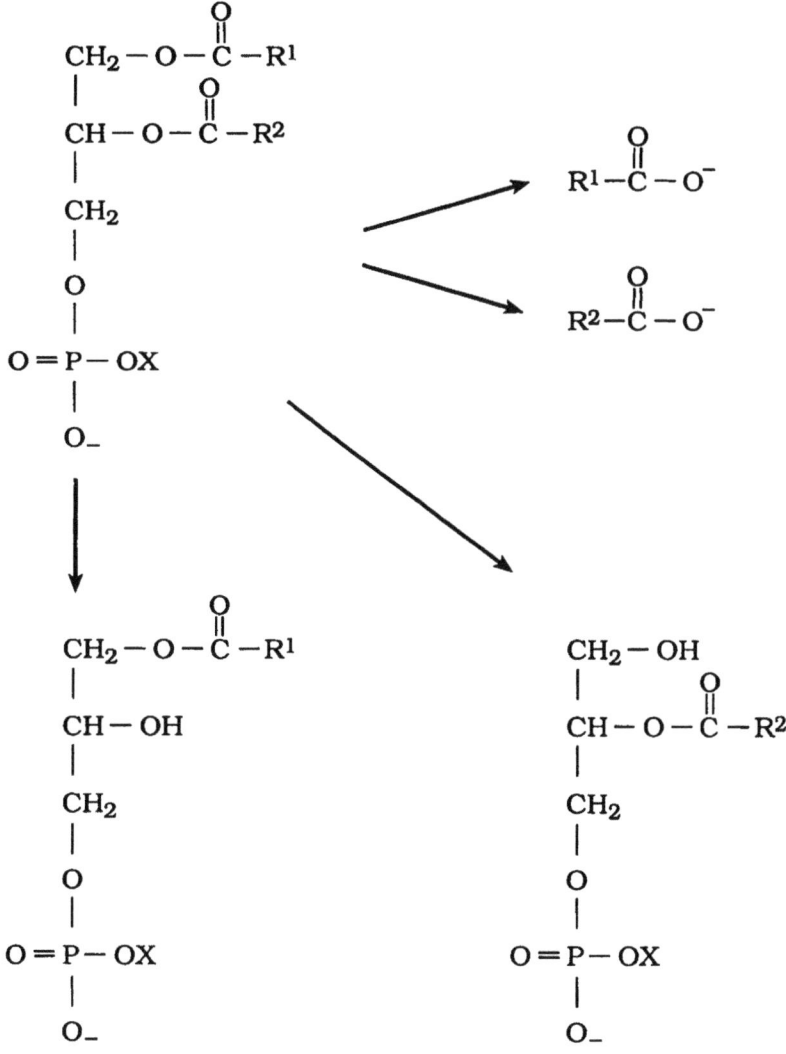

Figure 6.9 *Major, structurally informative ions in the negative ion FABMS spectra of phospholipids*

Tandem Mass Spectrometry of Phospholipids

Tandem mass spectrometry may be used to enhance the information content of FABMS spectra of phospholipids.[62,68–70] The MS/MS spectra can be used to identify the position of attachment of different acyl groups. The use of 9-fluorenylmethoxycarbonyl derivatives was claimed to enhance the information content of positive FABMS spectra.[69] Constant Neutral Loss scans have been used to characterise mixtures of phospholipids.[71] These approaches cannot be recommended for analysis of complex mixtures of phospholipids because of the propensity for FABMS to suffer from suppression effects. A strongly surface active component may be ionised preferentially to another component with lower surfactant properties. This could result in a failure to identify important mixture components. A more fruitful approach lies in the combination of tandem mass spectrometry and ESI MS or LC/MS (see below).

LC/MS of Phospholipids

Although TSP LC/MS has had limited success in the analysis of phospholipids,[72–75] detection limits were poor (10–100 ng), even in SIM mode. The TSP spectra yielded $[M + H]^+$ and $[M + Na]^+$ ions and structurally diagnostic fragment ions. More recent studies have shown conclusively that electrospray LC/MS is considerably more sensitive than TSP.[76–80] Conventional ESI spectra exhibited intense $[M + H]^+$ and $[M + Na]^+$ ions together with structurally diagnostic fragmentation that was dependent on the ESI capillary exit voltage.[76] Negative ion spectra contained intense $[M - H]^-$ and $[M - 2H]^{2-}$ ions and no fragment ions; a linear correlation between the amount of a standard phospholipid consumed and the intensity of these ions was linear over a 10 000-fold dynamic range.[78] The quantification limit in one study was approximately 0.5 pmol before flow splitting, 5 fmol after a 100:1 post-column split and sensitivity was found to be compound dependent.[76] Tandem mass spectrometry can be used to enhance the information content of the ES spectra.[77,79–81] Positive ion ESI MS/MS yields fragment ions that allow characterisation of the head group, whereas negative ion MS/MS produces RCOO$^-$ ions that may be used to identify the acyl substituents.[77] The same group noted that positional information (*sn*-1 or *sn*-2) could not be deduced from the MS/MS spectra. However, this was contradicted by the observation of a preferential loss of the acyl group from the *sn*-2 position in negative ion mode.[81] Another group also noted characteristic differences in the fragment ion spectra of *sn*-1 and *sn*-2 regioisomers when sodiated molecules were collisionally activated.[79] Sensitivity was excellent in MS/MS mode; product ion spectra were obtainable on ≤ 5 pmol of phospholipid.

Both reversed-phase[76] and normal-phase[80] LC/MS can be used to analyse phospholipids, although normal-phase techniques have advantages in their ability to separate phospholipids by class. The normal-phase separation of a standard mixture of phospholipids is shown in Figure 6.10.

APCI also yielded informative spectra, although this technique had the disadvantage (in contrast to ESI) that molecular weight information was not generated for all phospholipid classes. Increased cone voltage fragmentation was used to generate improved structural information. A recent comparison of particle beam and 'Ion Spray' (essentially a trademark form of APCI) LC/MS and LC/MS/MS of soybean phospholipids concluded that ion spray yielded more informative data than particle beam and was more sensitive.[82] The sensitivity of ion spray appeared to be inferior to that of ESI.

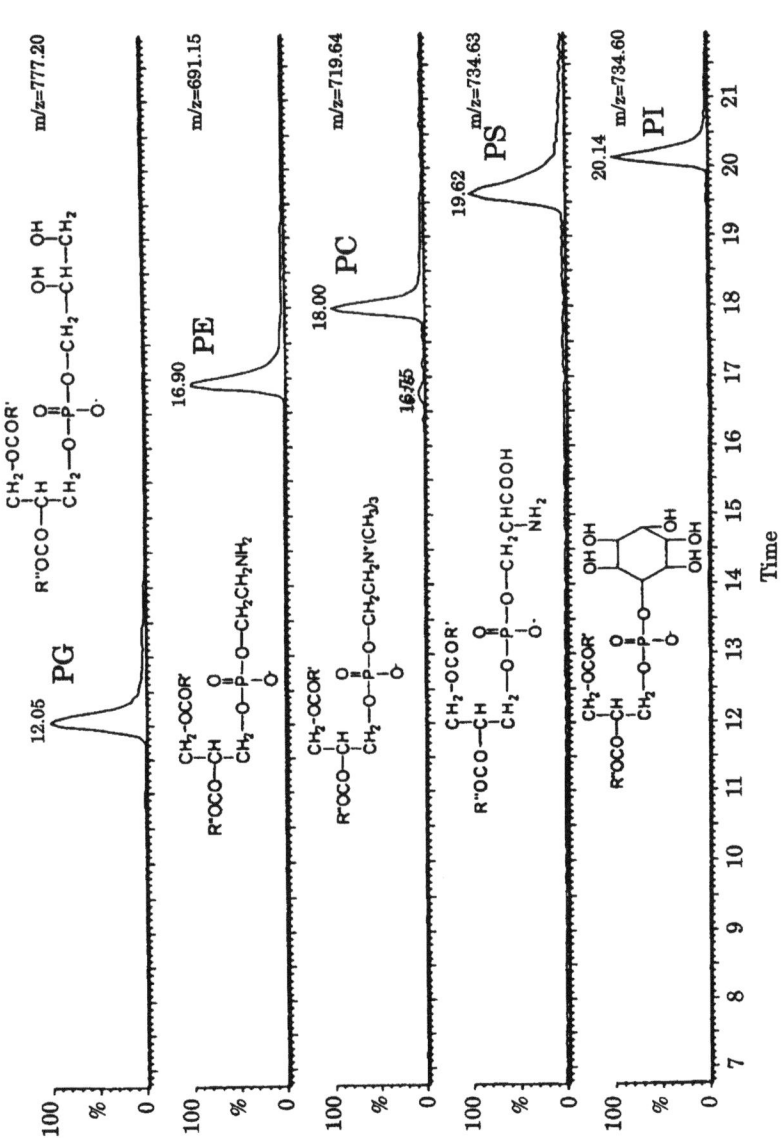

Figure 6.10 *Normal-phase negative ion ES LC/MS separation of a standard mixture of phospholipids, represented as the mass chromatogram of each phospholipid class*
Reproduced from A.A. Karlsson, P. Michelsen, A. Larsen and G. Odham, *Rapid Commun. Mass Spectrom.*, 1996, **10**, 775, with permission of the authors and publisher, © John Wiley & Sons Limited)

Recommended Mass Spectrometric Methods for Determining Compound Lipids

The sensitivity of positive and negative ion ESI LC/MS, allied with its ability to generate unequivocal molecular weight information, clearly indicates that this is likely to become the most useful mass spectrometric technique for analysing phospholipid mixtures. Additional structural information is available by applying MS/MS techniques and the combination of chromatography and mass spectrometry, when available, is likely to become the method of choice. Where these advanced techniques are unavailable, the more laborious approaches of chromatographic class separation, chemical degradation and analysis of fatty acid esters are capable of generating the required structural information.

5   Lipid Oxidation Products

The analysis of lipid oxidation products in foodstuffs and biological tissues is important for two major reasons. Firstly, their presence can influence food palatability and may be used as markers of spoilage; secondly, processes such as inflammation, ageing, atherosclerosis, cancer and ischaemia are believed to result from oxidative stress that can be related, at least partially, to the diet. Traditional (*i.e.* non-mass spectrometric) techniques for analysing lipid oxidation products suffer from several drawbacks, most notably poor inter-laboratory reproducibility, problems of interference from other components in the lipid matrix and insufficient sensitivity. For these reasons, mass spectrometric methods, with their characteristic high sensitivity and specificity, are becoming increasingly popular for determining lipid oxidation products.

GC/MS of Lipid Oxidation Products

Many foodstuffs contain significant concentrations of cholesterol. Cholesterol oxides (oxysterols) are also present in much lower concentrations, but at levels that may still be biologically significant. A GC/MS SIM method using 22-ketocholesterol as an internal standard has been developed for quantifying the oxysterols 7-hydroxycholesterol (α and β isomers), 20α-hydroxycholesterol, cholesterol-5,6-epoxides (α and β isomers), cholestanetriol, 25-hydroxycholesterol and 7-ketocholesterol.[83] The technique, which requires medium (3000) resolution (and therefore a magnetic sector mass spectrometer), was sensitive, robust and specific and had detection limits of approximately 2–6 ng oxysterol/g lipid. This represents a considerable improvement on existing, non-mass spectrometric methods for analysing oxysterols and can be recommended as a general method for quantifying these compounds in foods. It has been successfully applied to the determination of cholesterol oxidation products formed in butter and dairy spreads during storage.[84]

GC/MS methods are also suitable for determining cholesterol oxidation products in blood, as exemplified by the development of an IDMS method, again based on trimethylsilyl derivatives, for quantifying seven oxysterols in a single serum sample.[85] The technique was subsequently improved by using solid-phase extraction techniques to increase the speed of sample processing.[86] Intended as a general technique for quantifying these potentially adverse compounds, it has potential in relating dietary intake to levels of autoxidation products in human serum.

GC/MS has also contributed to an assessment of the possible health hazards of lipid oxidation products in repeatedly used frying oils.[87] Used frying oils from commercial food processors were found to contain a series of 4-oxoalde-hydes. Peroxidation of lipids yields unstable molecules that break down to lower molecular weight species. A GC/MS assay for simultaneously measuring several of these low molecular weight molecules, including malondialdehyde, 4-hydro-xynonenal and other aldehydes, has been developed.[88] The technique is based on reduction of the aldehydes to stable alcohols: these are subsequently derivatised to form *t*-butyldimethylsilyl ethers. Developed for determining lipid peroxidation markers in biological tissues, the method has potential as a sensitive technique for quantifying markers of lipid peroxidation in stored foods, or for relating dietary intake to oxidative damage. Alternative techniques, based on negative ion GC/MS of pentafluorobenzoyloxime-trimethyl-silyl derivatives, appear to be more sensitive and capable of measuring a wider range of peroxidation products at physiological levels.[89,90]

A quantitative GC/MS isomer-specific method for determining hydroxy fatty acids as markers of lipid peroxidation in food and biological samples is based on reduction of peroxidation products to hydroxy fatty acids.[91] Subsequent saponification and on-column methylation forms esters that yield two characteristic ions for each positional isomer. The method was shown to be robust and (when SIM techniques were used) capable of detecting 0.2 ng of a single isomer.

LC/MS of Lipid Oxidation Products

Although reported LC/MS methods for determining lipid oxidation/peroxidation products have all been developed for physiological studies, they have obvious potential for direct determination of incipient food spoilage or in quantifying possible diet-related markers of oxidative damage. Early efforts concentrated on TSP LC/MS of hydroperoxy polyunsaturated fatty acid derivatives[92,93] or epoxy fatty acids[93] and TSP or particle beam LC/MS of plasma cholesterol oxides.[94] Although these methods all showed promise, the TSP or particle beam methods can no longer be recommended because of the superiority of electrospray as a general LC/MS technique for labile compounds. Electrospray methods have now been developed for lipid ester ozonides and for core aldehydes formed by oxidation of model triacylglycerols and egg yolk phospholipids,[95] and for lipid hydroperoxides and long chain conjugated keto acids.[96] The first of these methods relies on negative and positive ion electrospray of ozonides, free aldehydes or dinitrophenylhydrazone derivatives. The second publication describes negative ion electrospray LC/MS and LC/MS/MS of the free oxidation products and reports detection limits in the low picogram range.

Recommended Mass Spectrometric Methods for Determining Lipid Oxidation Products

Any recommendation of mass spectrometric techniques for analysing lipid oxidation products is dependent on the type of information sought. Where determination of low molecular weight markers of oxide and peroxide decomposition products is desired, GC/MS of the free oxidation products (or of suitable derivatives of their reduction products) is sensitive and specific. These methods are also suitable for quantifying higher molecular weight molecules, especially oxysterols. If, on the other hand, it is desirable to assay the immediate products of oxidation, and provided these products are sufficiently stable to be isolated, ESI LC/MS has shown considerable promise and is ripe for further development.

6   References

1 J.J. Myher and A. Kuksis, *J. Chromatogr. B – Biomed. Appl.*, 1995, **671**, 3.
2 M.H. Gordon, in *The Encyclopaedia of Food Science, Food Technology and Nutrition*, ed. R. Macrae, R.K. Robinson and M.J. Sadler, Academic Press, London, 1993, pp. 2738–2747.
3 A. Kuksis and J.J. Myher, in *Mass Spectrometry*, ed. A.M. Lawson, Walter de Gruyter, Berlin, 1989, pp. 265–351.
4 R.P. Evershed, in *Developments in the Analysis of Lipids*, ed. J.H.P. Tyman and M.H. Gordon, Special Publication No. 160, Royal Society of Chemistry, Cambridge, 1994, pp. 123–160.
5 H.Y. Kim and N. Salem, *Progr. Lipid Res.*, 1993, **32**, 221.
6 M. Careri, A. Mangia and M. Musci, *J. Chromatogr. A*, 1998, **794**, 263.
7 A.D. Jones, *Chromatogr. Sci. Ser.*, 1994, **65**, 347.
8 G. Odham, in *Biochemical Applications of Mass Spectrometry*, ed. G.R. Waller and O. Dermer, Wiley-Interscience, New York, 1980, pp. 153–171.
9 S. Banni, B.W. Day, R.W. Evans, F.P. Corongiu and B. Lombardi, *J. Am. Oil Chem. Soc.*, 1994, **71**, 1321.
10 D.E. Minnikin, *Chem. Phys. Lipids*, 1978, **21**, 313.
11 B.A. Anderson and R.T. Holman, *Lipids*, 1974, **9**, 185.
12 B.A. Anderson, *Progr. Chem. Fats Lipids*, 1978, **16**, 279.
13 D.J. Harvey, *Biomed. Mass Spectrom.*, 1982, **9**, 33.
14 W. Vetter, W. Meister and G. Osterhelt, *Org. Mass Spectrom.*, 1988, **23**, 566.
15 N.J. Jensen and M.L. Gross, *Mass Spectrom. Rev.*, 1987, **6**, 497.
16 V.G. Voinov, V.M. Boguslavskiy and Y.N. Elkin, *Org. Mass Spectrom.*, 1994, **29**, 641.
17 J.Y. Zhang, Q.T. Yu and Z.H. Huang, *Shitsuryo Bunseki*, 1987, **35**, 23.
18 J.Y. Zhang, Q.T. Yu, B.N. Liu and Z.H. Huang, *Biomed. Environ. Mass Spectrom.*, 1988, **15**, 33.
19 K.B. Tomer, F.W. Crow and M.L. Gross, *J. Am. Chem. Soc.*, 1983, **105**, 5487.
20 N.J. Jensen, K.B. Tomer and M.L. Gross, *Anal. Chem.*, 1985, **57**, 2018.
21 J. Adams and M.L. Gross, *Anal. Chem.*, 1987, **59**, 1576.
22 J.-C. Promé, H. Aurelle, F. Couderc and A. Savagnac, *Rapid Commun. Mass Spectrom.*, 1987, **1**, 50.
23 F. Couderc, H. Aurelle, D. Promé, A. Savagnac and J.-C. Promé, *Biomed. Environ. Mass Spectrom.*, 1989, **16**, 317.
24 F. Couderc, *Lipids*, 1995, **30**, 691.
25 T. Murata, *Anal. Chem.*, 1977, **49**, 728.

26 T. Murata, *Anal. Chem.*, 1977, **49**, 2209.
27 E. Schulte, M. Hohn and U. Rapp, *Fresenius' Z. Anal. Chem.*, 1981, **307**, 115.
28 M. Barber, D. Bell, M. Eckersley, M. Morris and L. Tetler, *Rapid Commun. Mass Spectrom.*, 1988, **2**, 18.
29 C. Evans, P. Traldi, M. Bambagiotti-Alberti, V. Gianellini, S.A. Coran and F.F. Vincieri, *Biol. Mass Spectrom.*, 1991, **20**, 351.
30 M. Lamberto and M. Saitta, *J. Am. Oil Chem. Soc.*, 1995, **72**, 867.
31 J.J. Myher, A. Kuksis, L. Marai and S.F.K. Yeung, *Anal. Chem.*, 1978, **50**, 557.
32 P. Mareš, *Progr. Lipid Res.*, 1988, **27**, 107.
33 J.J. Myher, A. Kuksis, L. Marai and P. Sandra, *J. Chromatogr.*, 1988, **452**, 93.
34 R.P. Evershed, *J. Am. Soc. Mass Spectrom.*, 1996, **7**, 350.
35 J.E. Biller and K. Biemann, *Anal. Lett.*, 1974, **7**, 515.
36 P. Zöllner, E. Lorbeer and G. Remberg, *Org. Mass Spectrom.*, 1994, **29**, 253.
37 P. Zöllner and E. Lorbeer, *J. Mass Spectrom.*, 1995, **30**, 432.
38 K.L. Duffin, J.D. Henion and J.J. Schich, *Anal. Chem.*, 1991, **63**, 1781.
39 P.J.W. Schuyl, T. de Joode, M.A. Vasconcellos and G.S.M.J.E. Duchateau, *J. Chromatogr. A*, 1998, **810**, 53.
40 W.E. Neff and W.C. Byrdwell. *J. Liquid Chromatogr.*, 1995, **18**, 4165.
41 W.C. Byrdwell and E.A. Emken, *Lipids*, 1995, **30**, 173.
42 W.C. Byrdwell, E.A. Emken, W.E. Neff and R.O. Adlof, *Lipids*, 1996, **31**, 919.
43 W.E. Neff and W.C. Byrdwell, *J. Am. Oil Chem. Soc.*, 1995, **72**, 1185.
44 H. Kallio, P. Laakso, R. Huopalahti, R.R. Linko and P. Oksman, *Anal. Chem.*, 1989, **61**, 698.
45 P. Laakso and P. Manninen, *Lipids*, 1997, **32**, 1285.
46 P. Laakso and P. Voutilainen, *Lipids*, 1996, **31**, 1311.
47 H. Kallio and G. Currie, *Lipids*, 1993, **28**, 207.
48 H. Kallio and G. Currie, in *CRC Handbook of Chromatography: Analysis of Lipids*, ed. K.D. Mukherjee and N. Weber, CRC Press, Boca Raton, 1993, pp. 435–457.
49 G.J. Currie and H. Kallio, *Lipids*, 1993, **28**, 217.
50 P. Laakso and H. Kallio, *J. Am Oil Chem. Soc.*, 1993, **70**, 1161.
51 P. Laakso and H. Kallio, *J. Am Oil Chem. Soc.*, 1993, **70**, 1173.
52 P. Laakso and H. Kallio, *Lipids*, 1996, **31**, 34.
53 H. Kallio and P. Rua, *J. Am. Oil Chem. Soc.*, 1994, **71**, 985.
54 H. Kallio, G. Currie, R. Gibson and S. Kallio, *Ann. Chim.*, 1997, **87**, 187.
55 H. Kallio, J.-P. Kurvinen, O. Sjövall and A. Johansson, *ACS Symposium Ser.*, 1998, **702**, 221.
56 B.S. Alam and S.Q. Alam, *J. Nutr. Biochem.*, 1992, **3**, 609.
57 R.C. Murphy and K.A. Harrison, *Mass Spectrom. Rev.*, 1994, **13**, 57
58 G.R. Fenwick, J. Eagles and R. Self, *Biomed. Environ. Mass Spectrom.*, 1983, **10**, 382.
59 E. Ayanoglu, A. Wegmann, O. Pilet, G.D. Marbury, J.R. Hass and C. Djerassi, *J. Am. Chem. Soc.*, 1984, **106**, 5246.
60 D.N. Heller, R.J. Cotter, C. Fenselau and O.M. Uy, *Anal. Chem.*, 1987, **59**, 2806.
61 M.J.I. Mattina, S.D. Richardson, M. Wood, Q.Z. Zhou, M.J. Contado, F.M. Menger and L.E. Abbey, *Org. Mass Spectrom.*, 1988, **23**, 292.
62 H. Munster and H. Budzikiewicz, *Biol. Chem. Hoppe-Seyler*, 1988, **369**, 303.
63 B.N. Pramanik, J.M. Zechman, P.R. Das and P.L. Bartner, *Biomed. Environ. Mass Spectrom.*, 1990, **19**, 164.
64 H. Munster, J. Stein and H. Budzikiewicz, *Biomed. Environ. Mass Spectrom.*, 1986, **13**, 423.
65 S. Chen, E. Benfati, R. Fanelli, G. Kirschner and F. Pregnolato, *Biomed. Environ. Mass Spectrom.*, 1989, **18**, 1051.
66 M.-Y. Zhang, X.-Y. Liang, Y.-Y. Chen and X.-G. Liang, *Anal. Chem.*, 1984, **56**, 2288.
67 M.C. Whal, H.S. Kim, T.D. Wood, S. Guan and A.G. Marshall, *Anal. Chem.*, 1993, **65**, 3669.

68 N.J. Jensen, K.B. Tomer and M.L. Gross, *Lipids*, 1986, **21**, 580.
69 S. Chen, G. Menon and P. Traldi, *Org. Mass Spectrom.*, 1992, **27**, 215.
70 S. Chen, O. Curcuruto, S. Catinella, P. Traldi and G. Menon, *Biol. Mass Spectrom.*, 1992, **21**, 655.
71 D.N. Heller, C.M. Murphy, R.J. Cotter, C. Fenselau and O.M. Uy, *Anal. Chem.*, 1988, **60**, 2787.
72 H.-Y. Kim and N. Salem, *Anal. Chem.*, 1986, **58**, 9.
73 H.-Y. Kim and N. Salem, *Anal. Chem.*, 1987, **59**, 722.
74 H.-Y. Kim and N. Salem, *J. Chromatogr.*, 1987, **394**, 155.
75 N. Salem, H.-Y. Kim and J.A. Yergey, *Colloquie INSERM*, 1987, **152**, 151.
76 H.-Y. Kim, T.-C.L. Wang and Y.-C. Ma, *Anal. Chem.*, 1994, **66**, 3977.
77 J.L. Kerwin, A.R. Tuninga and L.H. Ericsson, *J. Lipid Res.*, 1994, **35**, 1102.
78 X.L. Han and R.W. Gross, *Proc. Natl. Acad. Sci. USA*, 1994, **91**, 10 635.
79 X.L. Han and R.W. Gross, *J. Am. Soc. Mass Spectrom.*, 1995, **6**, 1202.
80 A.A. Karlsson, P. Michelsen, A. Larsen and G. Odham, *Rapid Commun. Mass Spectrom.*, 1996, **10**, 775.
81 P.B.W. Smith, A.P. Snyder and C.S. Harden, *Anal. Chem.*, 1995, **67**, 1824.
82 M. Careri, M. Dieci, A. Mangia, P. Manini and A. Raffaelli, *Rapid Commun. Mass Spectrom.*, 1996, **10**, 707.
83 J.H. Nielsen, C.E. Olsen, C. Duedahl and L.H. Skibsted, *J. Dairy Res.*, 1995, **62**, 101.
84 J.H. Nielsen, C.E. Olsen, C. Jensen and L.H. Skibsted, *J. Dairy Res.*, 1996, **63**, 159.
85 O. Breuer and I. Björkhem, *Steroids*, 1990, **55**, 185.
86 S. Dzeletovic, O. Breuer, E. Lund and U. Diczfalusy, *Anal. Biochem.*, 1995, **225**, 73.
87 G.R. Takeoka, R.G. Buttery and C.T. Perrino, *J. Agric. Food Chem.*, 1995, **43**, 22.
88 C. Des Rosiers, M.-J. Rivest, M.-J. Boily, M. Jetté, A. Carrobeé-Cohen and A. Kumar, *Anal. Biochem.*, 1993, **208**, 161.
89 X.P. Luo, M. Yazdanpanah, N. Bhooi and D.C. Lehotay, *Anal. Biochem.*, 1995, **228**, 294.
90 A. Lodl-Stahlhofen, W. Kern and G. Spiteller, *J. Chrom.*, 1995, **673**, 1.
91 R. Wilson, R. Smith, P. Wilson, M.J. Shepherd and R.A. Riemersma, *Anal. Biochem.*, 1997, **248**, 76.
92 M. Yamane, A. Abe, S. Yamane and F. Ishikawa, *J. Chromatogr.*, 1992, **579**, 25.
93 M. Yamane, A. Abe and S. Yamane, *J. Chromatogr. B*, 1994, **652**, 123.
94 A. Sevanian, R. Seraglia, P. Traldi, P. Rossato, F. Ursini and H. Hodis, *Free Radical Biol. Med.*, 1994, **17**, 397.
95 A. Ravandi, A. Kuksis, J.J. Myher and L. Marai, *J. Biochem. Biophys. Meth.*, 1995, **30**, 271.
96 D.K. MacMillan and R.C. Murphy, *J. Am. Soc. Mass Spectrom.*, 1995, **6**, 1190.

Sugars and Carbohydrates

1 Introduction

The application of EIMS to structural studies on carbohydrates started at the same time as work on other biopolymers,[1] but the similarity of monosaccharide residues meant that the mass spectra were not as useful as those for peptides. Furthermore, neither the complex linkage of the sugar residues in oligosaccharides nor the differences between anomers of the common sugars could be determined from EI mass spectra of non-specific derivatives. Nevertheless, the fragmentation processes experienced under EI conditions were tabulated and a nomenclature established. Special sample derivatisation procedures were required for linkage determination and isomer differentiation.[2] However, chemical derivatisation was traditional in the routine preparation of volatile compounds from involatile monosaccharides and oligosaccharides and thus EIMS was adapted for primary structural studies. Oligosaccharides with four or five residues could be volatilised, but the blocking groups, trimethylsilyl for example, often increased the molecular mass to values beyond the range of the EI mass spectrometers in use at that time. GC was already used extensively in the analysis of free sugars and hydrolysed carbohydrate extracts from natural sources, using derivatives chosen to contribute to efficient separation.[3] The GC retention times alone were often sufficient to recognise the components of simple mixtures because many of the common anomeric forms produced individual peaks. However, the additional advantages of high resolution and high sensitivity detection, especially in quantitative SIM mode, and structure identification afforded by GC/MS for the analysis of complex mixtures of carbohydrate derivatives were obvious. The need to combine the three functions of the derivative into one chemical reaction: volatility, efficient GC separation and characteristic, structurally informative mass spectrometric fragmentation patterns, unified the efforts of biochemists, chemists, chromatographers and mass spectroscopists. A useful and comprehensive review of mass spectrometry in carbohydrate research was published in 1989 and summarises the state of the art at a time before LC/MS procedures were widespread.[4]

Of particular relevance to GC/MS was the application of the partially-methylated alditol acetate (PMAA) derivatives. These combined the attributes of volatility and structure-specific fragmentation so that the linkages of the monosaccharide residues in the polysaccharide chain could be deduced from the EI mass spectrum.[4-6] The advent of desorption ionisation methods in the 1980s (*e.g.* FABMS[7]) opened up many new areas of carbohydrate research because involatile samples could be ionised directly from the solid phase. The application of desorption ionisation mass spectrometry to the analysis of intact carbohydrates was limited, but it was possible to record molecular weight information on oligosaccharides,[8] and with chemical derivatisation some quite remarkable analyses could be performed.[9,10] The ionic nature of many glycosylated intermediates enhanced their ionisation in LSIMS experiments and some of the first examples of FABMS used glucosinolates and glycoalkaloids, which produced exemplary mass spectra.[11] The application of FABMS to glycoconjugates was reviewed in 1987,[12] and Domon and Costello published a systematic nomenclature for the fragmentation of glycoconjugates using FABMS/MS,[13] along the lines established for peptides. The mass spectrometry of food glycosides is discussed in more detail in Chapter 10.

A very useful review of the contemporary methods of analysis of plant cell wall polysaccharides was published in 1990.[14] The routine use of GC/MS in the analysis of alditol acetate derivatives, and the role of FABMS and LC/MS in relation to the analysis of oligoglycosyl alditols derived from the hydrolysis of pectic polysaccharides of primary cell walls was reviewed in 1990.[15] [M − H]⁻ ions could, at that time, be obtained from underivatised oligogalacturonides up to DP 11 by FABMS. Larger oligomers, from DP 2–14, could be analysed after derivatisation to their pentafluorobenzyl oximes and as their ammonium salts.

Although it was not essential to derivatise oligosaccharide samples for analysis by FABMS, it was realised that the mass spectrum could be improved, either qualitatively or quantitatively (greater sensitivity) by making a suitable chemical derivative. For example, *n*-alkyl *p*-aminobenzoates have been prepared for high-sensitivity assays of nicotinic acetylcholine receptor from Torpedo californica.[16] The designer derivative *N,N*-(2,4-dinitrophenyl)octylamine, which provided a UV chromophore (2,4-dinitrophenyl group), at the same time lends charge stability and incorporates a hydrophobic tail to enhance FAB ionisation.[17] The octyl group also enhances surface activity, making analytes amenable to separation by reversed-phase chromatography using C_{18} bonded phase columns. Thioglycerol or thioglycerol–TFA matrices were used in LSIMS analyses of peptides, glycopeptides, reducing oligosaccharides, silylated oligosaccharides and oligosaccharide alditols after HPLC separation on a porous graphitised carbon column.[18]

Recently, LC/MS has been applied to the analysis of carbohydrates and ESI and MALDI mass spectrometry have extended the mass range available for the analysis of intact oligosaccharides. A wide-ranging review of mass spectrometry includes a comprehensive discussion of the analysis of carbohydrates and glycoconjugates.[19] The use of ESI-MS and CID for molecular weight profiling, sequence, linkage and branching determination in carbohydrates has been reviewed,[20] as has MALDI-MS of oligosaccharides.[21] Some examples of further methodological developments, made since the major reviews, will be used to update mass spectrometric approaches to carbohydrate analysis and research. The application of mass spectrometry to the analysis of carbohydrates in foods will be discussed under the appropriate headings.

A comparison has been made between the current state of the art of GC/MS and GC/MS/MS, and the newer LC/MS and LC/MS/MS methods, for the analysis of sugar monomers.[22] The authors concluded that the GC methods retained higher sensitivities (in 1996) than the LC methods. The review also discussed GC/LC for complex matrix cleanup and MS/MS for the analysis of underivatised carbohydrates.

2 Analysis of Sugar Residues: GC/MS of Derivatised Sugars

GC/MS methods applied to methylation analysis generally follow normal procedures, *i.e.* permethylation, mineral acid hydrolysis of glycosidic linkages and preparation of suitable derivatives of the components of the hydrolysate. The mixtures are separated by GC and characterised by GC/MS.[4] A wide range of chemical derivatives have been used in the formation of products for analysis including alditol acetates, trimethylsilylated alditols, trifluoroacetylated alditols, aldononitrile acetates, oximes and pertrimethylsilyl, peracetyl and pertrifluoroacetyl derivatives.[23] However, permethylation analysis using partially methylated alditol acetates (PMAAs, see below) is still one of the most popular and successful techniques.

Alternative techniques include methanolysis with methanolic HCl, which generates mixtures of volatile anomers readily resolved by GC.[23] Much later, trimethylsilylation and methanolysis were employed to prepare neutral sugars and uronic acids for GC/MS analysis.[24]

High-temperature GC and GC/MS have been used for the quantitative determination of methylated DP 2-12 oligosaccharides in a range foods.[25]

Absolute configurations of monosaccharides were obtained from GC/MS analyses of the products of glycosidation with (−)-2-butanol or (+)-2-octanol.[26] GC/MS has also been used in carbohydrate profiling to distinguish between *Bacillus anthracis* and *B. cereus*, closely related pathogens that are difficult to recognise by pheno- or genotyping.[27] Different levels of galactose and galactosamine were found to differentiate vegetative cultures of *B. anthracis* from *B. cereus*, while fucose was found in spore cultures of *B. cereus* but not in *B. anthracis*.

The traditional methods of sugar analysis have been compared to modern methods such as GC, HPLC and ion chromatography (IC).[28] An example of this is the rapid HPLC analysis of doughs and baking products when fructose, glucose, sucrose, maltose and lactose were separated in 12 minutes.[29] Microdialysis was used to selectively sample the same sugars from milk and milk-based products prior to HPLC analysis.[30] LC/MS is becoming commonplace in the analytical laboratory and methods currently using HPLC may be found to benefit from the additional resolution of the combined techniques.

Similarly, since CE/MS became available, reports of the use of CE for sugar analysis indicate the advantages of speed of analysis, small sample size and higher resolution than HPLC. CE has been applied to the analysis of sugars in wine[31] and to the analysis of derivatised xylose, arabinose, glucose and galactose from a sorghum hydrolysate.[32] Novel 2-aminoacridone derivatives of sugars were separated by MECC.[33]

3 Partially Methylated Alditol Acetate (PMAA) Analysis of Linkage and Sugar Type

Figure 7.1 illustrates the procedure for permethylation, hydrolysis and acetylation of oligosaccharides diagrammatically. It has been emphasised that there is

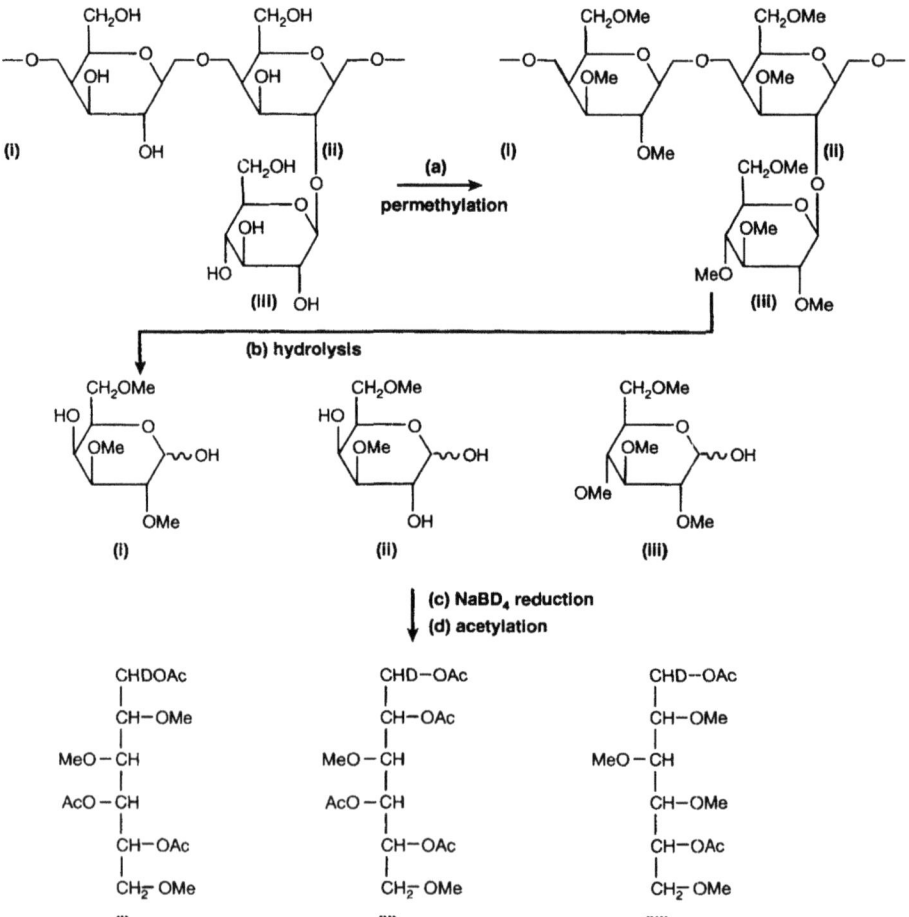

Figure 7.1   *Reaction pathways in the preparation of partially methylated alditol acetates (PMAAs) for linkage analysis of oligosaccharides*

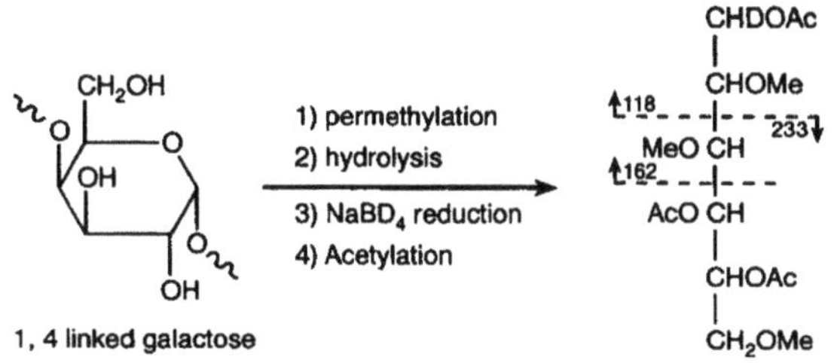

Figure 7.2 *Mass spectrometric characteristics of permethylated and acetylated (PMAA)*
derivative of 1,4-linked galactose derived from an oligosaccharide

no 'standard' version of linkage analysis:[34] particular types of analysis often require 'fine tuning'. However, the procedure outlined in the diagram is general.

The main mass spectrometric fragment ions of the PMAA derivative of 1,4-linked galactose and its 70 eV EI mass spectra are shown in Figures 7.2 and 7.3 respectively.

Ion chromatography and GC/MS methylation analysis confirmed the presence of sugars and reductones in complex peanut flavour precursor extracts.[35] Pyridine CIMS was used to enhance EIMS for methylation analysis.[36] Adducts formed between protonated pyridine and PMAAs are very stable and the mass spectra contain $[M + PyrH]^+$ ions.

4 Recent Methods for the Determination of Molecular Size

Oligosaccharides

ESI-MS is becoming useful for derivatised and underivatised oligosaccharides. A method for the ESI-MS and Tandem MS of trimethyl(*p*-aminophenyl)-ammonium derivatised oligosaccharides has been described[37] and the use of ESI-MS for underivatised glucopyranosyl disaccharides has been reported.[38] Instrumental methods for the analysis of oligosaccharides have been compared.[39]

Carbohydrates

One the earliest reports of the application of MALDI-MS to molecular weight determination of oligosaccharides was made in 1991.[40] The technique provided 1 pmol sensitivity with rapid sampling (5 min) and good mass spectrometric resolution, within 0.5 Da. Several other good examples of the value of MALDI-MS were given. For example, a comparison was made between a gel filtration chromatogram (500 µg, analysis time 650 min) and the MALDI spectrum (10 ng, 5 min analysis time) of a mixture of 1–11-mer glucose homopolymers (Figure 7.4).

Thermal damage to milk was visualised using MALDI-MS with sinapic acid as the matrix.[41] The MALDI mass spectra from 10 000 to 50 000 Th of

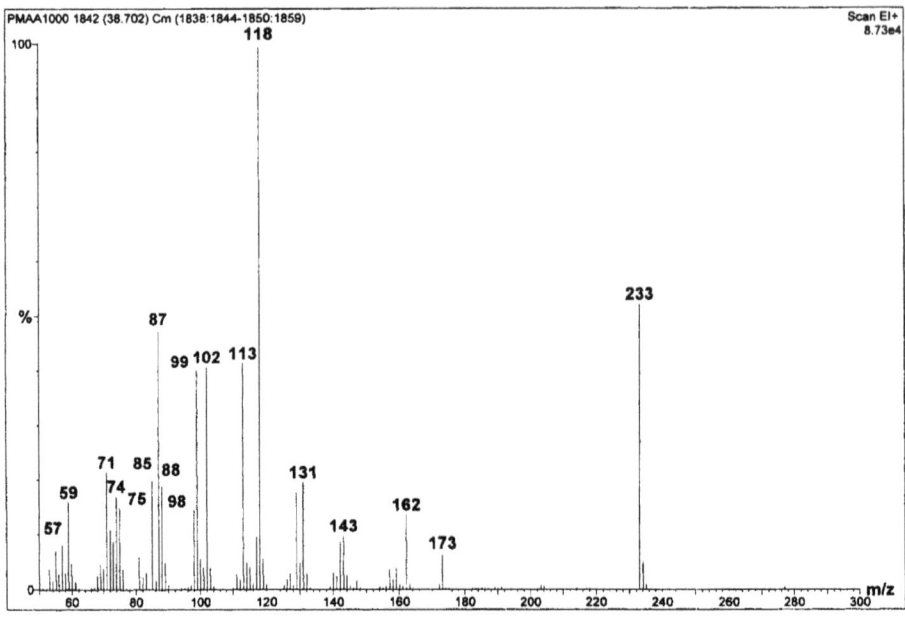

Figure 7.3   *The 70 eV EI mass spectrum of the PMAA derivative of 1,4-linked galactose*

pasteurised, UHT, raw milk and milk whey were presented and interpreted in terms of the protonation of caseins and other known constituents.

An example of the use of MALDI-ToF-MS for carbohydrate MW determination was given by the work on the allergenicity of hen egg-white proteins.[42] Deglycosylated ovomucoid consisted of five isoforms with MWs between 20.7 and 21.5 kDa compared to native ovomucoid that had a MW of 28 kDa. LC/ESI-MS/MS has been used to study carbohydrates utilising the post-column addition of metal chlorides.[43]

Glycoconjugates

Biologically active food glycosides are covered in more detail in Chapter 4 and only a small number of illustrative examples are therefore given here.

An on-line system of HPLC/UV-visible photodiode array/MS, employing both TSP and ESI, was used to determine the glycosides of saffron (*Crocus sativus* L.), including carotenoid glycosides.[44] The combined technique provided information about *cis* and *trans* isomers, molecular masses and glucose residue sequences. TSP was found to be suitable for the analysis of glycosidic carotenoids with up to three glucose residues and ESI was useful with glycosidic carotenoids with up to five glucoses.

The structure of a secretory proteinase inhibitor isolated from the latex of green fruits of papaya (*Carica papaya*) has been elucidated with the aid of mass

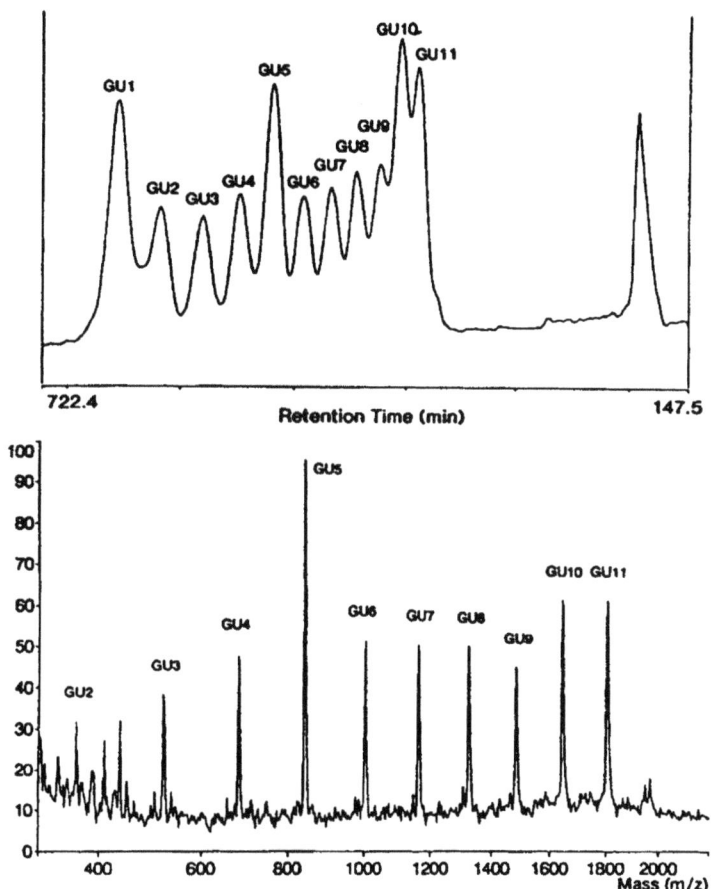

Figure 7.4 *Comparison of the MALDI mass spectrum (lower trace) of glucose homopolymers derived from a Dextran hydrolysate with the P4 gel filtration
chromatogram of the same material*
(Reproduced from K.K. Mock, M. Davey and J.S. Cottrell, *Biochem.
Biophys. Res. Commun.*, 1991, **177**, 644, with the permission of the authors
and publishers)

spectrometry.[45] 184 amino acids were sequenced by classical methods. Two
carbohydrate chains, each probably composed of (mannose) (5), (xylose) (1),
(fucose) (0–2), and (*N*-acetylglucosamine) (2) residues, were attached to Asp-84
and -90. Mass spectrometric and compositional analysis suggested that they
might represent a new class of plant xylose-containing carbohydrate chains with
five mannose residues.

FABMS and MS/MS have been used to study (−)-menthol-β-(D)-glucopyr-
anoside, a precursor of menthol in foods, and neohesperidin dihydrochalcone,
an alternative sweetener to sucrose, and their non-covalent association with β-

cyclodextrin as examples of 'the interaction of a carbohydrate host with glycoconjugate guests'.[46] The FAB matrix was thioglycerol. The CID spectra of β-cyclodextrin complexes were compared with those of the native compounds and the differences discussed. Mass spectrometric techniques were also used to study the secodammarane glycosides, potential sweeteners of plant origin from *Pterocarya paliurus*.[47]

5   The Separation of Oligosaccharides by LC/MS

Along with ^{1}H NMR spectroscopy, ESI-MS has been used to establish the structures of DP 2–11 oligosaccharides generated from rhamnogalacturonan I by enzyme action.[48] Major ions in the positive ion mass spectra were $[M + Na]^+$ and $[M + Na - 18]^+$ and a minor ion appeared at $[M + H - 18]^+$. The two ions formed by the loss of 18 mass units were thought to originate from the lactone form of the aldonic acid located at the reducing end of each oligosaccharide. Fragment ions were observed in the spectra of oligosaccharides above DP 6 which were shown to be produced in the ESI inlet and not during the preparation of the sample by hydrolysis of the glycosidic bonds. The authors therefore caution others when interpreting the spectra of unknown compounds. Further work with ESI-MS was reported in 1995.[49]

TSP-MS has been employed to analyse charged oligosaccharides from pectic region hydrolysates of apple.[50] TSP and ESI have been compared for the analysis of flavocoenzymes from sugar beet.[51] TSP spectra contained ions showing fragmentation around the flavin ring for riboflavin and riboflavin 5′-monophosphate while the flavin adenine dinucleotide spectrum showed cleavage at the adenine group. In contrast, ESI spectra contained mainly MW information. Two new compounds (riboflavin 3′-monosulfate and riboflavin 5′-monosulfate) were identified as major flavins in roots.

6   Sequence Determination of Oligosaccharides

The food industry *inter alia* uses pectins as, for example, protein stabilisers and in aqueous gel formation. Thus, oligomer sequence determination by mass spectrometry is an active area of development. The number and location of methyl-esterified residues is important in the understanding of the structure–function relationships of pectins and the specificity of pectolytic enzymes. The development of mass spectrometric methods for the sequence determination of oligosaccharides has been reviewed and the use of quadrupole ion traps for sequencing partially methyl-esterified oligogalacturonates described.[52] The standard ESI source was replaced by a laboratory-built nanoelectrospray source, to provide increased signal stability and decreased pressures in the interface. CID pathways in positive ion and negative ion modes were studied for an unsaturated pentamer of galacturonic acid. Negative ion ESI was found to be more sensitive (5–10 pmol μl^{-1}) than positive ion (10–50 pmol μl^{-1}). The fragmentation pathway is shown in Figure 7.5 using the nomenclature of Domon and Costello.[13] Cationised molecules were detected in the positive ion

a)   b)

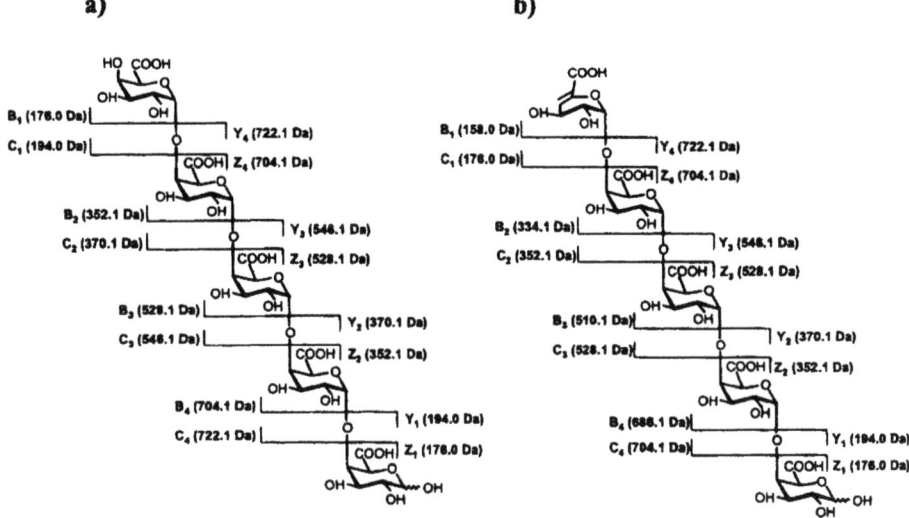

Figure 7.5   *Carbohydrate fragmentation nomenclature for saturated and unsaturated pentamers of galacturonic acid*
(Reproduced with permission from R. Korner, G. Limberg, T.M.I..E. Christensen, J.D. Mikkelsen and P. Roepstorff, *Anal. Chem.*, 1999, **71**, 1421, © 1999 American Chemical Society)

form and singly and doubly deprotonated species in negative ion mode. ^{18}O-Labelling of the reducing ends of the oligomers was effective in differentiating them from the isomeric non-reducing ends.

Interpretation of the mass spectra is based on two series of ions originating from the non-reducing end (B and C), and two series from the reducing end (Y and Z) representing glycosidic bond cleavages. Experience gained with the interpretation of the fragmentation pattern of the unsaturated pentamer (Figure 7.6) was used in the elucidation of the mass spectra of higher oligomers. For example, a similarly substituted decamer was obtained from an unseparated digest of partially methyl-esterified (DE 31%) pectin using exopolygalacturonase II enzyme (Figure 7.7).

The MS/MS analysis of oligomers bearing methyl-esterified residues required knowledge of the enzyme activity to reduce the number of possible isomers in contention. Models with two methyl-esterified sites in digests using different enzymes were studied. It was concluded that the fragmentation spectra of pentamers containing two methyl-esterified residues produced with three different enzymes all show clear differences explained by the distribution of methyl ester groups. However, there are still problems because CID pathways were found to vary with the number and distribution of methyl-esterified residues. Nevertheless, this application illustrates both the flexibility of mass spectrometry in its modern form and its ability to tackle real problems in food analysis.

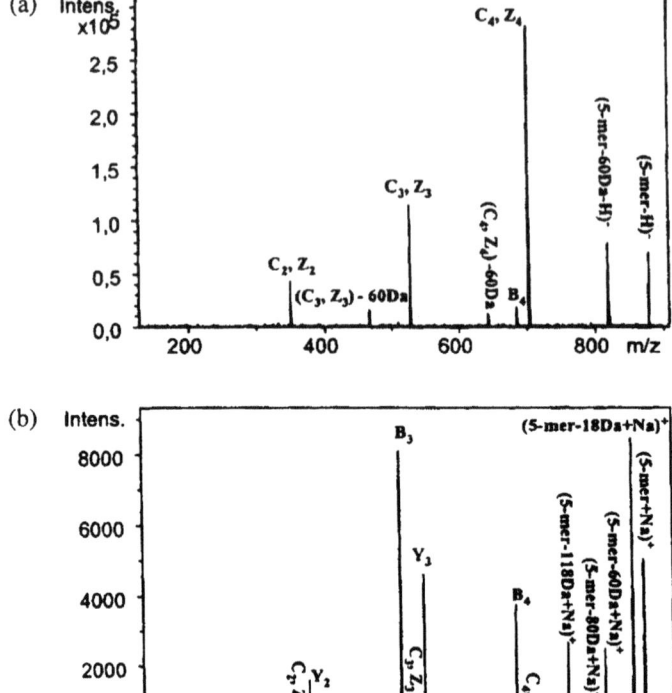

Figure 7.6   *MS/MS spectra of* (a) *the singly deprotonated and* (b) *the singly sodiated pentamer of galacturonic acid*
(Reproduced by permission from R. Korner, G. Limberg, T.M.I.E. Christensen, J.D. Mikkelsen and P. Roepstorff, *Anal. Chem.*, 1999, **71**, 1421, © 1999 American Chemical Society)

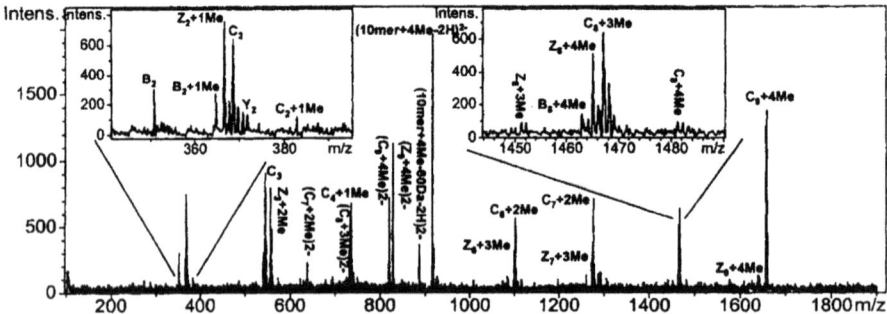

Figure 7.7   *Negative ion MS/MS spectrum of the doubly deprotonated decamer with two methyl esterified residues and* ^{18}O*-labelled reducing end from an unseparated PG II digest of partially methyl esterified (31%) pectin*
(Reproduced by permission from R. Korner, G. Limberg, T.M.I.E. Christensen, J.D. Mikkelsen and P. Roepstorff, *Anal. Chem.*, 1999, **71**, 1421, © 1999 American Chemical Society)

7 References

1 N.K. Kochetkov and O.S. Chizhov, *Adv. Carbohydr. Chem.*, 1966, **21**, 39.
2 D.S. Robinson, J. Eagles and R. Self, *Carbohyd. Res.*, 1973, **26**, 204.
3 G.G.C. Dutton, *Adv. Carbohydr. Chem. Biochem.*, 1974, **30**, 9.
4 J.P. Kamerling and J.F.G. Vliegenthart, in *Mass Spectrometry*, ed. A.M. Lawson, Walter de Gruyter, Berlin, 1989, p. 176.
5 A. Fox, S.L. Morgan and J. Gilbart, in *Analysis of Carbohydrates by GLC and MS*, ed. C.J. Biermann and G.D. McGinnis, CRC Press, Boca Raton, FL, 1989, p. 87.
6 N.C. Carpita and E.M. Shea, in *Analysis of Carbohydrates by GLC and MS*, ed. C.J. Biermann and G.D. McGinnis, CRC Press, Boca Raton, FL, 1989, p. 157.
7 M. Barber, R.S. Bordoli, R.D. Sedgwick and A.N. Tyler, *J. Chem. Soc. Chem Commun.*, 1981, 325.
8 R. Self, J. Eagles, F.A. Mellon and G.R. Fenwick, in *Mass Spectrometry of Large Molecules*, ed. S. Facchetti, Elsevier, Amsterdam, 1985 p. 209.
9 A. Dell and J.E. Thomas-Oates, in *Analysis of Carbohydrates by GLC and MS*, ed. C.J. Biermann and G.D. McGinnis, CRC Press, Boca Raton, FL, 1989, p. 217.
10 K.-H. Khoo and A. Dell, *Glycobiology*, 1990, **1**, 83.
11 R. Self, L.C.E. Taylor, C.V. Bradley, S. Santikarn and D.H. Williams, in *Introduction to Mass Spectrometry of Large Molecules*, ed. S. Daolio, SCI Series, Padova, 1982, p. 227.
12 H. Egge and J. Peter-Katalinic, *Mass Spectrom. Rev.*, 1987, **6**, 331.
13 B. Domon and C.E. Costello, *Glycoconjugate J.*, 1988, **5**, 397.
14 R.R. Selvandran and P. Ryden, in *Methods in Plant Biochemistry – Vol 2*, ed. P.M. Dey, Academic Press, London, 1990, p. 549
15 M. O'Neill, P. Albersheim and A. Darvill, in *Methods in Plant Biochemistry – Vol 2*, ed. P.M. Dey, Academic Press, London, 1990, p. 415.
16 L. Poulter, R. Karrer and A.L. Burlingame, *Anal. Biochem.*, 1991, **195**, 1.
17 Y. Zhang, R.A. Cedergren, T.J. Nieuwenhuis and R.I. Hollingsworth, *Anal. Biochem.*, 1993, **208**, 363.
18 M.J. Davies, K.D. Smith, R. A. Carruthers, W. Chai, A.M. Lawson, and E.F. Hounsell, *J. Chromatogr.*, 1993, **646**, 317.
19 A.L. Burlingame, R.K. Boyd and S.J. Gaskell, *Anal. Chem.*, 1994, **66**, 634R.
20 V.N. Reinhold, B.B. Reinhold and C.E. Costello, *Anal. Chem.*, 1995, **67**, 1772.
21 D.J. Harvey, *J. Chromatogr. A*, 1996, **720**, 429.
22 G.E. Black and A. Fox, *J. Chromatogr. A*, 1996, **720**, 51.
23 G.G.C. Dutton, *Adv. Carbohydr. Chem. Biochem.*, 1974, **30**, 9.
24 J. Bleton, P. Mejanelle, J. Sansoulet, S. Goursaud, and A. Tchapla, *J. Chromatogr. A*, 1996, **720**, 27.
25 N.-G. Carlsson, H. Karlsson and A-S. Sandberg, *J. Agric. Food Chem.*, 1992, **40**, 2404.
26 G.J. Gerwig, J.P. Kamerling and J.F.G. Vliegenthart, *Carbohydr. Res.*, 1979, **77**, 1.
27 A. Fox, G.E. Black, K. Fox and S. Rostovtseva, *J. Clin. Microbiol.*, 1993, **31**, 887.
28 S.V. Vercellotti and M.A. Clarke, *Int. Sugar J.*, 1994, **96**, 437.
29 J.M. Langemeier and D.E. Rogers, *Cereal Chem.*, 1995, **72**, 349.
30 S.I. Mannino, M.S. Cosio and P. Zimei, *Electroanalysis*, 1996, **8**, 353.
31 T.S. Colllins and A.L. Waterhouse, *ACS Abstr.*, 1993, **206**, 53.
32 C. Delgardo, T. Talou and A. Gaset, *Analusis*, 1993, **21**, 281.
33 M. Greenaway, G.N. Okafo, P. Camilleri and D. Dhanak, *J. Chem. Soc. Chem. Commun.*, 1994, 1691.
34 A. Jay, *Carbohydrate Res.*, 1996, **15**, 897.
35 J.R. Vercellotti, L.L. Munchausen, T.H. Saunders, P.J. Garegg and P.Seffers, *Food Chem.*, 1994, **50**, 221.
36 V. Patoprsty, V. Kovacik and S. Karacsonyi, *Rapid Commun. Mass Spectrom.*, 1995, **9**, 840.

37 M. Okamoto, K.-I. Takahashi, and T. Doi, *Rapid Commun. Mass Spectrom.*, 1995, **9**, 641.

38 B. Mulroney, J.C. Traeger, and B.A. Stone, *J. Mass Spectrom.*, 1995, **30**, 1277.

39 D.J. Harvey, T.J.P. Naven, B. Kuster, R.H. Bateman, M.R. Green and G. Critchley, *Rapid Commun. Mass Spectrom.*, 1995, **9**, 1556.

40 K.K. Mock, M. Davey and J.S. Cottrell, *Biochem. Biophys. Res. Commun.*, 1991, **177**, 644.

41 R. Marsilio, S. Catinella, R. Seraglia and P. Traldi, *Rapid Commun. Mass Spectrom.*, 1995, **9**, 550.

42 M. Besler, H. Steinhart and A. Paschke, *Food Agric. Immunol.*, 1997, **9**, 277.

43 M. Kohler and J.A. Leary, *Anal. Chem.*, 1995, **67**, 3501.

44 P.A. Tarantilis, G. Tsoupras and M. Poissiou, *J. Chromatogr.*, 1995, **699**, 107.

45 S. Odani, Y. Yokokawa, H. Takeda, S. Abe and S. Odani, *Eur. J. Biochem.*, 1996, **241**, 77.

46 A. Mele, W. Panzeri and A. Selva, *Eur. J. Mass Spectrom.*, 1997, **3**, 347.

47 E.J. Kennelly, L. Cai, L. Long, L. Shamon, K. Zaw, B-N. Zhou, J.M. Pezzuto and A.D. Kinghorn, *J. Agric. Food Chem.*, 1995, **43**, 2602.

48 J. An, L. Zhang, M.A. O'Neill, P. Albersheim and A.G. Darvill, *Carbohydr. Res.*, 1994, **264**, 83.

49 A.J. Whitcombe, M.A. O'Neill, W. Steffan, P. Albersheim and A.G. Darvill, *Carbohydr. Res.*, 1995, **271**, 15.

50 H.A. Schols, M. Mutter, A.G.T. Voragen, W.M.A. Niessen, R.A.M. van der Hoeven, J. van der Greef and C. Bruggink, *Carbohydr. Res.*, 1994, **261**, 335.

51 J. Abian, S. Susin, J. Abadia and E. Gelpi, *Analyt. Chim. Acta*, 1995, **302**, 215.

52 R. Korner, G. Limberg, T.M.I.E. Christensen, J.D. Mikkelsen and P. Roepstorff, *Anal. Chem.*, 1999, **71**, 1421.

CHAPTER 8

Quantification and Metabolism of Inorganic Nutrients

1 Introduction

In 1987 a large proportion of a review of mass spectrometric methods for inorganic trace analysis was devoted to Spark Source Mass Spectrometry (SSMS).[1] The authors noted that the (then) novel technique of Inductively Coupled Plasma Mass Spectrometry (ICP-MS) held the greatest potential for food analysis. This prophecy has been fulfilled, to such an extent that within six years a major text devoted to trace element determination contained only a brief mention of SSMS: the mass spectrometric technique given most coverage was ICP-MS.[2]

Mass spectrometers are, of course, not the only instruments useful for determining inorganic elements. Spectrochemical methods, atomic absorption, emission and fluorescence spectrometry, Neutron Activation Analysis (NAA) and X-ray methods can also be used. However, mass spectrometers have advantages over these techniques in many inorganic applications because of their sensitivity, selectivity, dynamic range and, for ICP-MS, multi-element capabilities and speed of analysis. Quantitative mass spectrometry, in the form of Isotope Dilution Mass Spectrometry (IDMS), is a primary method of analysis.[3] It can provide 'gold standard' data. These data are matrix-independent, based on the measurement of ratios and may comprise a form of direct traceability of quantitative measurements to the 'Avogadro measurement procedure' in which isotope ratio measurements are conducted to redetermine the Avogadro constant.[4] Although IDMS methods are expensive, they do provide definitive data and may be used to validate more convenient, but less robust, methods.

The main types of mass spectrometer currently used in inorganic analysis of foods are ICP-MS and TIMS. The principles of these techniques have already been described in Chapter 1. ICP-MS can be applied quantitatively to multi-element determination, single element analysis, IDMS or metabolic studies using stable isotopes. TIMS is used exclusively for IDMS or metabolic, stable isotope tracer measurement. It is most commonly employed as a single element method, although oligo-element determinations are possible for some elements under appropriate conditions. Quantitatively, TIMS is mainly used as a reference technique, for certifying standard reference materials or where high precision IDMS measurements (< 0.1% RSD) are required.

2 Sample Collection and Treatment for Inorganic Mass Spectrometry

It is beyond the scope of this book to give a comprehensive account of sample collection and treatment techniques for inorganic mass spectrometry. The interested reader is referred to several publications that cover these matters in detail.[1,5,6] However, some general points will be made here. First, it is very important to ensure that a representative sample of a foodstuff is taken for analysis. Foods are notoriously inhomogeneous and it is essential to collect sufficient sample to ensure accurate measurement. Furthermore, the very wide ranges of materials eaten as foods (meats, cheeses, fruit, vegetables, soups, spreads, pulses, *etc.*) have very different chemical and physical structures. These structures establish how the elements of interest are bound to the food. This in turn determines the optimal methods for liberating the inorganic elements quantitatively, so that they can be measured accurately.[5]

Samples are normally dried and subsequently dry or wet ashing (*i.e.* heating at high temperature in air in a muffle furnace, or acid digestion, respectively) are the first chemical stage of sample processing. Dry ashed samples are subsequently dissolved in acid. Several techniques are used for wet ashing, but heated, oxidising solutions of acid mixtures and hydrogen peroxide are widely employed. Heat can be applied directly in open vessels. However, microwave digestion in closed vessels is becoming more popular.[2,6] Subsequent sample processing is highly dependent on the mass spectrometric technique used. TIMS requires the processed sample of the target analyte to be of high purity and free of organic material, whereas sample processing before ICP-MS measurement is generally much simpler.

The most important stage of sample processing for TIMS is normally the isolation of the target element from the sample matrix.[7] Popular and convenient methods for sample purification include anion or cation exchange chromatography, solvent or chelate extraction, precipitation, electrolytic deposition and volatilisation.[2,7]

Sample preparation methods for ICP-MS are very closely related to the sample introduction techniques used (see Chapter 1 for a brief account of ICP-MS sample introduction methods). Because ICP-MS does not have the stringent requirements for analyte purity that are essential for successful TIMS, preparation methods are often easier and swifter. Although preparation can be very straightforward, for example acid digestion and solvent dilution, some form of purification is often advisable to eliminate potential monatomic and polyatomic interferences.[8] These can occur through interaction of the sample matrix with or within the ionising plasma of the mass spectrometer ion source. Purification schemes for ICP-MS range from simple clean-up and pre-concentration of polyvalent metal ions in high salt concentration biological fluids[9] to the types of ion-exchange procedure used in TIMS.

3 Quantitative Applications of TIMS

Quantitative analysis by TIMS is conducted exclusively by IDMS. The principle of IDMS is the addition to the sample of an exact amount of an enriched spike isotope of known isotopic distribution. Spike isotopes are generally stable isotopes that occur at low natural abundance in isotopically unenriched elements. The sample and spike are then mixed intimately to form a homogenous solution (if the spike is added to a solid sample, chemical dissolution of the solid is essential). The isotope ratios of the analyte are altered quantitatively by this procedure. Any subsequent steps of sample preparation need not be quantitative, as sample and spike are lost in equivalent proportions. However, for trace elements it is very important to ensure that the sample is not contaminated by accidental addition of the element to be analysed. This can occur if dust particles enter the sample, if chemicals used in sample preparation are impure, or if the material comprising the sample vessels contains the target analyte. After the analyte has been isolated, the altered isotope ratios are measured mass spectrometrically. The isotope ratio of the spike, the amount of spike added and the natural abundances of the unenriched isotopes are all known quantities, and hence the quantity of analyte can be calculated. More detailed descriptions of the IDMS procedure have been given by several authors.[2,10-12]

Isotope dilution TIMS is one of the most accurate quantitative methods for determining metals (isotope ratios are typically measurable with a precision of better than 0.1% RSD). However, it has been noted that routine application of TIMS is impractical because of the extensive sample preparation required.[2] Additionally, the sample loading, filament programming and isotope ratio measurement procedures are lengthy, compared with the rapid sample introduction and data acquisition methods used in ICP-MS. Even with automated sample introduction and data acquisition, TIMS requires, typically, 60 min per sample, to which must be added sample preparation time. This compares unfavourably with the 5–10 min per sample required for ICP-MS measurements. Consequently, TIMS has fallen out of favour as a general quantitative technique for toxic elements in foods. An exception is the certification of reference materials, where the high precision and accuracy of TIMS IDMS are valuable.

Because of the preference of ICP-MS to TIMS for quantifying nutrient elements, this topic will not be covered in any detail here, but will be discussed in relation to metabolic and speciation studies, as the need arises. The main focus of this chapter is to introduce the mass spectrometric methods used to study the metabolism of food-derived inorganic nutrients in humans.

4   Quantitative Applications of ICP-MS

Quantitative applications of ICP-MS to analysis of foods and diets are now routine and will not be discussed in any detail here. ICP-MS functions more like a type of sophisticated and selective atomic absorption spectroscopy than as mass spectrometry in these types of application. The unique advantage of ICP-MS in these applications is its speed, specificity and multi-element capability. Many instructive examples of the technique can be found in government reports and are freely available on the UK Ministry of Agriculture Fisheries and Food Web Site.[13]

5   Metabolic Studies of Inorganic Nutrients

Several inorganic nutrients are essential to maintain human life because of their vital role in the control of body biochemistry.[14] Excepting iron-deficiency anaemia, chronic or acute deficiencies of inorganic elements are rare in developed countries. However, sub-clinical or marginal deficiencies may be widespread. Information concerning the efficiency with which inorganic elements are absorbed from foods, and on the amounts necessary to maintain optimal health, is generally lacking. Measurement of the inorganic nutrient content of the diet alone is insufficient to establish requirements because only a proportion of the nutrients is absorbed. Bioavailability, strictly defined as the proportion of that nutrient in a food absorbed and used by the body,[15] is dependent on physiological and dietary factors. Severe deficiencies of inorganic elements are generally diagnosed easily, but marginal deficiencies are more difficult to recognise. This is because dietary sufficiency cannot usually be determined from the concentrations of inorganic elements in different body fluids and tissues. The concentrations of inorganic elements in blood, for example, are often under strict homeostatic regulation. Blood, or more specifically plasma, concentrations are maintained at the expense of body tissue stores (for example liver or muscle tissue) that cannot be sampled so easily. Recommendations of dietary intakes thus require a sound scientific understanding of the bioavailability of inorganic nutrients. Additional information on the amounts of nutrient in different body pools, kinetics of absorption and exchange between body pools is highly desirable. Furthermore, knowledge of nutrient interactions, of enhancers and inhibitors of absorption in foods and of the effects of the physiological status of individuals on absorption and metabolism is also highly desirable for assessing dietary adequacy.

Although radioactive isotopes can be used to study these factors, enriched stable isotopes provide the only *safe*, scientifically rigorous tools for studying inorganic metabolism in detail. Stable isotopes have the disadvantage that their measurement is difficult, requiring sophisticated techniques of inorganic mass spectrometry for accurate quantification. Nevertheless, as inorganic mass spectrometers become more widely available, many more laboratories are now able to conduct these studies.[16–20] Neutron Activation Analysis is an alternative to mass spectrometry for measuring stable isotopes. However, NAA has several

Table 8.1 *Natural Abundances of Stable Isotopes of Selected Nutrient Elements*

Element	Mass number	Natural abundance (atom %)
Ca	40	96.941
	42	0.647
	43	0.135
	44	2.086
	46	0.004
	48	0.187
Fe	54	5.8
	56	91.72
	57	2.2
	58	0.28
Zn	64	48.6
	66	27.9
	67	4.1
	68	18.8
	70	0.6
Se	74	0.89
	76	9.36
	77	7.63
	78	23.78
	80	49.61
	82	8.73

disadvantages and has been largely replaced by mass spectrometry for determining the stable isotopes of nutrient elements.[21]

The natural abundances of the stable isotopes of Ca, Fe, Zn and Se are shown Table 8.1.[22] The least abundant isotope is the ideal tracer because administration of small amounts of enriched forms of these isotopes can yield measurable changes in isotope ratios in blood, urine or faeces without disturbing metabolic balance. These stable isotopes can be purchased from a variety of sources, enriched to levels of up to and above 90 atom %.* In practice, the least naturally abundant isotope may not be the best choice because of limitations of cost, or measurement problems. For example, ^{46}Ca is very expensive and ^{80}Se is difficult to measure by conventional ICP-MS because of an interference peak due to $^{40}Ar_2^+$.

The most obvious drawback of stable isotopes is that they are not true tracers of absorption because they are not massless: they must be measurable above the natural abundance of endogenous isotopes in any biological sample. The quantity of an isotope administered in an absorption study should not exceed

---

* A useful list of suppliers of enriched stable isotopes can be found at the University of Vermont's Geochemistry Web Site, http://geology.uvm.edu/geowww/suppliers.html

the recommended daily intake of the element (and should preferably not exceed the normal range of element concentrations found in a single meal). If the typical amount found in a single meal is exceeded greatly, absorption may be abnormal (usually lower than expected) because the metabolic balance is perturbed. Conversely, administration of a small amount of the label may produce an immeasurably small change in the isotope ratio. Thus, the accuracy and precision of the mass spectrometric technique used to determine isotopes in metabolic studies are very important. Isotope ratios measured with a precision of 1% RSD or better are considered sufficient for conducting most types of absorption study.[23,24] However, it has been suggested that as high a precision as possible should be aimed for, whatever the cost.[24] The smallest detectable change in isotope ratio is three times the SD or RSD of the basal (*i.e.* natural abundance) measurement. However, an average change in isotope ratio of at least ten times basal RSD is desirable. This is because of the large variation in absorption and metabolic response between individuals, and even within the same individual.[25,26] Isotope ratio measurements (R) are optimal if the ratio does not deviate markedly from unity ($0.1 < R < 1$).[27] The approximate quantity of enriched stable isotope required to achieve the desirable average change can be calculated by making reasonable assumptions about the expected absorption and retention of the isotope. A worked example for enriched isotopes of ^{57}Fe and ^{58}Fe has been given for a 5–6 month old infant and is a useful example of this type of calculation.[25]

Until recently, only thermal ionisation mass spectrometers were able to provide a sufficient precision for most nutritional elements and were therefore considered to be the best of the four mass spectrometric techniques then available.[19] The precision attainable by ICP-MS was inferior to the ultimate precision of TIMS, and some isotopes were difficult or impossible to measure by ICP-MS because of polyatomic interferences from the argon plasma. However, improvements in instrument design, including high resolution and multiple collector systems, have narrowed the gap in isotopic range and precision between the rival techniques. Furthermore, ICP-MS, unlike TIMS, is a continuous-flow method and can be coupled directly to separation techniques, including HPLC and SEC. This has great advantages in speciation studies (see below).

6   Sample Preparation for Inorganic Mass Spectrometry

Sample preparation techniques for quantification of inorganic nutrients in foods or for stable isotope studies of metabolism depend significantly on the mass spectrometric techniques to be used for determining these elements. However, the preliminary stages of sample processing, whether the sample is a food or of biological origin (commonly blood, plasma, urine or faeces), are similar.[2,6–8,28,29] A typical procedure involves controlled drying of the analyte followed by sample mineralisation using dry or wet ashing. Subsequently, samples may be analysed directly by measurement of isotope ratios in diluted acid solutions (usually only feasible by ICP-MS). Alternatively, additional sample purification may be necessary. This is often conducted by ion-exchange chromatography, although other methods, including electrodeposition and solvent extraction, may also be used. Whatever sample preparation and processing techniques are used avoiding contamination is essential. This applies to all stages of the sampling and sample handling process. For example, syringes used for collecting blood can be a source of chromium contamination and the rubber seals of blood collecting vessels can introduce traces of zinc. Ensuring a clean, dust-free working environment and minimising sample manipulations, as far as practical, is important. Apparatus should be of high-purity PTFE or plastic construction (although PVC, a source of zinc contamination, should be avoided) and solvents must be of high purity. Validating any quantitative or stable isotope tracer study by testing is highly advisable, as is evaluating the analytical procedure using appropriate certified reference materials. These matters have been considered in more detail elsewhere.[6,18,30] Losses of sample are acceptable from *homogeneous*, isotopically labelled samples, provided these are not so large that sensitivity is compromised. This is because sample loss does not change the isotope ratio of a homogeneous sample.

Quantification of the inorganic nutrient of interest is usually essential in any stable isotope metabolic study. These quantitative data are often needed in the final calculation of absorption, bioavailability, *etc.* Measurements can be made by non-mass spectrometric techniques, such as atomic absorption spectroscopy, or by quantitative ICP-MS without isotope dilution. However, these methods are less precise than IDMS: accuracy, precision and convenience are improved by simultaneous determination of a quantitative spike isotope and the tracer isotope(s).[31] Inorganic nutrients possessing more than two naturally abundant isotopes can be quantified in this way. ^{66}Zn or ^{68}Zn, for example, are useful for IDMS quantification of zinc in experiments where the lower abundance ^{67}Zn and ^{70}Zn isotopes are used as metabolic tracers. For inorganic nutrients that only possess two stable isotopes, Cu for example, IDMS quantification can be conducted after sub-sampling and spiking a sub-sample. IDMS of the spiked and unspiked samples then yields quantitative information, after the appropriate calculations.

7 Mass Spectrometric Techniques for Inorganic Nutrient Studies

Electron Ionisation Mass Spectrometry and Combined Gas Chromatography/Mass Spectrometry (GC/MS)

Although EIMS and GC/MS are widely used to identify and quantify organic compounds, they can also be used for quantitative and isotope ratio measurement of inorganic elements. Native inorganic nutrients cannot be analysed by EIMS or GC/MS as they have insufficient vapour pressure below 500 °C. Early approaches involving conversion to metal halides were abandoned because of practical problems and the limited number of elements amenable to this method. Instead, inorganic elements must first be converted to an organically bound form by reaction with a suitable chelate. This makes them sufficiently volatile for evaporation from a heatable probe or for GC. Several useful chelates are based on β-diketones, but other derivatives have also been used.[18,19,28] Measurements can be made using widely available, cheap and convenient quadrupole bench-top instruments and the analysis of many inorganic elements is theoretically possible. However, several practical problems specific to EIMS and GC/MS of inorganic chelates prevent widespread adoption of this methodology. Successful applications are largely confined to GC/MS of Se and Cr derivatives.[18,28] The drawbacks of the EIMS and GC/MS approaches will be described in more detail later. The types of chelate used for elemental analysis by mass spectrometry, and the range of isotope ratio measurement precisions attained, have been summarized.[19,28]

EIMS and GC/MS are more widely available than TIMS or ICP-MS instruments. They can be automated and are, in bench-top forms, inexpensive to purchase, run and maintain. Nevertheless, the inorganic chelate method of IDMS quantification or metabolic tracer measurement does have several major disadvantages. Preparation of the inorganic derivative can be laborious; not all chelates are commercially available and those that are must often be purified carefully to reduce interference problems. Furthermore, samples may also require extensive purification if they are to be measured by heatable probe EIMS. However, these disadvantages are relatively trivial compared with the major drawback: the susceptibility of this approach to problems caused by metal exchange and cross-contamination. Inorganic elements are prone to exchange with metals in the mass spectrometer ion source[32] and to memory effects resulting from retention of chelates in the ion source.[33] Additional problems can occur because of interactions with GC columns. Some chelated inorganic nutrients are less vulnerable to these problems than others and strategies have been devised to help overcome some of these difficulties (summarised by Ducros[28]). Limited and inconsistent precision is another obvious major disadvantage of EIMS and GC/MS when used to measure isotope ratio of inorganic chelates. Although RSDs of 1–2% have been reported for isotope ratio measurement (summarised by Yergey[19]), these are at the upper limit of the desired precision and ratios with RSDs of 4–5% are common. For these reasons, EIMS and GC/MS methods are not recommended for routine use. Two exceptions may be made to this general observation: analysis of chelates of Se and, possibly, Cr. A comparison of Se analysis by several different methods, including IDMS GC/MS,[34] has shown that this element can be determined accurately by GC/MS, suggesting that tracer studies should also meet with success. Several laboratories have also reported good results with chelates of Cr and these data have been summarized.[28] Cr, on the other hand, is a notoriously difficult element to determine because of its low concentration in biological samples. It is suggested here that ICP-MS be explored for determining Cr isotopes in tracer studies, as sample handling can be reduced significantly, thus reducing the risk of contamination. GC/MS should be considered a method of last resort for Cr, if dedicated inorganic mass spectrometers are not available.

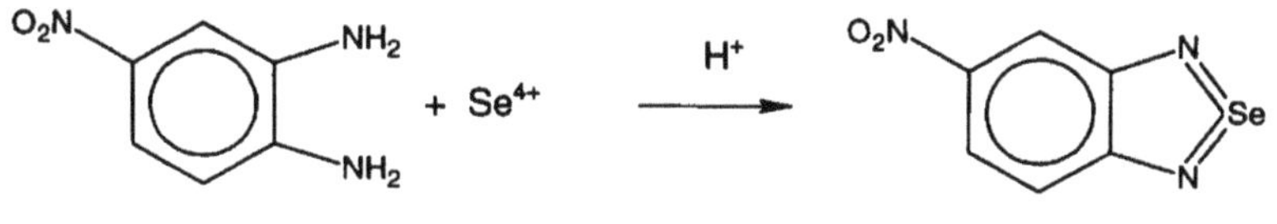

Figure 8.1 *Structure of the 4-nitro piazselenol derivative of selenium*

Se, although present at low concentrations, is more abundant in biological fluids than Cr and not as prone to contamination. GC/MS is therefore recommendable as a general method for quantifying this element by IDMS or for conducting metabolic studies using stable isotopes. The derivative of choice is the volatile piazselenol (Figure 8.1) generated by reaction with 4-nitrophenyl-enediamine.[35-39]

The full mass spectrum and an SIM trace of m/z 229 and 231 (containing the isotopes ^{80}Se and ^{82}Se) are shown in Figure 8.2 (a) and (b).

Isotope ratios for IDMS or metabolic tracer studies can be conducted on most modern quadrupole bench-top GC/MS instruments and may be automated with the aid of a GC auto-injector. The isotope ^{74}Se is considered the best for metabolic studies and ^{76}Se or ^{82}Se are also valuable in multiple tracer experiments or as quantitative spike isotopes. A detailed protocol for conducting GC/MS isotope ratio measurements and for performing IDMS calculations has been given.[28] Recently, negative ion GC/MS of piazselenol has been described and yielded improved detection limits over the positive ion technique.[40]

Fast Atom Bombardment Mass Spectrometry

FABMS and the very similar technique of secondary ion mass spectrometry (SIMS) were originally used for analysing involatile, polar organic molecules. These analytes must be dissolved in a low-boiling, viscous matrix, normally glycerol or 3-nitrobenzyl alcohol, to ensure the generation of an intense, persistent ion beam. In contrast, inorganic salts yield a continuous, stable ion beam under atom or ion bombardment without the addition of an organic matrix. Use of a liquid matrix is detrimental to inorganic isotope ratio measurement because interfering ions from the matrix reduce the accuracy and reproducibility of the data.

Isotope ratio measurements of inorganic samples are usually performed at medium or high resolution to ensure separation of the ions of interest from interfering ions.[41-43] However, successful measurements have been made on Fe at low resolution.[44]

Although good precision is obtainable (0.2–1% RSD for many isotope ratio measurements), accuracy and reproducibility are questionable.[19,29] A major source of potential error is caused by isotopic fractionation, which has been investigated systematically.[45] Although these errors can be corrected,[29] the overall effect is to reduce accuracy. Further uncertainty is caused by ions due to metal monohydrides; these can generate systematic errors.[45,46] It is possible to

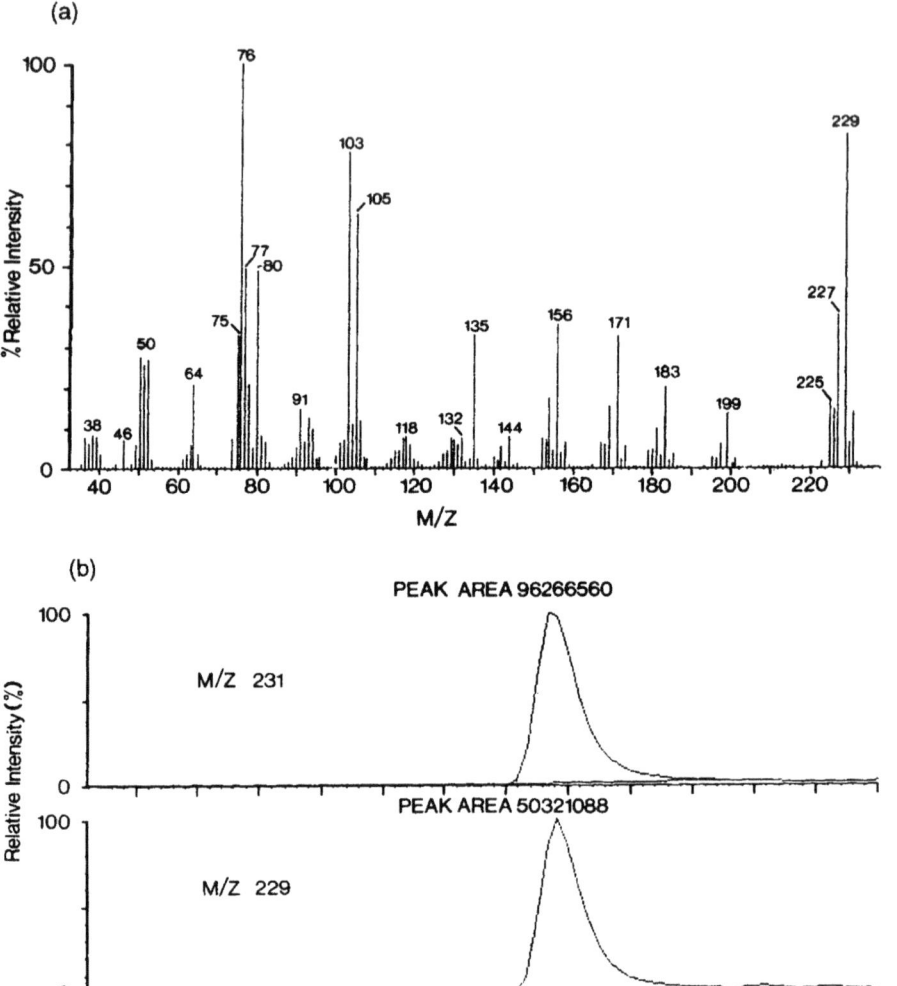

Figure 8.2   *Electron ionisation mass spectrum* (a) *and SIM GC/MS traces of m/z 229 and*
231 (b) *of the 4-nitro piazselenol derivative of selenium.*
(Reproduced from H.M. Crews, V. Ducros, J. Eagles, F.A. Mellon, P.
Kastenmayer, J.B. Luten and B.A. McGaw, *Analyst*, 1994, **119**, 2491, by
permission of the Royal Society of Chemistry)

choose stable isotope tracers that are not susceptible to hydride interference, or
to conduct measurements under conditions that allow correction of interference
effects.[46-48] However, this reduces the flexibility of the methodology and/or
lengthens analysis time.

Sample preparation techniques for FABMS isotope ratio measurement range
from very simple procedures (Ca measurements in plasma or urine samples[41]) to

more complex methods, including extraction, ion exchange or a combination of both.[44,49,50] A comprehensive summary of sample preparation techniques has been given.[29]

Applications of FABMS for determining stable isotopes are declining despite the accessibility the technique. This is partly because of the drawbacks and difficulties of the FABMS method and because dedicated inorganic mass spectrometers (TIMS and especially ICP-MS) are becoming more widely available. FABMS isotope ratio measurements are also difficult to automate, are labour-intensive and suffer from sensitivity and interference for some isotopes. Nevertheless, medium- or high-resolution FABMS has, in the past, provided a useful introduction to isotope ratio measurement. Guidelines have been given outlining recommended procedures for ensuring some degree of success with this technique.[29] However, it cannot be recommended as suitable for complex studies or those requiring high sensitivity and accuracy.

Thermal Ionisation Mass Spectrometry (TIMS)

TIMS is the mass spectrometric technique that is capable of yielding the most accurate and precise measurements of stable isotopes. Its use in geochemistry, geochronology, accurate atomic weight measurement and nuclear chemistry was already extensive before the first applications in the nutritional sciences. Consequently, a wealth of useful analytical and procedural information on the optimal use of TIMS was available to aid pioneers of the technique in nutrition studies.

The method of sample loading is very important in TIMS. Special loading rigs and heating cycles are necessary to ensure stable, intense ion beams. Loading techniques are dependent on quantity of a sample, its chemical form and the chemical nature of any ionisation enhancer used. The heating cycle used for sample ionisation is also decisive: every element has its own optimal ionisation temperature. This lies in the ranges 900–1500 °C for Cu, Fe and Zn, ionised by single filament techniques, and 1600–1900 °C for Ca and Mg that are determined using a double filament procedure.

Isotope ratios determined by TIMS differ from the correct values because isotopes evaporate from the filament at different rates, *i.e.* mass dependent isotope fractionation occurs. This can be corrected by several different techniques that recover the accuracy of the ratios. These corrections are sometimes unnecessary in nutritional studies, if isotopic enrichment is high.[7]

Samples for TIMS analysis require careful and conscientious preparation to remove all organic matter and any metals not under investigation. TIMS is particularly susceptible to contamination for several reasons. Firstly, organic material generates interfering ions throughout the mass scale; secondly, inorganic elements can reduce ionisation efficiency when present in large quantities and thirdly, in rarer cases, cause isobaric interferences with the element of interest. Separation of the nutritional element is achieved by cation or anion exchange, precipitation, extraction and volatilisation.[7] Ion exchange chromatography is the most popular, rapid and successful separation technique for many nutritional elements.

Precision and accuracy of isotope ratios are dependent on the type of mass analyser used: magnetic sector instruments equipped with multiple collectors yield RSDs mostly in the range 0.01–0.1%, whereas quadrupole mass analysers generate data with RSDs of 0.1–1%. High precision is unnecessary in many nutritional experiments. Exceptions are isotope ratios determined on large body pools or on populations that, because of developmental or environmental stresses, are metabolically very efficient at absorbing mineral nutrients. A good example of the first case is the determination of Fe absorption from foods by measuring uptake in red blood cells. The body pool of iron is large compared with intake, and absorption of this element is particularly inefficient. Increases in isotope ratio are therefore small, compared with natural abundance, and must be measured with high precision. Calcium excretion, by way of contrast, can be very low in populations that have a small dietary intake of this element. Measurements of Ca absorption by urinary monitoring are therefore difficult because such a small quantity of label is excreted and because the concentration of Ca is low.[51] Nonetheless, high-sensitivity magnetic sector TIMS instruments are capable of measuring the kinetics of calcium metabolism. Studies carried out on a normal subject, after dosing with sub-milligram amounts of enriched stable isotopes, indicated that robust data could be obtained from saliva samples for up to 170 h.[52]

Modern TIMS instruments are automated or semi-automated and are capable of measuring a maximum of about 13–20 samples per day. The alternative technique, ICP-MS, can yield a much higher throughput (40–60 samples per day). TIMS also suffers from the need for careful chemical separation of the nutrient element from the sample matrix. This increases the time required for a complete analysis. However, TIMS can measure *all* the isotopes of Ca, Fe or Se, whereas polyatomic interferences impose limitations on determining all these isotopes by ICP-MS (with the exception of a very recent generation of collision cell and/or cool plasma-equipped instruments). TIMS also yields consistently higher accuracy data than any of the other mass spectrometric techniques discussed here and is considered the most general technique for determining nutrient elements.[19] However, the pre-eminent position of TIMS is now being challenged by a new generation of ICP-MS instruments (see below)

Inductively Coupled Plasma Mass Spectrometry (ICP-MS)

ICP-MS is the most recent, dedicated technique of inorganic mass spectrometry used to determine nutritional elements and their isotope ratios. The technique is very sensitive, with detection limits in the μg l^{-1} range for many elements.[19] However, accurate isotope ratio measurements require higher concentrations (at least one and preferably two orders of magnitude greater). This is more than sufficient for the vast majority of nutritional applications. Exceptions occur for a small (but significant) number of isotopes that are vulnerable to isobaric interferences from monatomic or polyatomic ionic species generated in the argon plasma. The main isobaric interferences for nutritional elements have been listed.[8] The most notable nutritional isotopes to suffer from this problem are ^{40}Ca (isobaric with $^{40}Ar^+$), ^{80}Se (isobaric with $^{40}Ar_2^+$) and ^{56}Fe (isobaric with $^{40}Ar^{16}O^+$).

Optimisation techniques for ICP-MS measurement of isotope ratios have been developed[53,54] and summarised[8] and will not be discussed here. A strategy aimed at reducing plasma source noise has successfully yielded precisions of 0.05% RSD on isotope ratio measurements.[55] Isotope ratios of nutritional elements are determinable with RSDs of less than 2%, provided enough material is collected.[8] Although not up to the quality of the best TIMS data, this figure is satisfactory for many metabolic studies. A comparison of quadrupole TIMS, FABMS and ICP-MS isotope ratio measurements confirms that Zn isotopes can be determined with similar precision by all three methods.[46] However, ICP-MS has been described as 'only partially successful' in stable isotope metabolic studies.[19] This view may be overly pessimistic: continuing improvements in methodology and instrument design are making ICP-MS one of the most attractive methods for conducting metabolic studies. Faecal balance studies, where sample collection errors are usually greater than any errors in isotope measurement, are generally compatible with ICP-MS. Plasma source methods are also particularly appropriate in speciation studies (see below).

Sample introduction techniques for ICP-MS are very varied and are often closely linked to the sample preparation method used.[8] Biological fluids or solids are often simply acid digested, diluted and introduced into the ICP-MS ion source via a conventional nebuliser. The technique works well for simple analyses of nutrient elements, but isotope ratio measurements usually require some form of sample purification. This can reduce or eliminate many major polyatomic interferences formed by plasma reactions of matrix components, for example $^{40}Ar^{23}Na$, which interferes with ^{63}Cu, and $^{35}Cl^{16}O_2$, an interferent of ^{67}Zn. Purification methods include extraction, precipitation, ion-exchange chromatography, size-exclusion chromatography, hydride generation and electrothermal vaporisation (ETV).[8] Hydride generation and ETV are good examples of techniques that combine purification and sample introduction.

Alternative sample introduction techniques for ICP-MS have been described[56] and summarized.[8] Although the popular nebuliser/spray chamber method is still used for many samples, other approaches are possible. These include combined LC/ICP-MS[57] and hydride generation for sample purification and introduction of Se isotopes.[58]

ICP-MS is now confirmed as a very powerful and versatile method for studying nutrient metabolism. It is rapid, sensitive, specific and generally has the least demanding requirements for sample preparation of all the inorganic mass spectrometric methods discussed here. Practical disadvantages of the technique are lower precision than TIMS and susceptibility to isobaric interferences, although some interferences can be reduced by appropriate instrumental or sample preparation strategies.[8] High resolution magnetic sector ICP-MS also has the potential to improve data significantly, although at high capital cost.[59,60] The recent development of collision cell techniques for application to both quadrupole and magnetic sector ICP-MS has the potential to revolutionise ICP-MS technologies. Collision cells can reduce argon and argide interferences to negligible levels, allowing all isotopes of Fe and Se (for example) to be measured, even when limited amounts of sample are available (see Chapter 1, p. 18).

A particularly powerful feature of ICP-MS is its potential in speciation studies. It can be used to detect specific oxidation states (which may drastically modify nutritional value) and to aid identification of mineral-binding proteins.[57,61] It has been predicted that ICP-MS could well supersede alternative inorganic mass spectrometric techniques in metabolic studies in the near future.[8] The development of collision cell techniques makes this prediction even more likely to become fact.

Future Prospects for Inorganic Mass Spectrometry in Food and Nutrition Studies

A recent forecast regarding applications of FABMS and GC/MS to metabolic studies predicted that these would decline and that TIMS could be limited to studies requiring high precision: ICP-MS was seen as the technique of the future.[62] The coupling of LC or SEC to ICP-MS is expected to increase both in straightforward speciation studies and for studying the uptake and incorporation of stable isotopes into storage or transport proteins. The potential of laser desorption mass spectrometry has also been noted. Recent applications of LDMS to isotope ratio determination of the nutrient minerals Cu, Ca, Mg, Fe and Zn suggest that it is possible to attain accuracy and precision close to that of ICP-MS.[63] An advantage of LDMS was the lack of the usual interference problems. However, the most exciting recent development in the field has been collision cell technology. Several instruments that exploit this technique are now available commercially and could provide the basis for new, rapid, sensitive and versatile multielement methods for determining nutrient elements and for studying their metabolism.

Selected Applications of Inorganic Mass Spectrometry to Food and Nutrition Studies

Two selected examples displaying the application of inorganic mass spectrometry to food and nutrition science will be given. A wider range of applications is described in several books and reviews recommended to the interested reader.[16,18-20]

The first example is the measurement of the bioavailability of Fe in different weaning foods, a study conducted in parallel with measurements of the effect of ascorbic acid on Fe absorption.[64] Fe demand in infants of 6 months or more is high because body stores, normally sufficient to satisfy physiological requirements, have been depleted by this age. Despite this, little is known about the bioavailability of Fe in particular weaning foods or diets. Radioisotopes are

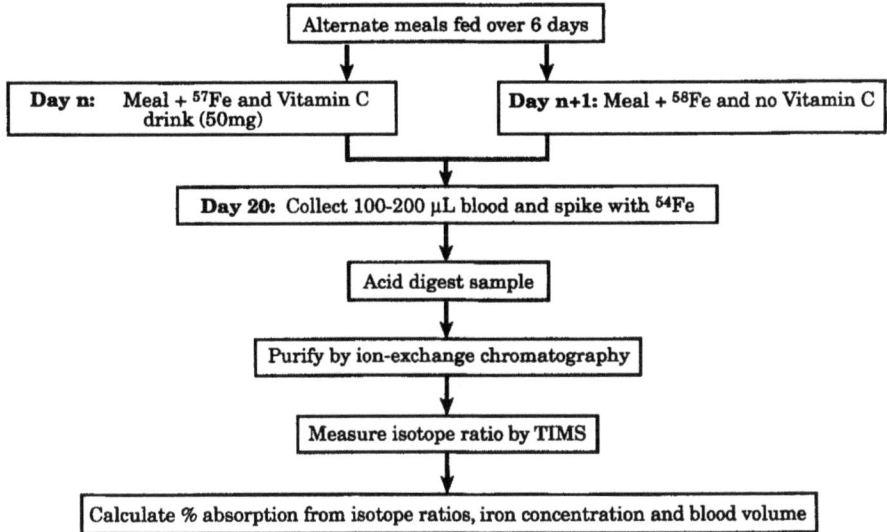

Figure 8.3 *Protocol for measuring the absorption, by erythrocyte incorporation, of the stable isotopes of iron*

particularly unsuitable for administration to healthy infants. Consequently, the use of stable isotopes is very appropriate in this type of study. The study design involved the use of the stable isotopes ^{57}Fe and ^{58}Fe to determine absorption and a third, spike isotope, ^{54}Fe, to measure the concentration of Fe in blood.

The method is based on the tissue retention.[26] Tissue retention can be used to measure absorption when most of the inorganic element of interest is incorporated into a specific tissue. Approximately 90% of absorbed Fe is incorporated into red blood cells 14 days after administration. Thus Fe absorption can be measured simply by administering an isotope with a meal, collecting a small blood sample (~ 250 μl, *i.e.* a heel prick) and determining the isotope ratios. Four of the weaning foods most commonly used in the UK, that contribute significantly to Fe intake, were chosen for labelling. These were a vegetable-based infant formula; a breakfast cereal, wholemeal bread and low-sugar tinned baked beans. A commercial apple juice drink containing 50 mg ascorbic acid or a placebo drink containing no ascorbate was administered with the meals. The study protocol is shown in diagrammatic form in Figure 8.3.

Blood samples were spiked with a known quantity of ^{54}Fe to allow measurement of total Fe content. Samples were processed by microwave assisted digestion in concentrated HNO_3 and H_2O_2 and Fe was isolated by ion-exchange chromatography.* Isotope ratios were then measured by quadrupole TIMS. Weights of isotopes of ^{57}Fe, ^{58}Fe and ^{54}Fe were calculated from the isotope ratios and the absorption of iron determined from these values. The results are presented in Figure 8.4.

---

* Ether extraction of $FeCl_3$ is now the preferred method in one of the authors' laboratories.

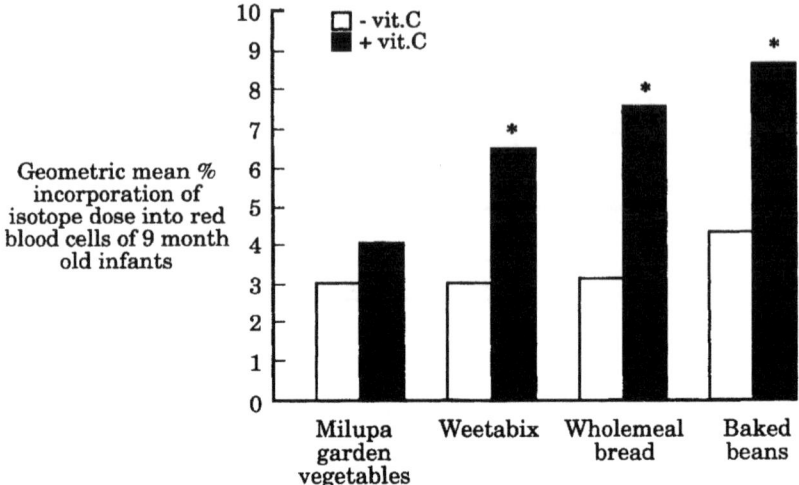

Figure 8.4 *Bioavailability (% incorporation into red blood cells) of selected weaning foods, given to 9 month old infants, with and without a vitamin C containing drink*
(Reproduced from S.J. Fairweather-Tait, T. Fox, S.G. Wharf and J. Eagles, *Pediatr. Res.*, 1995, **37**, 389, with permission of the authors and publisher)

Statistical analysis showed no significant differences in the bioavailability of Fe between the foods. However, ascorbic acid significantly increased Fe uptake from all the foods by a factor of two, except the infant formula (which already contained high levels of ascorbic acid). This type of study is very valuable in providing quantitative data on Fe absorption for a particularly vulnerable section of the population.

The second example is a study of metals in food and other proteins by combined SEC/ICP-MS and LC/ICP-MS.[57] Attention here is focused on the results for food proteins. Reversed-phase chromatography/ICP-MS was applied to the speciation of Zn in an *in vitro* simulated gastrointestinal digest of chicken meat samples labelled intrinsically with ^{67}Zn and extrinsically with ^{68}Zn. The Zn isotope profiles obtained by LC/ICP-MS are shown in Figure 8.5.

Isotope ratios determined from these profiles agreed with those measured by TIMS and conventional ICP-MS. This study displays the great potential of LC/ICP-MS in metabolic studies. Conventional stable isotope studies usually measure gross absorption and label retention in undifferentiated body pools or other biological samples. The combined chromatography/MS technique could enable studies of the uptake of stable isotopes by specific proteins and herald a 'molecular' approach to metabolic studies of minerals. Obtaining additional information by investigating the proteins in more detail should be possible, for example by ESI-MS. A simultaneous LC/ESI-MS, LC/ICP-MS experiment is conceivable, yielding information on the nature of the protein and the stable isotope incorporation in a single experiment. Although there may be significant practical difficulties with this proposal, the benefits obtainable from this

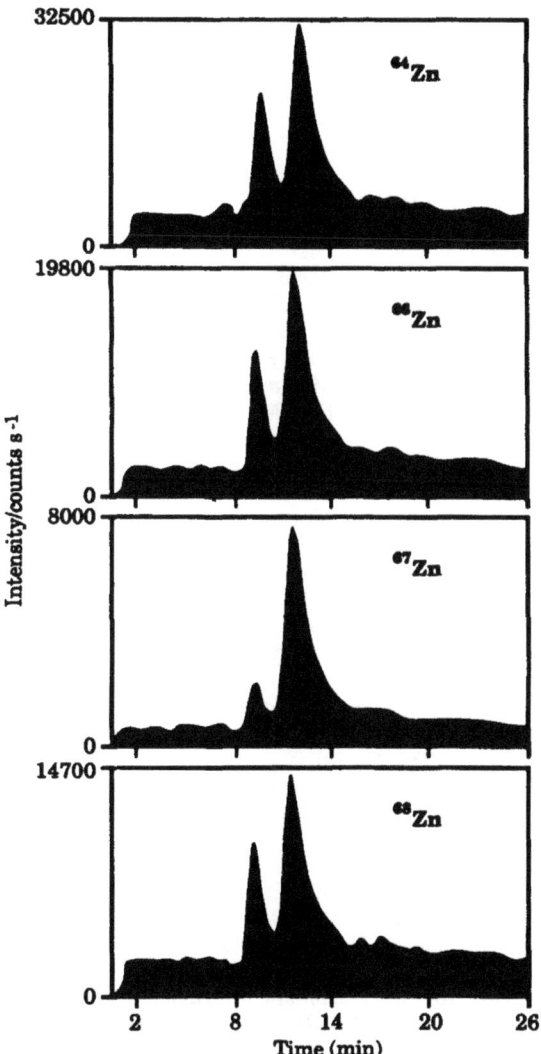

Figure 8.5 *SEC/ICP-MS elution profiles of isotopes of zinc bound to proteins in a proteolytic digest of chicken meat*
(Reproduced from L.M.W. Owen, H.M. Crews, R.C. Hutton and A. Walsh, *Analyst*, 1992, **117**, 649, by permission of the Royal Society of Chemistry)

methodology are potentially very considerable. Speciation by combined SEC/ICP-MS is also useful in general nutritional analysis, as demonstrated by a recent example analysing minerals in human milk whey and infant formulae.[65]

The combination of the ultra high-resolution separation technique capillary electrophoresis with ICP-MS (operated in electrospray or ion spray modes) also has great potential, explored in a recent paper.[66] Although several technical challenges remain to be overcome, the methodology is very promising.

8 References

1 A.M. Ure and J.R. Bacon, in *Applications of Mass Spectrometry in Food Science*, ed. J. Gilbert, Elsevier Applied Science, London, 1987, pp. 343–375.
2 C. Vandecasteele and C.B. Block, *Modern Methods for Trace Element Determination*, John Wiley & Sons, Chichester, 1993.
3 P. De Bièvre, *Anal. Proc.*, 1993, **30**, 328.
4 P. De Bièvre, *Fresenius' J. Anal. Chem.*, 1994, **350**, 277.
5 G.J. DeMenna, in *Spectroscopic Techniques for Food Analysis*, VCH Publishers Inc., New York, 1994, pp. 147–179.
6 B.A. McGaw, in *Stable Isotopes in Human Nutrition: Inorganic Nutrient Metabolism*, ed. F.A. Mellon and B. Sandström, Academic Press, London, 1996, pp. 47–53.
7 P. Kastenmayer, in *Stable Isotopes in Human Nutrition: Inorganic Nutrient Metabolism*, ed. F.A. Mellon and B. Sandström, Academic Press, London, 1996, pp. 81–96.
8 H.M. Crews, J.B. Luten and B.A. McGaw, in *Stable Isotopes in Human Nutrition: Inorganic Nutrient Metabolism*, ed. F.A. Mellon and B. Sandström, Academic Press, London, 1996, pp. 97–115.
9 K. Akatsuka, J.W. McLaren, J.W. Lam and S.S. Berman, *J. Anal. At. Spectrom.*, 1992, 7, 889.
10 K.G. Heumann, in *Inorganic Mass Spectrometry*, ed. F. Adams, R. Gijbels and R. Van Grieken, Wiley Interscience, 1988, pp. 301–376.
11 K.G. Heumann, *Int. J. Mass Spectrom. Ion Proc.*, 1992, **118/119**, 575.
12 H.P. Longerich, *At. Spectrosc.*, 1989, **10**, 112.
13 See, for example, http://www.maff.gov.uk/food/infsheet/index.htm, especially Report No. 156, Metals and other elements in dietary supplements and licensed medical products.
14 B. Sandström, in *Stable Isotopes in Human Nutrition: Inorganic Nutrient Metabolism*, ed. F.A. Mellon and B. Sandström, Academic press, London, 1996, pp. 3–9.
15 B.L. O'Dell, *Nutr. Rev.*, 1984, **42**, 301
16 J.R. Turnlund, *Crit. Rev. Food Sci. Nutr.*, 1991, **30**, 387.
17 S.K. Aggarwal, M. Kinter, R.L. Fitzgerald and D.A. Herold, *Crit. Rev. Clin. Lab. Sci.*, 1994, **31**, 35.
18 H.M. Crews, V. Ducros, J. Eagles, F.A. Mellon, P. Kastenmayer, J.B. Luten and B.A. McGaw, *Analyst*, 1994, **119**, 2491.
19 A.L. Yergey, *J. Nutr.*, 1996, **126**, 355S.
20 F.A. Mellon and B. Sandström (eds.), *Stable Isotopes in Human Nutrition: Inorganic Nutrient Metabolism*, Academic Press, London, 1996, 156 pp.
21 S.J. Fairweather-Tait, in *Stable Isotopes in Human Nutrition: Inorganic Nutrient Metabolism*, ed. F.A. Mellon and B. Sandström, Academic Press, London, 1996, pp. 57–58.
22 P. De Bièvre and P.D.P. Taylor, *Int. J. Mass Spectrom. Ion Proc.*, 1993, **123**, 149.
23 M. Janghorbani, *Progr. Food Sci. Nutr.*, 1984, **7**, 303.
24 B.T.G. Ting and M. Janghorbani, *Anal. Chem.*, 1986, **58**, 1334.
25 B. Sandström, S.J. Fairweather-Tait, R. Hurrell and W. van Dokkum, *Nutr. Res. Rev.*, 1993, **6**, 71.
26 W. van Dokkum, S.J. Fairweather-Tait, R. Hurrell and B. Sandström, in *Stable Isotopes in Human Nutrition: Inorganic Nutrient Metabolism*, ed. F.A. Mellon and B. Sandström, Academic Press, London, 1996, pp. 23–42.
27 A.A. van Heuzen, T. Hoekstra, and B. van Wingerden, *J. Anal. At. Spectrom.*, 1989, **4**, 483.
28 V. Ducros, in *Stable Isotopes in Human Nutrition: Inorganic Nutrient Metabolism*, ed. F.A. Mellon and B. Sandström, Academic Press, London, 1996, pp. 61–72.
29 J. Eagles and F.A. Mellon, in *Stable Isotopes in Human Nutrition: Inorganic Nutrient Metabolism*, ed. F.A. Mellon and B. Sandström, Academic Press, London, 1996, pp. 74–80.

30 J. Hoffmann, *Fresenius' Z. Anal. Chem.*, 1988, **331**, 220.
31 J.R. Turnlund and W.R. Keyes, *J. Micronutr. Anal.*, 1990, **7**, 117.
32 D.B. Christie, M. Hall, C.M. Moynihan, K.M. Hambidge and P.V. Fennessey in *Stable Isotopes in Nutrition*, ed. J.R. Turnlund and P.E. Johnson, American Chemical Society, Washington, 1984, pp. 127–138.
33 W.T. Buckley, S.N. Huckin, J.J. Budac and G.K. Egenford, *Anal. Chem.*, 1982, **54**, 504.
34 M. Ihnat, M.S. Wolynetz, Y. Thomassen and M. Verlinden, *Pure Appl. Chem.*, 1986, **58**, 1063.
35 D.C. Reamer and C. Veillon, *Anal. Chem.*, 1981, **53**, 2166.
36 D.C. Reamer and C. Veillon, *J. Nutr.*, 1983, **113**, 786.
37 W.R. Wolf, D.E. Lacroix and J. Kochansky, *J. Micronutr. Anal.*, 1988, **4**, 145.
38 S.K. Aggarwal, M. Kinter and D.A. Herold, *Anal. Biochem.*, 1992, **202**, 367.
39 V. Ducros and A. Favier, *J. Chromatogr.*, 1992, **583**, 35.
40 P. Van Dael, D. Barclay, K. Longet, S. Metairon and L.B. Fay, *J. Chromatogr B*, 1998, **715**, 341.
41 D.L. Smith, *Anal. Chem.*, 1983, **55**, 2391.
42 J. Eagles, S.J. Fairweather-Tait and R. Self, *Anal. Chem.*, 1985, **57**, 469.
43 P.L. Peirce, K.M. Hambidge, C.H. Goss, L.V. Miller and P.V. Fennessey, *Anal. Chem.*, 1987, **59**, 2034.
44 D.R. Flory, L.V. Miller and P.V. Fennessey, *Anal. Chem.*, 1993, **65**, 3501.
45 L.V. Miller, K.M. Hambidge and P.V. Fennessey, *Anal. Chim. Acta*, 1990, **241**, 249.
46 J. Eagles, S.J. Fairweather-Tait, F.A. Mellon, D.E. Portwood, R. Self, A. Götz, K.G. Heumann and H.M. Crews, *Rapid Commun. Mass Spectrom.*, 1989, **3**, 203.
47 D.E. Pratt, J. Eagles and S.J. Fairweather-Tait, *J. Micronutrient Anal.*, 1987, **3**, 107.
48 J. Eagles, S.J. Fairweather-Tait, D.E. Portwood, R. Self, A. Götz and K.G. Heumann, *Anal. Chem.*, 1989, **61**, 1023.
49 A.A.-R. Gharaibeh, J. Eagles and R. Self, *Biomed. Mass Spectrom.*, 1985, **12**, 344.
50 R. Self, S.J. Fairweather-Tait and J. Eagles, *Anal. Proc.*, 1985, **22**, 194.
51 S.J. Fairweather-Tait, A. Prentice, K.G. Heumann, L.M.A. Jarjou, D.M. Stirling, S.G. Wharf and J.R. Turnlund, *Am. J. Clin. Nutr.*, 1995, **62**, 1188.
52 S.M. Smith, M.E. Wastney, L.E. Nyquist, C.-Y. Shih, H. Wisemann, J.E. Nillen and H.W. Lane, *Rapid Commun. Mass Spectrom.*, 1996, **31**, 1263.
53 N. Furuta, *J. Anal. At. Spectrom.*, 1991, **6**, 199.
54 I.S. Begley and B.L. Sharp. *The 1994 Winter Conference on Plasma Spectrochemistry Handbook*, ed. R.M. Barnes, paper PhP-11, p. 229.
55 I.S. Begley and B.L. Sharp, *J. Anal. At. Spectrom.*, 1994, **9**, 171.
56 J.M. Carey, F.A. Byrdy and J.A. Caruso, *J. Chrom. Sci.*, 1993, **31**, 330.
57 L.M.W. Owen, H.M. Crews, R.C. Hutton and A. Walsh, *Analyst*, 1992, **117**, 649.
58 W.T. Buckley, J.J. Budac, D.V. Godfrey and K.M. Koenig, *Biol. Mass Spectrom.*, 1992, **21**, 473.
59 N. Bradshaw, E.F.H. Hall and N.E Sanderson, *J. Anal. At. Spectrom.*, 1989, **4**, 801.
60 A.J. Walder and P.A. Freedman, *J. Anal. At. Spectrom.*, 1992, **7**, 571.
61 L. Stuhne-Sekalec, S.X. Xu, J.G. Parkes, N.F. Olivieri and D.M. Templeton, *Anal. Biochem.*, 1992, **205**, 278.
62 F.A. Mellon and B. Sandström, in *Stable Isotopes in Human Nutrition: Inorganic Nutrient Metabolism*, ed. F.A. Mellon and B. Sandström, Academic Press, London, 1996, pp. 119–120.
63 I.L. Koumenis, M.L. Vestal, A.L. Yergey, S. Abrams, S.N. Deming and T.W. Hutchens, *Anal. Chem.*, 1995, **67**, 4557.
64 S.J. Fairweather-Tait, T. Fox, S.G. Wharf and J. Eagles, *Pediatr. Res.*, 1995, **37**, 389.
65 P. Brätter, I. Navarro Blasco, V.E. Negretti de Brätter and A. Raab, *Analyst*, 1998, **123**, 821.
66 J.W. Olesik, J.A. Kinzer, E.J. Grunwald, K.K. Thaxton and S.V. Olesik, *Spectrochim. Acta*, 1998, **53**, 239.

CHAPTER 9

Analysis and Metabolism of Vitamins in Foods

1 Introduction

Vitamins are micronutrients and comprise classes of organic natural food components that are distinct from macronutrients (proteins, fats, carbohydrates *etc.*). They are essential, usually in very small amounts, for the maintenance of normal body function and are not synthesised by the host in amounts sufficient to meet normal physiological needs (definition paraphrased from Combs).[1]

In 1993, it was noted that applications of mass spectrometric methods to vitamins, although increasing, retained significant scope for wider exploitation.[2] This reference contained a summary of the small number of mass spectrometric studies of vitamins published at that time. Developments in the intervening period, particularly the wider availability and increased ease of use of LC/MS, have ensured a rapid rise in the application of mass spectrometric techniques in vitamin research.

Although vitamins can be analysed qualitatively and quantitatively by mass spectrometry, this is, apart from a small number of notable exceptions, not a cost-effective use of the methodology. Attention is increasingly being focused on two areas: identifying the products of vitamin metabolism and the use of stable isotope methods to trace the absorption and metabolic fate of vitamins. These applications represent important methods for advancing knowledge of dietary requirements in the population as a whole. They are also capable of identifying 'at risk' groups and assessing whether supplementation is desirable. Finally, if supplementation is required, stable isotope methods are very useful for defining the optimal form of the supplement, and how well it is absorbed from manufactured foods, raw materials, or tablets.

Vitamins fall into two broad categories, fat and water-soluble. The structures of water-soluble vitamins are quite diverse; fat-soluble vitamins, in contrast, all contain isoprenoid units. Not all vitamins have been studied by mass spectrometry sufficiently to warrant extensive coverage. Attention will therefore be focused on those vitamins for which mass spectrometric data exist.

2 Water-soluble Vitamins

B Group Vitamins

Thiamin

Thiamin, previously known as thiamine or vitamin B_1, is the trivial name of the combined pyrimidine/thiazole molecule that has the chemical structure shown in Figure 9.1.

It is widely distributed in plant and animal foods, with yeasts and liver a particularly rich source. However, cereal grains are the most important dietary source for humans.[1] Thiamin occurs mainly as the free form in plants and in phosphorylated forms in animal tissues. The metabolically active form is the diphosphate ester. This is a cofactor for several coenzymes that cleave the C–C bond of α-ketoacids and has an essential role in carbohydrate metabolism and neural function.[3] The classical symptom of thiamin deficiency in humans is *beri-beri*.

Despite its importance and the prevalence of thiamin deficiency diseases in certain regions of the world, particularly Southeast Asia, attempts to investigate its metabolism by mass spectrometric techniques are rare. The first of three recent publications describes a mass spectrometric study of the thermal degradation of thiamin in various media.[4] Identification of HPLC fractions of the degradation products, which contain many organoleptically interesting flavour products, was accomplished by GC/MS. TSP LC/MS using acidic media has shown promise in the analysis of thiamin[5] and an APCI-based LC/MS method has been used to quantify thiamin in dried yeast, with a detection limit of 2 ng.[6] The APCI methodology appears to be superior to a particle beam (PB) LC/MS method (developed as part of a general method for analysing water-soluble vitamins), where the detection limit was 90 ng in positive ion SIM mode.[7] This result is unsurprising, as the sensitivity of APCI is generally known to be superior to that of PB LC/MS.

Riboflavin

Riboflavin, sometimes called vitamin B_2, is a substituted isoalloxazine with the structure shown in Figure 9.2.

It is widely distributed in foods, mainly in protein bound forms.[1] Meats and dairy products are the main source of riboflavin in developed countries, but green leafy vegetables also contain appreciable concentrations. It is an integral component of the coenzymes riboflavin-5′-phosphate and flavin adenine di-

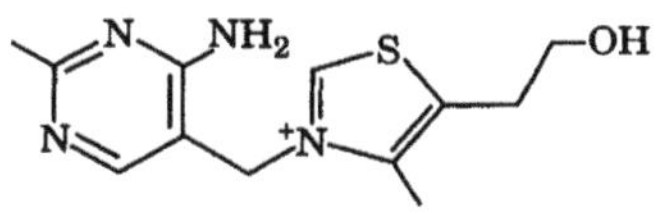

Figure 9.1 *The chemical structure of thiamin*

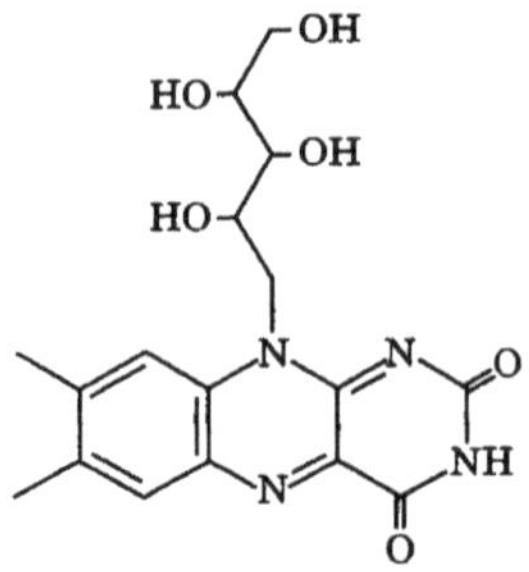

Figure 9.2   *The chemical structure of riboflavin*

nucleotide (these molecules are interconvertible through the action of the enzyme FAD-pyrophosphorylase). The coenzymes are essential for the functioning of a wide range of electron transfer enzymes. Severe riboflavin deficiency manifests itself in epithelial lesions and nervous disorders. Sub-clinical riboflavin deficiencies are not uncommon in developed countries.

Few mass spectrometric studies of riboflavin have been performed. However, the appearance of several recent papers suggests increased interest in the potential of mass spectrometry in riboflavin analysis. An LC/MS (frit FABMS) analysis of riboflavin and tocopherol is essentially a demonstration of the potential of column switching techniques used to overcome problems caused by HPLC buffer systems that are incompatible with this form of mass spectrometry.[8] PB LC/MS is unsuitable for analysing riboflavin as even microgram levels of this vitamin were undetectable, leading the authors to speculate that this highly polar compound was lost during transmission through the PB interface.[7] Riboflavin in milk decomposes readily when irradiated by light and identification of the main photolysis products, lumiflavin and lumichrome, is aided by mass spectrometry.[9] Riboflavin-derived coenzymes have been analysed by TSP and electrospray LC/MS.[10] Positive and negative ion TSP spectra exhibited intense fragment ions containing the flavin ring, whereas electrospray yielded mainly molecular weight information. The techniques were used to characterise riboflavin sulfates in sugar beet. Electrospray in particular shows considerable promise in the analysis of riboflavin and riboflavin-derived molecules and has obvious potential in stable isotope metabolic studies.

Pantothenic Acid

Pantothenic acid, very occasionally known as vitamin B_3, is formally a derivative of pantoic acid (2,4-dihydroxy-3,3-dimethylbutyric acid) and alanine, with the structure shown in Figure 9.3.

Meats, whole grains, avocado, broccoli, mushrooms and some yeasts are good sources of pantothenic acid and it occurs in these foods mainly in protein-bound forms or in CoA.[1] It functions in CoA and acyl-carrier protein as a source of acyl groups for fatty acid and cholesterol synthesis. Dietary deficiencies are rare and are usually only found when total food intake is inadequate.

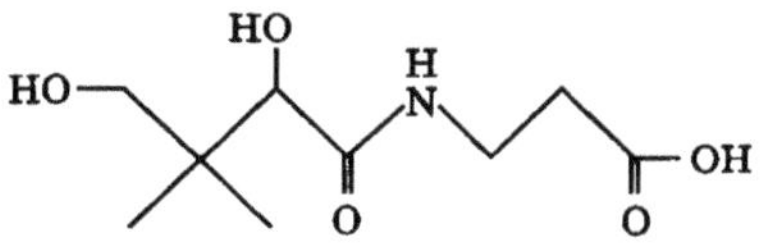

Figure 9.3 *The chemical structure of pantothenic acid*

Little use has been made of mass spectrometry in pantothenate analysis. However, GC/MS identification of the silylated urinary metabolites of a pantothenic acid homologue, calcium 4-(2,4-dihydroxy-3,3-dimethylbutyryl-amido)butyrate, shows the potential of mass spectrometry in metabolic studies of pantothenates.[11] The general PB LC/MS study of water-soluble vitamins (mentioned above) yielded detection limits of 100 ng in EI mode when the fragment ions at m/z 73, 113 and 131 were monitored.[7] ESI mass spectrometry has revealed evidence for the presence of covalently attached pantheine-4'-phosphate in a mammalian oxoreductase enzyme.[12] Alkaline hydrolysis of the protein resulted in a decrease in the mass spectrometrically determined molecular weight of the protein from 10751.6 to 10449.4. This reduction in mass corresponds to the removal of an acyl group attached by a thioester linkage.

Vitamin B₆

Vitamin B_6 is the generic descriptor of a group of seven metabolically interconverted 2-methyl-3,5-di(hydroxymethyl)pyridine derivatives (Figure 9.4). It is present in many foods, but nuts, vegetables, grains and meats are especially rich sources.

It is involved in many metabolic processes, including amino acid metabolism, biosynthesis of neurotransmitters, tryptophan-niacin conversion and synthesis of haem precursors.[1]

Classical methods for determining vitamin B_6 concentrations in biological

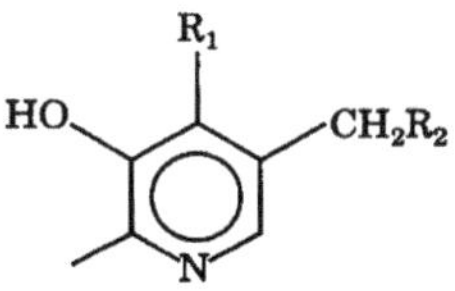

R_1	=	CH_2OH, R_2	=	OH	pyridoxine
	=	CHO,	R_2 =	OH	pyridoxal
	=	COOH,	R_2 =	OH	pyridoxic acid
	=	CH_2NH_2,R_2	=	OH	pyridoxamine
R_1	=	CHO,	R_2	= $-OPO_3^-$	pyridoxal 5'-phosphate
R_1	=	CH_2NH_2, R_2	= $-OPO_3^-$		pyridoxamine 5'-phosphate

Figure 9.4 *The chemical structure of vitamin B₆*

samples have several drawbacks. This, and the lack of definitive data on absorption and bioavailability, has spurred the development of stable isotope methods for quantitative and dietary studies of this vitamin.[12-14] The successful synthesis of several deuterium labelled B_6 vitamers[15] was followed by the development of an EI GC/MS IDMS method for quantifying these compounds in liver, milk, urine and faecal samples.[12] Individual B_6 vitamers were isolated as separate fractions by a complex HPLC procedure. Aldehydes were reduced to the alcohol form, acetylated and analysed by GC/MS. Unique mass spectra containing low-intensity molecular ions and intense $[M - 42]^+$ ions, due to loss of the elements of ketene from the acetates, were obtained. These fragment ions were used for quantitative purposes and yielded good data down to levels of 0.02 nmol ml^{-1}.

The authors noted that the cost of the analysis in staff time and equipment precludes quantitative IDMS of B_6 vitamers from being adopted as a general method. It is best used as a reference technique for verifying analytical data acquired by other methods. However, it also has considerable potential for conducting metabolic studies in human subjects. A greatly simplified technique, involving IDMS GC/MS of a *t*-butyldimethylsilyl derivative, has been developed for urinalysis of 4-pyridoxic acid, the main non-biologically active metabolite of vitamin B_6.[14] The relative bioavailability of deuterium-labelled pyridoxine and pyridoxine-5'-β-D-glucoside has been compared using a stable isotope GC/MS method.[13] The glucoside is a major form of vitamin B_6 in plant-derived foods and its bioavailability was greater in humans than in animal models. This type of study helps to clarify the role of intestinal mucosa and microflora in the absorption of micronutrients from foods. It also emphasises the importance of basing dietary information on human studies that, if they are to provide definitive data safely, must be conducted using stable isotope methods.

Mass spectrometric data have been obtained on the B group vitamins thiamin, nicotinic acid, pyridoxine, nicotinamide and pantothenic acid in food and drinks.[5] Solvent-mediated TSP LC/MS yielded detection limits of 10–100 pg in SIM mode and was used, with no sample pre-treatment, to analyse water-soluble vitamins in a commercial health drink. Determination of trace amounts of thiamin in fermented soyabean paste was also possible. PB LC/MS has not shown the same promise; the detection limits for pyridoxal, pyridoxamine and pyridoxine were 6, 400 and 225 ng respectively when positive and negative ion SIM were applied to the determination of these vitamers.[7] The promise displayed by the relatively insensitive (and unpredictable) TSP method suggests that the newer techniques of electrospray and APCI have the potential to determine B_6 and other N-containing vitamins. This proposition has been supported by a brief conference report.[16]

Folates

Folate is a generic term for a group of compounds related to folic acid (pteroylmonoglutamic acid). The structures of folic acid and polyglutamyltetrahydrofolic acid are shown in Figure 9.5(a) and 9.5(b) respectively.

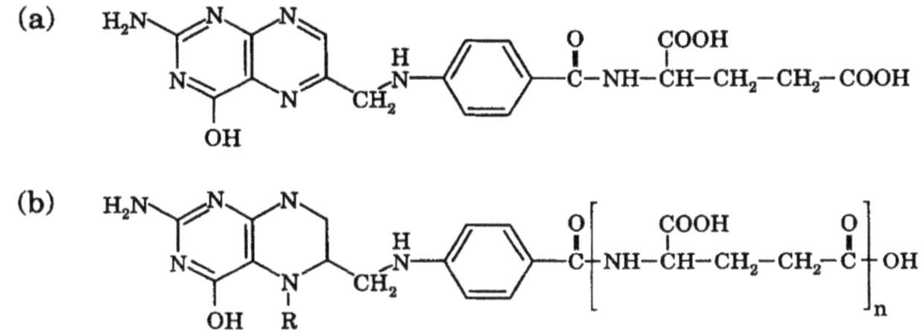

Figure 9.5 *The chemical structures of (a) folic acid and (b) polyglutamyl tetrahydrofolic acid (R = H). Other common forms are polyglutamyl 5-methyltetrahydrofolic acid (R = CH₃) and polyglutamyl-5-formyltetrahydrofolic acid (R = CHO)*

Folic acid is not itself present in living cells and is only found in foodstuffs when added as a supplement. Most naturally occurring folates have a single carbon unit at N-5 (and/or N-10) and contain 2–8 glutamyl residues; they occur widely in foods, and liver, mushrooms and green leafy vegetables are particularly rich sources in the human diet.[1] The monoglutamates are the biologically active forms and food polyglutamates must be enzymically deconjugated by intestinal enzymes before they can be absorbed.[17] Folate is essential for the one-carbon transfer reactions required in metabolic pathways, including amino acid interconversions and purine and pyrimidine biosynthesis.[18] Deficiencies are manifest by a variety of symptoms, including anaemia, skin lesions and poor growth. Perhaps more importantly, folates can reduce plasma homocysteine levels (elevated plasma homocysteine concentration is a known risk factor for cardiovascular disease). Elderly people may be at increased risk of folate deficiency. Folate supplements are currently advised to women in the UK and other countries, at the time of conception, as a possible measure to reduce the incidence of neural tube defects (NTDs). The association of folates with reduced risk of chronic disease and with prevention of NTDs has generated greater interest in studies of their metabolism in humans, and consequently in improved mass spectrometric methods for determining these members of the B-vitamin group.

Folates are polar, involatile molecules and early applications of EI mass spectrometry to the characterisation of these compounds was only successful after methylation or permethylation to increase volatility.[19,20] Good EI spectra, containing both molecular weight and structural information, can be obtained after derivatisation. However, the method has major drawbacks because multiple products are formed with some folates and it cannot be recommended as a general analytical technique. The state of the art in mass spectrometric analysis of folates was last reviewed in 1990.[21] Condensed-phase ionisation techniques were then considered promising in the analysis of folates. These included field desorption mass spectrometry[21–23] and FABMS.[21,24] FDMS fell out of favour because of technical difficulties with the method and was superseded by FABMS and, latterly, ESI, as a routine method for analysing polar, involatile molecules. FABMS was successful in identifying synthetic folates and folic acid and their metabolites after dissolution in DMSO and addition of a thioglycerol matrix.[24] Nevertheless, this technique is not recommendable as a general method of folate analysis: the folates do not yield intense FAB spectra and the ionisation method is insufficiently sensitive for this class of vitamin.

The metabolism of folates can be studied by GC/MS of extracts of biological fluids, after controlled chemical degradation and derivatisation.[25] It is possible to introduce deuterium labels in either the 3,3,4,4 positions of the glutamate,[26] or the 3',5' positions of the pteroyl[27] moieties of the molecule. If comparison of the relative bioavailability of mono- and poly-glutamyl folates in humans is to be undertaken successfully, labelling of both the pteroyl and glutamyl portions of the molecule is essential.[25] These studies, designed to measure the influence of food composition on folate absorption, are highly desirable. Foods (other than those supplemented with folic acid) contain folates in the polyglutamate form and dietary composition can affect the rates at which these molecules are deconjugated and absorbed. The protocols used for measuring folate bioavailability in humans typically involve administration of an oral dose of the $^2H_2$-labelled polyglutamate and intravenous injection of the corresponding mono-glutamate.[28] Urinary collections are then performed for 48 h and the isotope labelling of urinary folates is determined by GC/MS. Similar techniques may also be used to compare the bioavailability of different tetrahydrofolates with folic acid.[29] The mass spectrometric analysis is conducted after oxidative cleavage of the C-9–N-10 bond, following isolation of the urinary folates by affinity chromatography.[26] GC/MS analysis is performed on the trifluoro-acetylated lactam (Figure 9.6) by electron capture negative ion (ECNI) mass spectrometry.[30]

Selected ion monitoring chromatograms of the ions at m/z 426 (d_0), 428 (d_2) and 430 (d_4) from this derivative are shown in Figure 9.7.

Integration of the areas of these peaks allows calculation of the molar ratios of the different labelled urinary folates. Experiments designed to measure the kinetics and pool sizes of folates in humans[31] and to determine urinary excretion of folate metabolites[32] have been conducted using similar methodology. It is possible to obtain adequate isotope ratio measurements on 200 fmol injections of the 4-aminobenzoylglutamate lactam.[25] These detection limits were sufficient for use in subjects when protocols that resulted in excretion of 10–100 nmol of labelled urinary folate were used. A drawback of the methodology is that it is necessary to place volunteers on a 'saturation regime' of folate to attain these levels of excretion. Such conditions may lead to abnormal folate metabolism

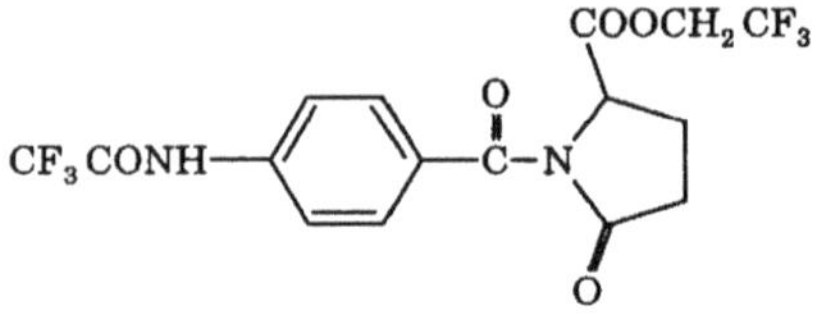

Figure 9.6 *The chemical structure of the trifluoroacetylated β-lactam derivative of folate*

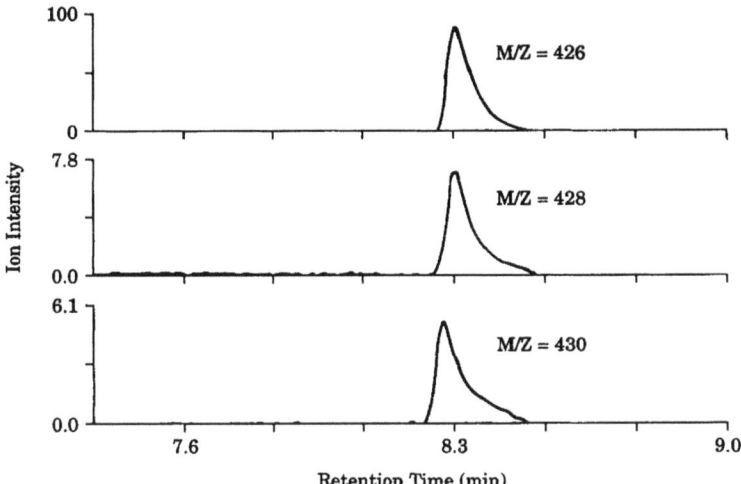

Figure 9.7 *Selected ion monitoring chromatograms of the ions at m/z 426 (d_0), 428 (d_2) and 430 (d_4) from the folate derivative shown in Figure 9.6*
(Reproduced with permission from J.F. Gregory and J.P. Toth, in *Food Technol.*, 1988, **42**, 230, © 1988 The Institute of Food Technologists)

and care is needed in the interpretation of studies conducted under these conditions.

These experiments indicate differences in the bioavailability of monoglutamyl folates.[29] They also show that the bioavailability of hexaglutamyl folate is approximately half that of monoglutamyl folate.[28]

Improved mass spectrometric methods (*i.e.* increased sensitivity, selectivity and speed) for measuring isotopically labelled folates are desirable. This should allow urinary excretion experiments without the necessity of maintaining volunteers on a saturation regime (*i.e.* non-physiological) of folate. Biovailability studies could also be conducted by collecting plasma, rather than urine, samples. A reduction in sample processing is also desirable, to speed up analysis time. The achievement of both improved sensitivity and selectivity and speed of analysis may be possible by using alternative ionisation and sample introduction techniques. ESI, APCI, or possibly GC/MS/MS of different volatile derivatives of folates deserve detailed investigation. A brief report of the ESI of labelled folates in human plasma has already appeared.[33] Negative ion ESI yields the most sensitive data for folate but there are practical difficulties in introducing real folate samples into the ESI ion source, as these molecules are very alkali sensitive.[34] Unfortunately, alkaline conditions are required to release folates from column traps and ionise them with the greatest efficiency. Overcoming these practical problems may be possible, in which case ESI becomes a very attractive proposition. Alternatively, APCI methods may be suitable if ionisation is sufficiently soft that the fragile folate molecules remain intact. A sensitive APCI (negative or positive ion) method could overcome most of the practical difficulties outlined above.

Isotope methods have also been developed to quantify red blood cell folates by IDMS.[35] The blood cell samples were spiked with a folate standard in which the benzene ring has been labelled with ^{13}C on all six carbons. Folates were isolated using bovine folate binding protein, and then chemically cleaved to 4-aminobenzoic acid, pteridines and glutamic acids. GC/MS isotope ratio measurements on derivatives of the 4-aminobenzoic acid were then used to generate quantitative data. The method was sensitive and specific and was used to generate reference data on a group of normal individuals. Because IDMS methods are generally expensive and time-consuming (compared with immunoassay methods, for example), this application represents the type of experiment best suited to isotope quantification techniques. IDMS is generally costly and protracted for routine analysis, but is an excellent method for generating reference values, for calibrating less robust techniques, or in 'needle in a haystack' applications where more traditional methods run into difficulties.

GC/MS methods are also useful for diagnosing folate and vitamin B_{12} deficiencies in 'at-risk' populations. Homocysteine accumulates because of folate and/or vitamin B_{12} (the two vitamins are closely interrelated) deficiencies and sensitive isotope dilution GC/MS assays have been developed for this metabolite.[36,37] GC/MS of *t*-butyldimethylsilyl derivatives of homocysteine yielded data that showed elevated homocysteine levels in subjects suffering from folate and cobalamin deficiency, even those free of haematological and neurological abnormalities.

Vitamin B_{12}

Vitamin B_{12} is a generic term for all compounds containing the corrin nucleus that possess the biological activity of cyanocobalamin (Figure 9.8).[1] The corrin ring is cobalt-centred and analogues with different groups (OH$^-$, CH$_3$, 5'-deoxyadenosyl, H$_2$O and nitrito) at the cyano ligand position are also of dietary importance. Because its biological synthesis is restricted to bacteria, its distribution is confined to foods formed by bacterial fermentation, or to the tissues of food animals that have absorbed vitamin B_{12} from their intestinal microflora, or from their diet.[1] It is especially concentrated in liver (the main storage organ). The biological activity of vitamin B_{12} is associated with two coenzyme forms that are essential to the functioning of three enzymes involved in propionate, amino acid and single carbon metabolism. Its metabolism is intimately linked with that of folates.[38] Vitamin B_{12} deficiencies result in anaemia and neurological changes. Lacto-ovo vegetarians are at particular risk from B_{12} deficiency, although this may take some time to develop (and can be avoided by consumption of yeast-derived products). Malabsorption of vitamin B_{12} is found occasionally and is manifest by the condition pernicious anaemia.

Several mass spectrometric techniques are suitable for the analysis of intact cobalamins and cobalamin derivatives. These include FABMS,[39,40] FDMS[41,42] and LDMS.[43] All techniques yielded useful positive and negative ion spectra

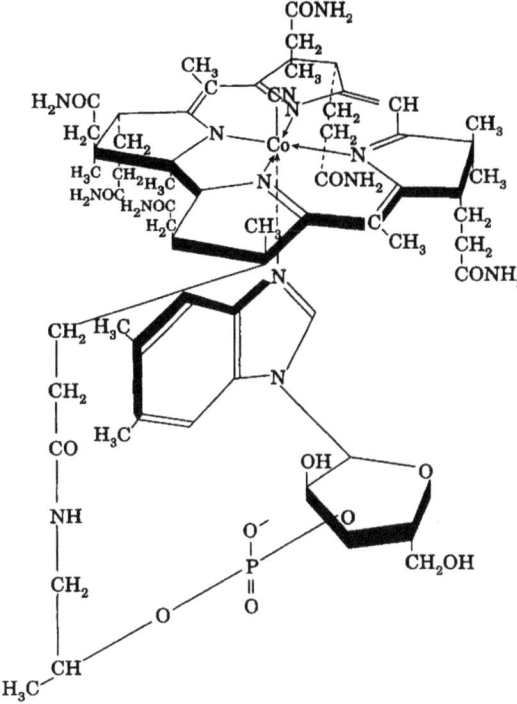

Figure 9.8 *The chemical structure of vitamin B$_{12}$*

that contained $[M + H]^+$ and $[M - H]^-$ ions, and structurally informative fragments (thermolytically generated, in the case of FDMS). Tandem mass spectrometry of $[M + H]^+$ ions generated by FABMS has also been investigated.[44] These methods are useful for characterising synthetic cobalamins but are not of particular interest in food or nutritional studies. A possible exception would be their application to stable isotope metabolic studies. However, none have been reported at the time of writing and alternative mass spectrometric techniques, particularly electrospray, probably have potential if this type of study is contemplated.

Electrospray mass spectrometry can detect intact, non-covalently bound protein–cobalamin complexes.[45] The cobalamin-binding domain of methionine synthase isolated from *E. coli* was identified and characterised in this way. This demonstrates the potential of ESI in identifying non-covalent vitamin–enzyme complexes.

The microorganism-produced cobalamins differ from native cobalamin only in the nucleotide loop portion of the molecule. A GC/MS method, based on *t*-butyldimethylsilyl derivatisation, has been developed for analysing these moieties.[46] A deuterium label was incorporated biosynthetically into the lower axial ligand of cobalamin for quantitative measurements. This methodology also has potential in human metabolic studies.

Elevated urinary and serum methylmalonic acid (MMA) levels are a useful index of cobalamin deficiency in various populations. Isotope dilution GC/MS methods, based on *t*-butyldimethylsilyl derivatives, have been developed for quantifying this metabolite in urine[47] and serum.[48] Dicyclohexyl esters have been investigated as alternative GC/MS derivatives.[49] However, improved silylation procedures have enhanced the sensitivity and specificity of the technique, yielding detection limits of 17 fmol and a linear range of 0.04–1.7 μmol l^{-1}.[50] GC/MS methods were concluded to have distinct advantages over conventional techniques in a recent review of metabolite assays in cobalamin and folate deficiencies.[48]

LC/MS methods that have produced good mass spectrometric data for vitamin B_{12} include the (now obsolete) Direct Liquid Introduction (DLI) method in negative ion mode[51] and atmospheric pressure spray (APS).[52] Ionisation in the APS source was effected by using ammonium acetate in the solvent in a manner similar to solvent-mediated ionisation in TSP ion sources. However, the spectral characteristics of APS and TSP were very different. The APS technique (based on a modified atmospheric pressure ion source) was evaluated against a variety of molecules, including sugars, glycosides and peptides. It showed some promise for vitamin B_{12} and indicates that the newer techniques of APCI and ESI are ripe for further exploitation in the analysis of this molecule.

Niacin

Niacin is the generic term for pyridine-3-carboxylic acid (nicotinic acid, Figure 9.9) and derivatives, particularly nicotinamide.

It is found in particularly high concentration in yeasts and meats, but is quite widely distributed (although in uneven concentration) in plant foods.[1] It is mainly protein-bound in plants and in animal tissues exists mostly as nicotinamide adenine dinucleotide (NAD) and its phosphate, NADP.

Several forms of niacin are bound in ways that prevent its absorption. Many animal species, humans included, can synthesise NAD and NADP from tryptophan.[1] It functions metabolically as an essential component of the electron transport mechanism of cells. Deficiencies of niacin are usually linked with deficiencies of other essential nutrients, particularly vitamin B_6. Overt deficiencies are characterised by skin lesions (particularly in areas of skin exposed to sunlight) and gastrointestinal and neurological disorders.

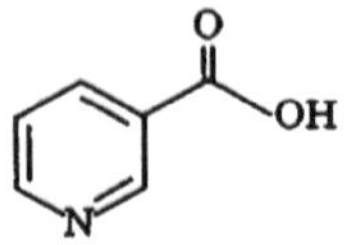

Figure 9.9 *The chemical structure of nicotinic acid, a member of the niacin family of vitamins*

Mass spectrometric applications to niacin analysis are quite rare. Exceptions are the development of GC/MS methods to identify trace impurities (2,3-, 2,5- and 2,6-pyridinedicarboxylic acids) in bulk niacin[53] and to determine plasma levels of nicotinic acid and nicotinamide.[54] The first application has potential as a general method for quality control of niacin supplements in foods. The methodology for quantifying niacin in human plasma samples has potential for assessing the importance of niacin metabolism in cancer prevention. This could be accomplished by increasing the understanding of the relationship between dietary niacin and circulating levels of NAD-precursors (reduced NAD levels are predicted to enhance carcinogenesis). It is also possible that a more rapid and sensitive means of determining niacin in plasma will be provided by particle beam, electrospray, or APCI LC/MS.

Biotin

Biotin is the molecule *cis*-hexahydro-2-oxo-1*H*-thieno[3,4-*d*]imidazole-4-penta-noic acid. The structures of biotin (top) and that of its enzyme-bound form (bottom) are shown in Figure 9.10.

It is widely distributed in foods in low concentration. The main dietary sources for humans are milk, liver, eggs and some vegetables.[1] It is generally found in protein-bound forms and its bioavailability is very variable, being highly dependent upon the ease with which the binding proteins are hydrolysed. Biotin acts as a carboxyl carrier in four carboxylase enzymes in animals and is important in lipid, glucose, energy and amino acid metabolism.[1] Deficiencies are rare because of its widespread distribution. However, its absorption is antagonised by certain proteins, particularly the glycoprotein avidin that is found in egg whites.

Little interest has been shown in mass spectrometric studies of biotin related

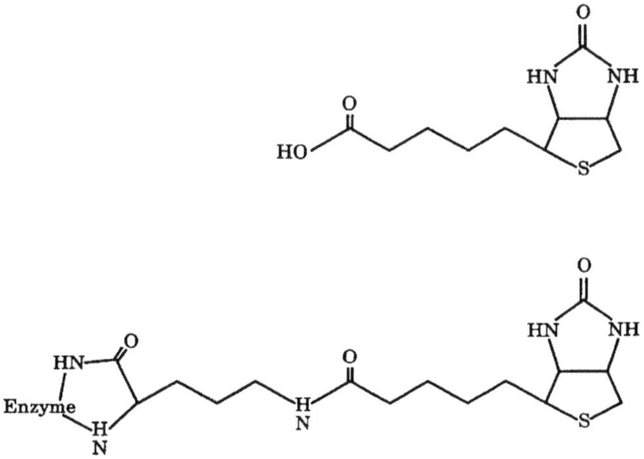

Figure 9.10 *The chemical structure of biotin*

to its dietary importance, although a (now obsolete) form of LC/MS has quantified biotin and dethiobiotin.[55] Particle beam LC/MS has been used to characterise *N*-acyl-D-biotinols[56] and has also been applied to the analysis of biotin and other water-soluble vitamins. However, the sensitivity, as is often the case with PB LC/MS, was poor (detection limit of 250 ng).[7]

Biotinylated hydroxy amino acid residues in peptides have also been identified successfully by mass spectrometry.[57] Amino acid sequence ions in the FABMS spectra of peptides reacted with N-OH succinimide esters of biotin revealed the location of *O*-biotinylation. These experiments showed that biotinylation was not confined to primary amino groups: certain amino acids containing highly reactive hydroxyl groups were also available for *O*-acylation. The methodology has potential in food science applications as it could be used to aid the characterisation of naturally occurring biotinylated peptides.

Most reported mass spectrometric data are confined to clinical and metabolic studies. Examples include determination of the metabolic origins of the S and 3'-N atoms in biotin by stable isotope mass spectrometry[58] or identification of inherited vitamin responsive disorders.[59]

An interesting and novel application of mass spectrometry to biotin is the study of complexes of this molecule with its antagonist avidin and related glycoproteins.[60–62] ESI was used to identify intact, non-covalent avidin–biotin complexes and the results showed that this form of mass spectrometry could detect, in the gas phase, complexes observed in the solution phase. The ability to detect these complexes was related, at least qualitatively, to their stability in solution.[61] This demonstration of the ability of ESI to detect large biomolecular non-covalent complexes that appear to reflect solution phase stoichiometry is, if generally applicable, very important. The technique has potential as a sensitive and specific method for investigating non-covalent molecular interactions *in vitro* and could be used to quantify significant effects on the absorption and utilisation of micronutrients *in vivo*.

Peroxidation of NAD(+) *in vitro* has been examined using the state-of-the-art mass spectrometric technique, combined CE/MS.[63] Using this technique, it was possible to monitor reaction products and detect unstable intermediates. This underlines the great potential of ESI for conducting fundamental metabolic studies, especially when it is combined with a rapid, high-resolution separation technique.

Other Water-soluble Vitamins

Vitamin C

Vitamin C is a dibasic acid, known also as ascorbic acid or L-ascorbic acid, which forms a redox system with the free radical monodehydro-L-ascorbic acid and dehydro-L-ascorbic acid (DHA). The structure of ascorbic acid is shown in Figure 9.11.

It is found in many foods, including fruits, vegetables, liver and kidneys. Its concentration can decrease drastically during storage and cooking.[1] Vitamin C

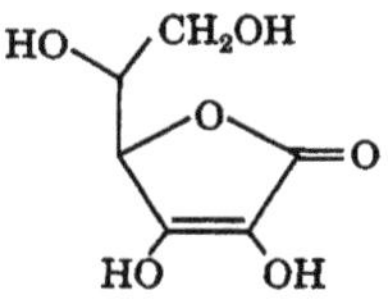

Figure 9.11 *The chemical structure of ascorbic acid*

has a variety of biological functions, all related to its redox properties. These functions include involvement in the synthesis of hormones, neurotransmitters, collagen and carnitine, and in the absorption of iron and other substances.[64] The role of vitamin C in iron absorption may be as an enhancer of bioavailability, by reducing Fe^{3+} to the more readily absorbed Fe^{2+}.

Acute vitamin C deficiency results in scurvy. This is unusual in developed countries, but mild vitamin C deficiency is more common and can be a consequence of smoking, stress, some diseases and ageing. This can increase the risk of heart disease and hypertension.[1]

Several mass spectrometric methods have been used to characterise vitamin C. These include electron ionisation,[65] LDMS[66] and GC/MS.[67,68] However, the characterisation of ascorbic acid and dehydroascorbic acid by GC/MS of *t*-butyldimethylsilyl derivatives, and the development of an isotope dilution assay for ascorbate and dehydroascorbate, represents an important advance in mass spectrometric studies of vitamin C.[69] Labels used were [$^{13}C_6$]ascorbate and [$^{13}C_6$]- and [6,6-$^2H_2$]-dehydroascorbate. The IDMS assay yielded good sensitivity (detection limits of 9 pg ascorbic acid) and linearity and was used to monitor ascorbic acid–dehydroascorbate interconversion in aqueous solutions and in human plasma. The method is satisfactory for determining both molecules in biological and aqueous fluids, even when oxidation processes occur. The same group used GC/MS methods to study variations in ascorbic acid oxidation routes in H_2O_2 and cupric ion solutions.[70] These *in vitro* experiments showed that different oxidative stresses generate dissimilar ascorbic acid oxidation products.

An IDMS GC/MS method for determining vitamin C in non-fat milk powder uses *t*-butyldimethylsilyl derivatives and ^{13}C-labelled ascorbic acid as the spike isotope.[71] There was some discrepancy between the IDMS methods and HPLC quantification. Because the L-ascorbic acid content of milk can alter rapidly with time, a certified reference value was not assigned.

A recent study of the feasibility of determining ascorbic acid kinetics in humans has concluded that conventional GC/MS of ^{13}C-labelled trimethylsilyl derivatives is suitable for metabolic studies.[72] In contrast to earlier (non-metabolic) investigations, the trimethylsilyl derivative was superior to the *t*-butyldimethylsilyl derivative for both conventional GC/MS and for GC/combustion/MS. This was because (i) the TBDMS derivatisation reagent yielded a mixture of tri- and tetra-substituted products and (ii) the TBDMS derivatives add a greater carbon load than TMS, reducing the measurable enrichment if GC/C/MS is used. Although both conventional GC/MS and GC/

C/MS were suitable for measuring plasma vitamin C after administration of an oral dose, the simpler GC/MS method may be preferable (and more accessible). The technique was used to measure plasma kinetics in an adult volunteer and the metabolism was shown to fit a three-compartment model.

3   Fat-soluble Vitamins

Vitamin A

Vitamin A is the generic descriptor for compounds with the biological activity of retinol.[1] This includes retinoids and some carotenoids (pro-vitamin A). The structures of retinol and of the main pro-vitamin A, β-carotene, are shown in Figure 9.12.

Vitamin A exists in several forms in animal products, mainly as long-chain fatty acid esters. Carotenoids are present in plant and animal products (in the latter because of dietary exposure). Foods rich in pro-vitamin A include green, leafy and yellow or orange vegetables (*e.g.* spinach, carrots and red peppers), eggs, liver and oily fish. Some processed foods, margarines in particular, are high in vitamin A because of fortification.

Vitamin A is important in the visual process as a component of rhodopsin and the iodopsin photosensitive visual pigments. This vitamin is also necessary for reproduction, immune function, bone metabolism, dermatological health and cancer prevention. Severe vitamin A deficiency leads to blindness, infection and stunted growth.

It was shown as early as 1970 that retinol and retinoids yield characteristic EI spectra.[73] Most subsequent reports of the mass spectrometry of retinol are focused on the determination of body stores of this vitamin by isotope ratio GC/MS. This is done by oral administration of deuterium-labelled retinol, with subsequent measurement of the ratio of deuterated and undeuterated forms in a blood sample taken after a suitable time (typically 2–7 weeks in humans) has elapsed. Early work on animal models was based on measurement of fragment

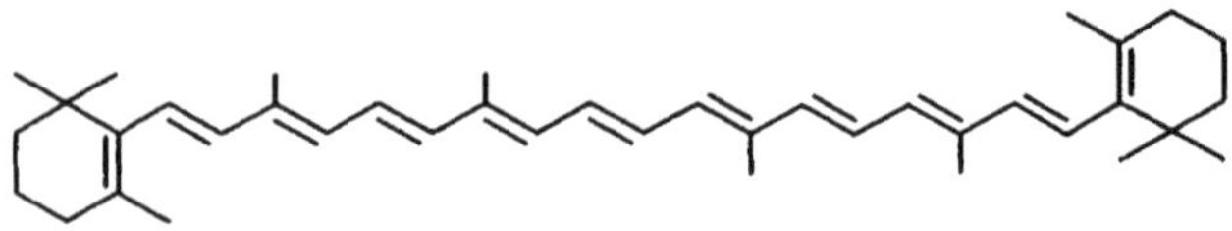

β-carotene

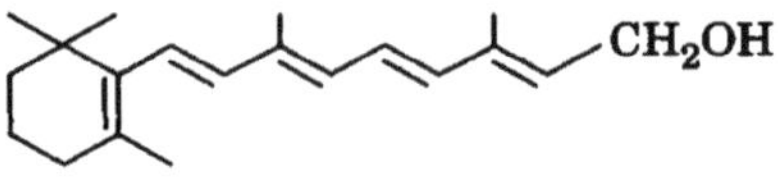

retinol

Figure 9.12   *The chemical structures of retinol and β-carotene*

ions ($[M - CH_2O - TMS]^+$) in trimethylsilylated derivatives of retinol.[74] Free retinol tends to dehydrate under GC conditions and an alternative approach advocated GC/MS determination of anhydroretinol rather than retinol or retinol derivatives.[75,76] Although the method was useful, extensive *cis*-isomerisation occurred during the dehydration step. Some subsequent investigations relied on GC/MS measurement of intact retinol[77] and yielded reasonable correlation between isotopic and biopsy measurements of retinol liver stores.[78] However, derivatisation does have considerable advantages and an improved protocol for determining ratios of [^{2}H$_4$]retinol to retinol has been developed by conversion of the analytes to their *t*-butyldimethylsilyl derivatives.[79] This method yielded improved sensitivity, precision and reliability and subsequent improvements in retinol isolation procedures based on solid-phase extraction have introduced additional refinements.[80] The stable isotope method is highly recommendable for determining the vitamin A stores of individuals and for observing the effects of diet on vitamin A status. Further improvements in the sensitivity of the method are still desirable, so that measurements can be performed on smaller blood samples. This is especially important in serial infant studies, or if sampling is conducted under non-ideal field conditions. Negative ion electron capture GC/MS (of the trimethylsilyl derivative) has been developed and exhibits an order of magnitude improvement in sensitivity over previous methods.[81] The negative ion spectrum exhibits an abundant ion at m/z 268 in unlabelled retinol, formed by loss of the silyl group. The improved methodology allows detection of [^{2}H$_8$]retinol for 32 days; even after this length of time the enrichment was 0.07%, well above the detection limit of 0.01%. An alternative technique utilising EI ion trap mass spectrometry of *t*-butyldimethylsilyl derivatives of retinol has been described.[82] The method, in which fragments outside the mass range of interest were ejected, yielded increased dynamic range (compared with conventional, quadrupole mass spectrometers). Standards required dilution to minimise mass shifts due to ion-coupling phenomena in the trap. The method yielded linear calibration curves over molar concentration ratios of 0.0023–1.0000 ([^{2}H$_4$]retinol:retinol) and yielded values for natural abundance close to theoretical.

Several fat-soluble vitamins (A, D$_3$, E, K$_1$ and K$_2$) have been determined by the now obsolete 'Direct Liquid Introduction' LC/MS technique. TSP LC/MS has been used to characterise the oxidised and glucuronidated metabolites of retinol in monkey plasma.[83] Microbore (0.32 mm i.d.) HPLC columns have been coupled to the CI source of a quadrupole mass spectrometer *via* a frit interface and used to analyse *cis*- and *trans*-retinoic acids.[84] These molecules are important vitamin A analogues and possess a wide range of physiological functions. However, better LC/MS methods are now available, exemplified by a report of ESI LC/MS of retinoids that indicates that this technique may be suitable for metabolic studies.[85] Detection limits in SIM mode were 23 pg for retinoic acid (negative ion mode, monitoring $[M - H]^-$) and 0.5 ng for retinol in positive ion mode (when monitoring the $[M + H]^+$ ion). More recently, an APCI LC/MS for quantitation of retinol and retinyl palmitate in human serum has been developed in the same laboratory.[86] Detection limits in SIM mode were 0.670 and 0.720 pmol injected on-column for all-*trans*-retinol and all-*trans*-retinyl palmitate respectively. The combination of high dynamic range and good sensitivity indicates that the technique may also be useful for human bioavailability studies.

Particle beam LC/MS has also shown promise in the analysis of retinol and retinol esters (as well as other fat-soluble vitamins).[87] Detection limits using a narrow bore (2.00 mm) HPLC column were in the low nanogram range, sufficient for general analysis (determination of vitamins in a powdered milk infant formula). However, these values are insufficient for general metabolic studies.

Carotenoids

Interest in the carotenoids has increased rapidly in the past fifteen years. This interest does not stem so much from their pro-vitamin A activity as from epidemiological evidence of their putative role in cancer prevention.[88,89] This has stimulated interest in the development of better qualitative and quantitative analytical procedures for determining carotenoids in foods and in human plasma. There are several problems associated with the analysis of carotenoids and methods are still imprecise.[90-92] It follows that the sensitivity and specificity of mass spectrometry make it a particularly attractive technique for solving some of the more intractable problems in carotenoid analysis.

Carotenoids usually yield intense EI spectra containing molecular ions and abundant, often structurally informative, fragment ions.[93-96] A more recent, comprehensive investigation into the EI and positive and negative ion CI mass spectrometry of carotenoids has been conducted.[97] Samples were introduced into the mass spectrometer ion source using a 'direct exposure probe', when the sample is heated on a metal wire located very close to the ionising medium. This increased absolute sensitivity by at least an order of magnitude. Electron capture NICI was more than 100-fold more sensitive than EI, although all the ionisation techniques used generated useful, often complementary, information. Positive and negative CI methods were particularly suitable for identifying the acyl groups of carotenoid esters. Use of deuterated ammonia as a positive or negative ion CI reagent gas allowed the number of exchangeable hydrogens to be determined in OH containing carotenoids such as lutein. NICI mass spectrometry[98] and NICI tandem mass spectrometry[99] have been used to identify the *in vitro* oxidation products of β-carotene and α-tocopherol. The *in vitro* experiment modelled a biological membrane under peroxyl radical attack and oxidation products detected included endoperoxides and aldehydes. The methodology is suitable for studying model systems but may be unsuitable for biological studies. This is because a wide range of carotenoids and more complex mixtures of oxidation products are encountered in biological systems. These systems therefore require LC/MS for definitive identification of all components.[100] However, tandem mass spectrometry is still potentially very useful in carotenoid analysis. It can either amplify LC/MS data (see below) or aid characterisation of novel carotenoids isolated from plant sources. This was recently exemplified by the structural characterisation of novel or rare *seco*-carotenoids and carotenones.[101,102]

Electron capture NICI particle beam LC/MS was one of a range of techniques used for identifying carotenoids and their oxidation products in human plasma.[100] Although the particle beam interface may not be optimal for LC/MS analysis of carotenoids because of its limited sensitivity, this is a landmark paper that provides objective evidence for the appearance of diet-derived carotenoids in the human blood stream. The absence of carotenoid epoxides (these are abundant in the human diet) from plasma samples, despite the presence of their parent carotenoids, indicates that their absorption and metabolism may be very different from that of hydroxy and hydrocarbon derivatives. The authors rightly pointed out the need for stable isotope studies of the absorption and metabolism of carotenoids in humans, to provide definitive data. A later study by the same group used similar techniques to identify, quantify and determine relative concentrations of carotenoids and their metabolites in human milk and serum.[103] Novel oxidative metabolites of lycopene (epimeric 2,6-cyclolycopene-1,5-diols) were identified in the course of this study. A more recent particle beam LC/MS study has shown that addition of the eluent modifiers ammonium acetate and triethylamine to the mobile phase yielded better sensitivity. Detection limits were improved by factors of between 3 and 15 times when negative ion techniques were used to ionise the analytes.[104] Under these circumstances, a detection limit of 200 pg was obtained for β-carotene in SIM mode.

FABMS and MS/MS and continuous flow FAB LC/MS[105–107] and electrospray LC/MS[108] have also been used to identify carotenoids in extracts of fresh and processed foods. The isomers α-carotene, β-carotene and lycopene yielded distinct, characteristic MS/MS spectra when ionised by static or continuous flow FABMS. LC/MS yielded molecular weight and structural information and good detection limits for full scan spectra (5 ng for lutein and 15 ng for α-carotene). However, the most promising of the above techniques (until the development of APCI methods), judged in terms of versatility and sensitivity, was electrospray.[108] Hydrocarbon carotenoids form $M^{+\bullet}$ ions rather than $[M + H]^+$ under positive ion FABMS and ESI conditions. This is a consequence of the chemical properties of carotenoids and the effect was exploited, in the case of electrospray, by using solution-phase chemistry to enhance the ionisation process.[108] Post-column addition of the oxidising agent 2,2,3,3,4,4,4-heptafluoro-1-butanol at a concentration of 0.1% (v/v) to the HPLC solvent resulted in greatly increased ion yields. Limits of detection for lutein and β-carotene were 1 and 2 pmol respectively in SIM mode. This is not only 100-fold lower than for photodiode array detection, it is sufficient to allow metabolic studies to be conducted if suitable labelled compounds are available. A recent interesting innovation has been the use of silver ions to enhance ionisation of carotenoids under ESI conditions.[109] The technique was demonstrated to be viable in the analysis of lutein, zeaxanthin, canthaxanthin, echinenon, β-carotene and lycopene (as well as tocopherols, see below). Argentation was achieved by post-column addition of a silver perchlorate solution at a concentration of 50 μg ml^{-1} and flow of 50 μl min^{-1}. Sensitivity and selectivity were increased by monitoring loss of elemental silver from the carotenoid extracts in MS/MS (MRM) mode. The method was used to analyse extracts of tomato juice, carrot and vegetable juices. This innovative work may represent a useful advance in the analysis and quantification of carotenoids, if generally applicable. The only possible drawback is the use of a perchlorate salt; perchlorates and hot mass spectrometer ion sources are a potentially explosive combination (if perchlorate salts are allowed to accumulate). It is possible that the methodology may prove hazardous if used on instruments equipped with ESI sources that require higher operating temperatures than the one used in the above study.

ESI LC/MS can be conducted at higher flow rates than FAB LC/MS and is more robust. However, APCI LC/MS can accommodate still higher flow rates and is yet more robust. This may be the technique of choice for LC/MS of carotenoids. Detection limits are similar to ESI LC/MS and an oxidising solvent is not required.[110-112] Figure 9.13 shows the HPLC/UV and HPLC/APCI/MS traces of a mixture of carotenoid standards, with representative mass spectra of β-carotene and zeaxanthin.

Additional structural information may be obtainable by LC/MS/MS, if required. However, in this case it may be advisable to use ESI LC/MS, or to find some means of enhancing the production of M$^{+\cdot}$ ions under APCI conditions. This is because M$^{+\cdot}$ ions of β-carotene form a characteristic and abundant fragment ion due to loss of the elements of toluene under CID MS/MS conditions. In contrast, the [M + H]$^+$ ions formed in APCI mode do not yield informative fragment ions.[113] The APCI LC/MS method also has great potential in stable isotope studies of carotenoid metabolism, besides its potential as a method for qualitative and quantitative analysis of carotenoids in foods. Innovations in carotenoid analysis using LC/MS have been reviewed.[111]

A small number of studies of carotenoid metabolism have been conducted, but not, so far, by LC/MS methods. High precision isotope ratio GC/combustion/MS has been used in human studies of the metabolism of ^{13}C-labelled β-carotene.[114] The unsaturated carotene must first be reduced to its fully saturated form before GC/MS measurements can be made. Although this is inconvenient, the high precision of the measurements enables studies to be conducted using small amounts of administered label (1.87 μmol). This technique requires careful application because reduction of small quantities of carotene is difficult without suffering sample losses or destroying the sample.

Larger doses must be employed if conventional organic mass spectrometry is used to quantify isotopes. For example, a 73 μmol oral dose of ^{2}H$_8$-labelled β-carotene was administered when tandem MS techniques were employed to measure isotope ratios.[115,116] These early reports indicate the great promise of stable isotope techniques in measuring the dietary metabolism of carotenoids. As discussed above, LC/MS techniques may yet be the methods of choice for tackling the complex and difficult problems posed by the need to obtain reliable data on the metabolism of these molecules in humans.

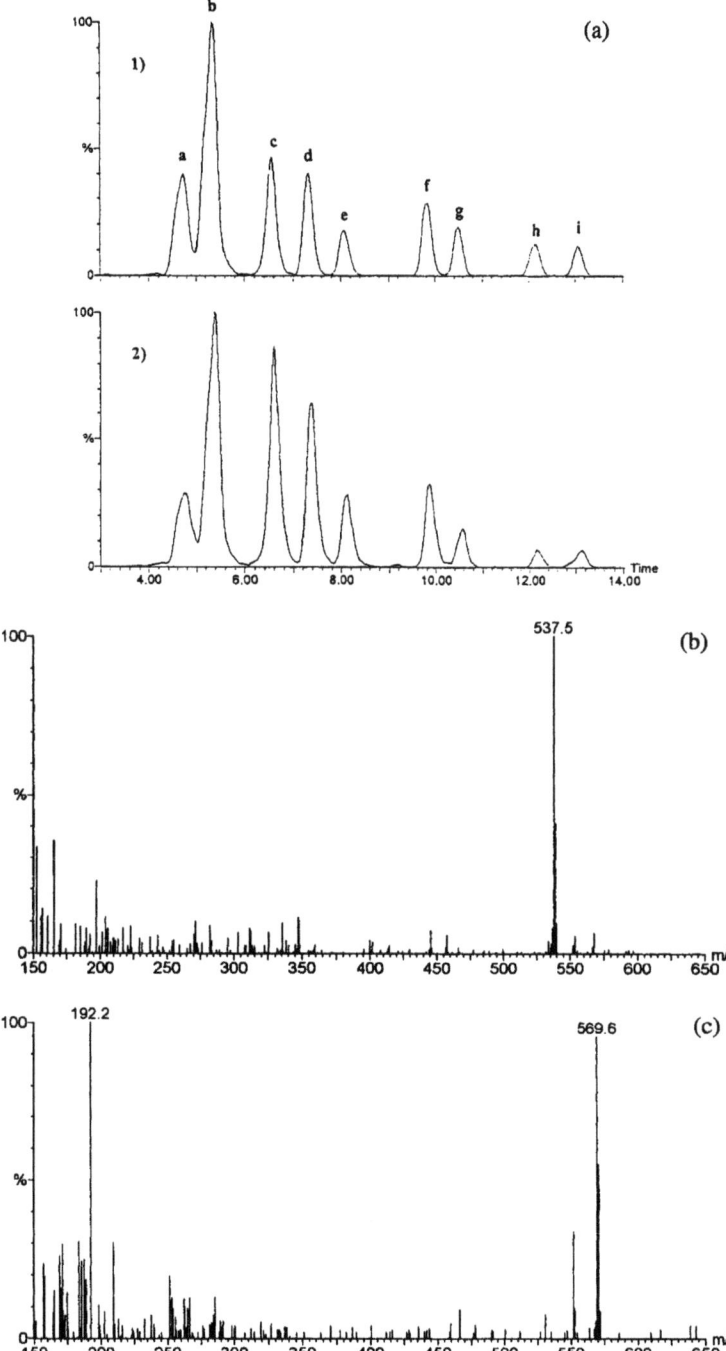

Figure 9.13 (a) *The HPLC/UV (1) and HPLC/APCI-MS (2) traces for a mixture of carotenoid standards, with representative mass spectra of β-carotene (b) and zeaxanthin (c)*
(Data supplied by P.A. Clarke, K.A. Barnes, J.R. Startin F.I. Ibe and M.J. Shepherd and reproduced here with permission)

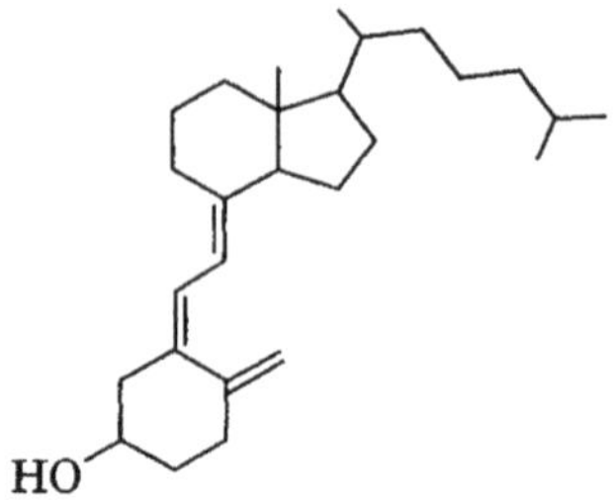

Figure 9.14   *The chemical structure of vitamin D₃*

Vitamin D

Vitamin D is the generic name for a group of closely related secosteroids that exhibit the qualitative biological activity of cholecalciferol, vitamin D_3 (Figure 9.14).[117]

Vitamin D is not required in the diets of populations exposed regularly to sunlight, because of endogenous epidermal photolytic transformation of 7-dehydrocholesterol. It is found in nature mainly in 'pro-vitamin D' forms ergosterol and ergocalciferol.[1] Fish oils are particularly rich in vitamin D but it is usually found in low concentration in other foods. Consequently, it is often used as a fortificant in commonly consumed foods (bread, milk products and infant foods for example). Vitamin D deficiency results in the skeletal diseases of rickets in children and osteomalacia in adults.

Mass spectrometry has been used extensively in the characterisation of Vitamin D and, particularly, its metabolites, more so than any other vitamin. This is because the products of vitamin D metabolism are those molecules that express the biological activity of vitamin D in its most potent form.[118] Mass spectrometric methods provide the safest and most sensitive techniques, both for conducting metabolic studies in humans and for quantifying a wide range of vitamin D metabolites in plasma. Several different mass spectrometric methods have been used to characterise these metabolites. These include heavy ion-induced desorption,[119] Desorption Chemical Ionisation,[120] FABMS,[121,122] TSP[123–125] and electrospray.[126,127] However, the principal method of choice, at least until very recently, has been SIM GC/MS of volatile, usually silylated, derivatives. Methods for studying single metabolites of vitamin D were developed in the 1970s.[128–131] These methods eventually progressed to multiple metabolite assays.[132,133] The methodology has improved steadily with the introduction of capillary columns, increased mass spectrometer sensitivity and the use of a stable isotope labelled internal standards for IDMS quantification.[134] A recent report of note describes a GC/MS method for assaying *underivatised* metabolites of vitamin D_2 and D_3.[135] This is achieved by quantitative dehydration of the metabolites in a high-temperature injector and pre-column packed with aluminium powder. The assay sensitivity for 25-OH vitamin D was approximately 1 ng ml^{-1}.

Although most quantitative and metabolic studies of vitamin D and its metabolites have been conducted by GC/MS, there is increasing interest in the application of LC/MS techniques, often with the aid of tandem mass spectrometry to increase selectivity. A drawback of the LC/MS methods described to date is the need for chemical derivatisation of the metabolites. Flow injection continuous flow FAB LC/MS/MS of metabolites derivatised by cycloaddition with 4-phenyl-1,2,4-triazoline-3,5-dione yields detection limits in the nanogram–picogram range. These values improve to low picogram levels when capillary LC is used.[122] Diels–Alder derivatives of vitamin D_3 and its metabolites can be detected at low picomolar levels by TSP LC/MS.[123] Use of amino or crown ether tagging reagents to derivative neutral vitamin D molecules renders these molecules amenable to electrospray ionisation.[125,126] Electrospray LC/MS/MS has been used to investigate the metabolism of a synthetic vitamin D analogue.[124] Particle beam LC/MS can be used to determine vitamin D, although the reported sensitivity may be insufficient for most metabolic studies.[136] The role of mass spectrometry in vitamin D research has been reviewed.[137]

Vitamin E

Vitamin E is the generic descriptor for all tocopherol and tocotrienol derivatives, the most biologically active being α-tocopherol.[138] The structure of α-tocopherol is shown in Figure 9.15.

Plants are the primary source of vitamin E and its main sources in the human diet are vegetable oils, cereal grains and seeds. It is a biological antioxidant with the capacity to stabilise polyunsaturated lipids and reduce lipid peroxidation.[1] Vitamin E deficiency has deleterious effects on the neuromuscular, vascular and reproductive systems. Severe vitamin E deficiency is uncommon and attention is mainly focused on the effects of sub-clinical deficiencies on optimal health.

Vitamin E molecules are very amenable to mass spectrometric analysis. GC/MS is particularly useful for determining vitamin E if these molecules are first derivatised by trimethylsilylation. IDMS GC/MS methods for measuring α-tocopherol with picogram sensitivity were described more than 15 years ago.[139] A more recent example describes the analysis of vitamin E and its oxidation products by isotope dilution GC/MS.[140] The 'Simon metabolites' of α-tocopherol may also be analysed successfully by GC/MS.[141]

Several studies of vitamin E metabolism have also been conducted with the aid of GC/MS of isotopically labelled substrates; these publications were

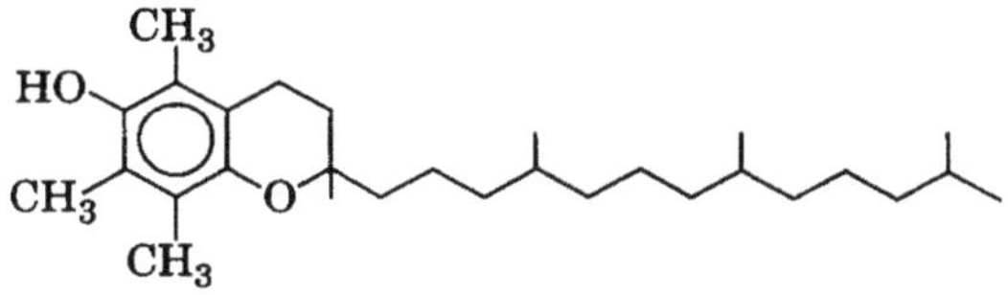

Figure 9.15 *The chemical structure of α-tocopherol*

reviewed in 1990.[142] A recent study uses GC/MS methods to compare the relative bioavailabilities of *RRR* (*i.e.* nature-identical) and all-racemic α-tocopheryl acetate (the commonest supplemental form).[143] This showed that the *RRR* form was more bioavailable than the all-racemic form. High sensitivity LC/combustion/IRMS is suitable for determining fat-soluble vitamins, including α-tocopherol.[144]

Early applications of LC/MS utilised the now obsolete belt LC/MS technique to determine tocopherol isomers in wheat germ oil.[145] More recently, frit-FAB LC/MS was successful in determining tocopherol.[8] Capillary SFC/MS is also suitable for characterising isomeric, underivatised tocopherols.[146] Particle beam LC/MS of vitamin E and its acetate yielded detection limits of 0.7 and 0.5 ng respectively. The technique was used to analyse these vitamers in an infant formula.[87] Modern LC/MS techniques have recently been used to determine vitamin E (and carotenoids, see above) using APCI LC/MS.[109] A novel feature of this application was the formation of tocopherol–Ag^+ adducts by post-column addition of silver perchlorate.

Tandem mass spectrometry is useful for profiling α-tocopherol and its oxidative products (see above under carotenoids)[98,99] and is suitable for characterising tocochromanols with vitamin E activity unambiguously.[147]

Vitamin K

Vitamin K is the generic name for 2-methyl-1,4-naphthoquinone with the biological activity of phylloquinone (Figure 9.16).

Vitamin K has anti-haemorrhagic properties because of its involvement in the biosynthesis of plasma clotting factors. It is found in highest concentration in green leafy vegetables. However, because dietary requirements are low most foods contribute significantly to intake; indeed synthesis by intestinal microflora is sufficient to meet needs.[1] Human cases of vitamin K deficiency are rare and are usually associated with lipid malabsorption diseases. Deficiency can result in haemorrhagic disease. However, newborn infants and the very young are at higher risk than adults because of poor placental transfer and lack of vitamin K in breast milk.

GC/MS methods for determining vitamin K have been developed.[148,149] The most recent of these references describes an IDMS GC/MS method for determining vitamin $K_{1(20)}$ in plasma using perfluoroacyl derivatives of the reduced molecule. Although this vitamer is amenable to GC/MS without derivatisation, sensitivity is believed to be insufficient for measurement in

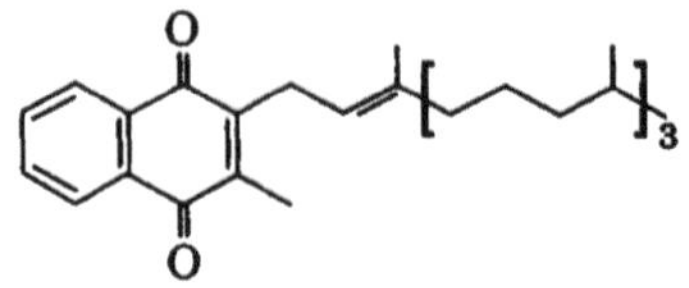

Figure 9.16   *The chemical structure of vitamin K*

biological samples. The limit of quantification of the heptafluorobutyryl derivative (chosen as the best because it was least prone to interference) was 2 pg ml^{-1}. Positive ion EI was preferred to ECNI because these molecules did not yield molecular ions under electron capture conditions.

LC/MS, based on the particle beam interface, can be used to determine vitamin K_1 in vegetables.[136] Both EI and positive and negative ion CI were used. Detection limits of 2 ng were obtained by ENCI. The EI, positive CI and negative CI mass spectra of vitamin K_1, obtained by particle beam mass spectrometry sample introduction, are shown in Figure 9.17. Although these data are encouraging, particle beam LC/MS lacks sensitivity and APCI LC/MS is likely to provide increased sensitivity.

Mass spectrometry is also very useful for studying vitamin K metabolism.

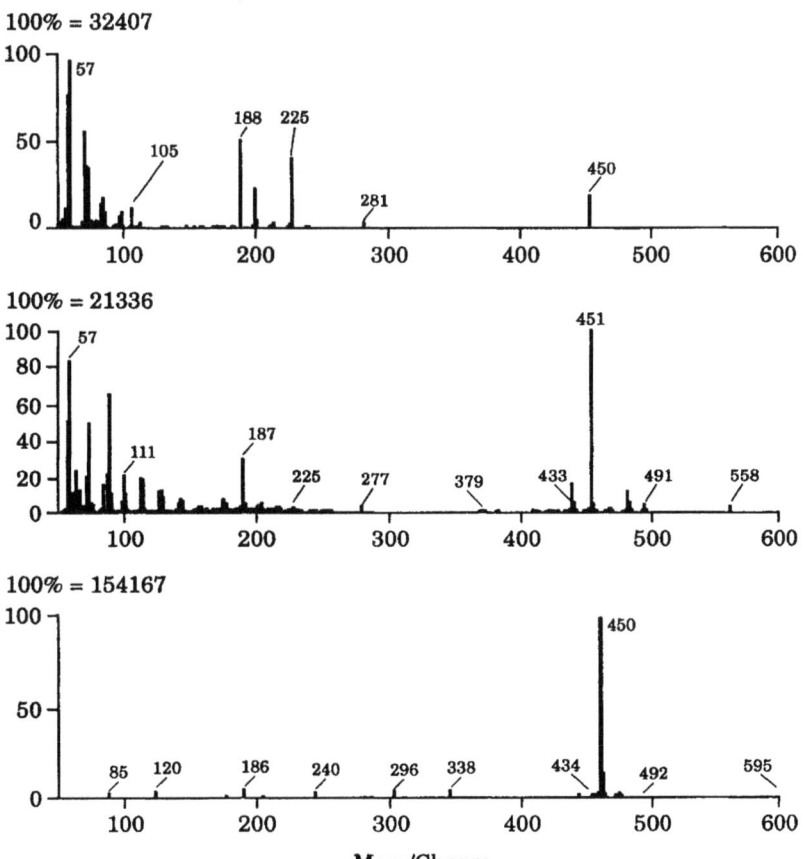

Figure 9.17 *The EI (top), positive (middle), and negative CI (bottom) mass spectra of vitamin K_1, obtained by particle beam mass spectrometry sample introduction* (Reproduced from M. Careri, A. Mangia, P. Manini and N. Taboni, *Fresenius' J. Anal. Chem.*, 1996, **355**, 48, with permission of the authors and publisher)

Dioxygen transfer during vitamin K-dependent carboxylase catalysis has been investigated using ^{18}O labelled molecular oxygen.[150] These experiments were conducted with the aid of EI mass spectrometry and negative ion CI SFC/MS. It was shown that 0.95 mol atoms of oxygen were incorporated into the epoxide product of vitamin K and 0.05 mol atom into the quinone oxygen.

4 General Conclusions and Recommendations

A variety of mass spectrometric techniques has been used to identify, quantify and study the metabolism of water- and fat-soluble vitamins. GC/MS (where the vitamin is suitably volatile or a volatile derivative can be prepared simply) is still a popular and useful method that can be recommended for many vitamins. It has obvious advantages where unambiguous structural characterisation is required. However, the possibilities of the newer ionisation techniques, particularly APCI and ESI, have not been exploited fully in vitamin studies. LC/MS using these ionisation modes has obvious potential. Furthermore, the development of CE/MS and microspray techniques may yield significant improvements in sensitivity, speed and separating power and are ripe for exploitation.

5 References

1 G.F. Combs Jr., *The Vitamins: Fundamental Aspects of Nutrition and Health*, Academic Press Inc., San Diego, 1992, 528 pp.
2 F.A. Mellon, in *Spectroscopic Techniques for Food Analysis*, ed. R.H. Wilson, VCH Publishers Inc., New York, 1993, pp. 177–213.
3 R.E. Davis and G.C. Icke, in *Adv. Clin. Chem.*, 1983, **23**, 93.
4 M. Guntert, H.J. Bertram, R. Emberger, R. Hopp, H. Sommer and P. Werkhoff, *ACS Symp. Ser.*, 1994, **564**, 199.
5 J. Iida and H. Murata, *Anal. Sci.*, 1990, **6**, 273.
6 K. Yamanaka, S. Horimoto, M. Matsuoka and K. Banno, *Chromatographia*, 1994, **39**, 91.
7 M. Careri, R. Cilloni, M.T. Lugari and P. Manini, *Anal. Commun.*, 1996, **33**, 159.
8 N. Asakawa, H. Ohe, M. Tsuno, Y. Nezu, Y. Yoshida and T. Sato, *J. Chromatogr.*, 1991, **541**, 231.
9 T. Toyosaki and A. Hayashi, *Milchwiss.-Milk Sci. Int.*, 1993, **48**, 607.
10 J. Abian, S. Susin, J. Abadia and E. Gelpi, *Anal. Chim. Acta*, 1995, **302**, 215.
11 M. Matsumoto, T. Kuhara, Y. Inoue, T. Shinka and I. Matsumoto, *J. Chromatogr.-Biomed. Appl.*, 1991, **562**, 139.
12 D.L. Hachey, S.P. Coburn, L.T. Brown, W.F. Erbelding, B. DeMark and P.D. Klein, *Anal. Biochem.*, 1992, **151**, 159.
13 J.F. Gregory, P.R. Trumbo, L.B. Bailey, J.P. Toth, T.G. Baumgartner and J.J. Cerda, *J. Nutr.*, 1991, **121**, 177.
14 D.M. Leyland, R.P. Evershed, R.H.T. Edwards and R.J. Beynon, *J. Chromatogr.*, 1992, **581**, 179.
15 S.P. Coburn, C.C. Lin, W.E. Schaltenbrand and J.D. Mahureen, *J. Labelled Compd. Radiopharm.*, 1982, **19**, 703.
16 M. Careri, R. Cilloni, A. Mangia, P. Manini and A. Raffaelli, *XII National Symposium of Analytical Chemistry*, Florence, Italy, 1995, paper p70.
17 J.F. Gregory, S.D. Bhandari, L.B. Bailey, J.P. Toth, T.G. Baumgartner and J.J. Cerda, *Am. J. Clin. Nutr.*, 1991, **53**, 736.
18 M. Jagerstad and K. Pietrzik, *Int. J. Vit. Nutr. Res.*, 1993, **63**, 285.

19 C.E. Hignite and D.L. Azarnoff, *Biomed. Mass Spectrom.*, 1978, **5**, 161.
20 R.G. Smith, J.C. Pegues, D. Farquhar, T.L. Loo and Y.-M. Ming, *Biomed. Mass Spectrom.*, 1981, **8**, 144.
21 M. Przybylski, in *Chemistry and Biology of Pteridines*, ed. H.-Ch. Curtius, S. Ghisla and N. Blau, Walter de Gruyter, Berlin, 1990, pp. 140–146.
22 M.C. Kirk, W.C. Coburn Jr. and J.R. Piper, *Biomed. Mass Spectrom.*, 1976, **3**, 245.
23 M. Anbar and G.A. St. John, *Anal. Chem.*, 1976, **48**, 198.
24 M.G. Nair, A. Abraham and S. Weintraub, in *Chemistry and Biology of Pteridines*, ed. H.-Ch. Curtius, S. Ghisla and N. Blau, Walter de Gruyter, Berlin, 1990, pp. 181–187.
25 J.F. Gregory and J.P. Toth, in *Evaluation of Folic Acid Metabolism in Health and Disease*, ed. M.F. Picciano, E.L.R. Stoksted and J.F. Gregory, Alan R. Liss Inc. New York, 1990, pp. 151–169.
26 J.F. Gregory and J.P. Toth, *Anal. Biochem.*, 1988, **170**, 94.
27 J.F. Gregory and J.P. Toth, *J. Labelled Comp. Radiopharm.*, 1988, **25**, 1349.
28 J.F. Gregory, S.D. Bhandari, L.B. Bailey, J.P. Toth, T.G. Baumgartner and J.J. Cerda, *Am. J. Clin. Nutr.*, 1991, **53**, 736.
29 J.F. Gregory, S.D. Bhandari, L.B. Bailey, J.P. Toth, T.G. Baumgartner and J.J. Cerda, *Am. J. Clin. Nutr.*, 1992, **55**, 1147.
30 J.P. Toth and J.F. Gregory, *Biomed. Environ. Mass Spectrom.*, 1988, **17**, 73.
31 A.E. Vonderporten, J.F. Gregory, J.P. Toth, J.J. Cerda, S.H. Curry and L.B. Bailey, *J. Nutr.*, 1992, **122**, 1293.
32 P.A. Kownackibrown, C.Z. Wang, L.B. Bailey, J.P. Toth and J.F. Gregory, *J. Nutr.*, 1993, **123**, 1101.
33 S.R. Dueker, A.D. Jones, A. Arjomand, Y.K. Ho, Y.Y. Chen and A.J. Clifford, *FASEB J.*, 1996, **10**, 1133.
34 A.I. Mallet, personal communication.
35 C.R. Santhosh-Kumar, J.C. Deutsch, K.L. Hassell, N.M. Kolhouse and J.F. Kolhouse, *Anal. Biochem.*, 1995, **51**, 311.
36 S.P. Stabler, P.D. Marcell, E.R. Podell and R.H. Allen, *Anal. Biochem.*, 1987, **162**, 185.
37 S.P. Stabler, P.D. Marcell, E.R. Podell, R.H. Allen, D.G. Savage and J. Lindenbaum, *J. Clin. Invest.*, 1988, **81**, 466.
38 H. van den Berg, *Int. J. Vit. Nutr. Res.*, 1993, **63**, 282.
39 M. Barber, R.S. Bordoli, R.D. Sedgwick and A.N. Tyler, *Biomed. Mass Spectrom.*, 1981, **8**, 492.
40 K. Kurumaya, T. Sakamoto, Y. Okada and M. Kajiwara, *J. Chromatogr.*, 1988, **435**, 235.
41 H.-R. Schulten and H.M. Schiebel, *Naturwissenschaften*, 1978, **65**, 223.
42 H.M. Schiebel and H.-R. Schulten, *Biomed. Mass Spectrom.*, 1982, **9**, 354.
43 S.W. Graham, P. Dowd and D.M. Hercules, *Anal. Chem.*, 1982, **54**, 649.
44 I.J. Amster and F.W. McLafferty, *Anal. Chem.*, 1985, **57**, 1208.
45 J.T. Drummond, R.R. Ogorzalek Loo and R.G. Matthews, *Biochemistry*, 1993, **32**, 9282.
46 D.P. Sundin and R.H. Allen, *Arch. Biochem. Biophys.*, 1992, **298**, 658.
47 P.D. Marcell, S.P. Stabler and R.H. Allen, *Anal. Biochem.*, 1985, **150**, 58.
48 S.P Stabler, P.D. Marcell, E.R. Podell, R.H. Allen and J. Lindenbaum, *J. Clin. Invest.*, 1986, **77**, 1606.
49 K. Rasmussen, *Clin. Chem.*, 1989, **35**, 260.
50 J.S. Straczek, F. Felden, B. Dousset, J.L. Gueant and F. Belleville, *J. Chromatogr. A*, 1993, **620**, 1.
51 M. Dedieu, C. Juin, P.J. Arpino and G. Guichon, *Anal. Chem.*, 1982, **54**, 2372.
52 M. Sakairi and H. Kambara, *Anal. Chem.*, 1989, **61**, 1159.
53 R.D. Kirchhoefer, *J. AOAC Int.*, 1994, **77**, 117.
54 E.L. Jacobson, A.J. Dame, J.S. Pyrek and M.K. Jacobson, *Biochimie*, 1995, **77**, 394.

55 M. Azoulay, P.L. Desbene and F. Frappier, *J. Chromatogr.*, 1984, **303**, 272.
56 G.R. Chipman and K.A. Cruickshank, *J. Chromatogr.*, 1991, **554**, 141.
57 B.T. Miller, M.E. Rogers, J.S. Smith and A. Kurosky, *Anal. Biochem.*, 1994, **219**, 240.
58 E. Demoll, R.H. White and W. Shive, *Biochemistry*, 1984, **23**, 558.
59 M. Tuchman, M.T. McCann, P.E. Johnson and B. Lemieux, *Pediatr. Res.*, 1991, **30**, 315.
60 B.L. Schwartz, K.J. Lightwahl and R.D. Smith, *J. Am. Soc. Mass Spectrom.*, 1994, **5**, 201.
61 B.L. Schwartz, D.C. Gale, R.D. Smith, A. Chilkoti and P.S. Stayton, *J. Mass Spectrom.*, 1995, **30**, 1095.
62 K. Eckart and J. Spiess, *J. Am. Soc. Mass Spectrom.*, 1995, **6**, 912.
63 Z.X. Zhao, H.R. Udseth and R.D. Smith, *J. Mass Spectrom.*, 1996, **31**, 193.
64 G. Maiani, E. Azzini and A. Ferro-Luzzi, *Int. J. Vit. Nutr. Res.*, 1993, **63**, 289.
65 Y.C. Ng, T. Akera, C.S. Han, W.E. Brasselton, R.H. Kennedy, K. Temma, T.M. Brody and P.H. Sato, *Biochem. Pharmacol.*, 1985, **34**, 2525.
66 J.M. McMahon, *Anal. Biochem.*, 1985, **147**, 535.
67 D. Knaeck and T. Podleski, *Proc. Natl. Acad. USA*, 1985, **82**, 575.
68 C.G. Honegger, W. Krenger, H. Langemann and A.J. Kempf, *J. Chromatogr.*, 1986, **381**, 249.
69 J.C. Deutsch and J.F. Kolhouse, *Anal. Chem.*, 1993, **65**, 321.
70 J.C. Deutsch, C.R. Santhosh-Kumar, K.L. Hassell and J.F. Kolhouse, *Anal. Chem.*, 1994, **66**, 345.
71 P. Ellerbe, L.T. Sniegoski, J.M. Miller and E.V. White, *J. Res. Nat. Bureau Stand.*, 1988, **93**, 367.
72 P. Ellerbe, L.T. Sniegoski, J.M. Miller and E.V. White, *J. Res. Nat. Bureau Stand.*, 1988, **93**, 367.
73 R.L. Lin, G.R. Waller, E.D. Mitchell, K.S. Yang and E.C. Nelson, *Anal. Biochem.*, 1970, **35**, 435.
74 D.R. Hughes, P. Rietz, W. Vetter and G.A.J. Parr, *Int. J. Vit. Nutr. Res.*, 1976, **46**, 231.
75 M.E. Cullum, J.A. Olson and S.W. Veysey, *Int. J. Vit. Nutr. Res.*, 1983, **53**, 3.
76 M.E. Cullum, M.H. Zile and S.W. Veysey, *Int. J. Vit. Nutr. Res.*, 1984, **54**, 11.
77 A.J. Clifford, A.D. Jones and H.C. Furr, *Meth. Enzymol.*, 1990, **189**, 94.
78 H.C. Furr, O. Amedee-Manesme, A.J. Clifford, H.R. Bergen, A.D. Jones, D.P. Anderson and J.A. Olson, *Am. J. Clin. Nutr.*, 1989, **49**, 713.
79 G.J. Handelman, M.J. Haskell, A.D. Jones and A.J. Clifford, *Anal. Chem.*, 1993, **65**, 2024.
80 S.R. Dueker, J.M. Lunetta, A.D. Jones and A.J. Clifford, *Clin. Chem.*, 1993, **39**, 2318.
81 G. Tang, J. Qin and G.G. Dolinikowski, *J. Nutr. Biochem.*, 1998, **9**, 408.
82 S.R. Dueker, R.S. Mercer, A.D. Jones and A.J. Clifford, *Anal. Chem.*, 1998, **70**, 1369.
83 C. Eckhoff, W. Wittfoht, H. Nau and W. Slikker, *Biomed. Environ. Mass Spectrom.*, 1990, **19**, 428.
84 U.B. Ranalder, B.B. Lausecker and C. Huselton, *J. Chromatogr. Biomed. Appl.*, 1993, **617**, 129.
85 R.B. van Breemen and C.-H. Huang, *FASEB J.*, 1996, **10**, 1098.
86 R.B. van Breemen, D. Nikolic, X. Xu, Y. Xiong, M. van Lieshout, C.E. West and A.B. Schilling, *J. Chromatogr. A*, 1998, **794**, 245.
87 R. Andreoli, M. Careri, P. Manini, G. Mori and M. Musci, *Chromatographia*, 1997, **44**, 605.
88 R. Peto, R. Doll, J.D. Buckley and M.B. Sporn, *Nature*, 1981, **290**, 201.
89 R.B. Shekelle, M. Lepper, S. Liu, C. Maliza, W. Raynor, A.H. Rossof, P. Oglesby, A.M. Shyrock and J. Stammler, *Lancet*, 1981, 1185.

90 E. Rojas-Hidalgo and B. Olmedilla, *Int. J. Vit. Nutr. Res*, 1993, **63**, 265.
91 K.J. Scott, *Food Chem.*, 1992, **45**, 357.
92 K.J. Scott and D.J. Hart, *Food Chem.*, 1993, **47**, 403.
93 C.R. Enzell, G.W. Francis and S. Liaanen-Jensen, *Acta Chem. Scand.*, 1969, **23**, 727.
94 U. Schwieter, G. Englert, N. Rigassi and W. Vetter, *Pure Appl. Chem.*, 1969, **20**, 365.
95 W. Vetter, G. Englert, N. Rigassi and U. Schwieter, in *Carotenoids*, ed. O. Isler, Birkhäuser, Basel, 1971, pp. 243.
96 C.R. Enzell and I. Wahlberg, in *Biochemical Applications of Mass Spectrometry*, ed. G.R. Waller and O.C. Dermer, Wiley, New York 1980, pp. 407–438.
97 W.R. Lusby, F. Khachik, G.R. Beecher and J. Lau, *Meth. Enzymol.*, 1992, **213**, 111.
98 T.D. McClure and D.C. Liebler, *Chem. Res. Toxicol.*, 1995, **8**, 128.
99 T.D. McClure and D.C. Liebler, *J. Mass Spectrom.*, 1995, **30**, 1480.
100 F. Khachik, G.R. Beecher, M.B. Goli, W.R. Lusby and J.C. Smith, *Anal. Chem.*, 1992, **64**, 2111.
101 A. Selva and F. Cardini, *Org. Mass Spectrom.*, 1993, **28**, 570.
102 A. Selva, E. Cardini and M. Chelli, *Org. Mass Spectrom.*, 1994, **29**, 695.
103 F. Khachik, C.J. Spengler, J.C. Smith, L.M. Canfield, A. Steck and H. Pfander, *Anal. Chem.*, 1997, **69**, 1873.
104 M. Careri, P. Lombardi, C. Mucchino and E. Cantoni, *Rapid Commun. Mass Spectrom.*, 1999, **13**, 118.
105 H.H. Schmitz, R.B. van Breemen and S.J. Schwartz, *Meth. Enzymol.*, 1992, **213**, 322.
106 R.B. van Breemen, H.H. Schmitz and S.J. Schwartz, *Anal. Chem.*, 1993, **65**, 965.
107 R.B. van Breemen, H.H. Schmitz and S.J. Schwartz, *J. Agric. Food Chem.*, 1995, **43**, 384.
108 R.B. van Breemen, *Anal. Chem.*, 1995, **67**, 2004.
109 C. Rentel, S. Strohschein, K. Albert and E. Bayer, *Anal. Chem.*, 1998, **70**, 4394.
110 R.B. van Breemen, C.-R. Huang, Y. Tan, L.C. Sander and A.B. Schilling, *J. Mass Spectrom.*, 1996, **31**, 975.
111 R.B. van Breemen, *Anal. Chem.*, 1996, **68**, 299A.
112 P.A. Clarke, K.A. Barnes, J.R. Startin and M.J. Shepherd, *Rapid Commun. Mass Spectrom.*, 1996, **10**, 1781.
113 F.A. Mellon, previously unpublished observation.
114 R.S. Parker, J.E. Swanson, B. Marmor, K.J. Goodman, A.B. Spielman, J.T. Brenna and S.M. Viereck, *Ann. NY Acad. Sci.*, 1994, **691**, 86.
115 S.R. Dueker, A.D. Jones, G.M. Smith and A.J. Clifford, *Anal. Chem.*, 1994, **66**, 4177.
116 J.A. Novotny, S.R. Dueker, L.A. Zech and A.J. Clifford, *J. Lipid Res.*, 1995, **36**, 1825.
117 H. van den Berg, *Int. J. Vit. Nutr. Res.*, 1993, **63**, 257.
118 H.F. DeLuca, *J. Steroid Biochem.*, 1979, **11**, 35.
119 J. Fohlman, P.A. Peterson, I. Kamensky, P. Hakansson and B. Sundqvist, *Nucl. Instr. Meth. Phys. Res.*, 1982, **198**, 169.
120 N. Leboulch, L. Cancela, L. Miravet and C. Lange, *Biomed. Environ. Mass Spectrom.*, 1986, **13**, 53.
121 I. Jardine, Q.F. Scanlan, V.R. Mattox and R. Kumar, *Biomed. Environ. Mass Spectrom.*, 1984, **11**, 4.
122 B. Yeung, P. Vouros and G.S. Reddy, *J. Chromatogr.*, 1993, **645**, 115.
123 D. Watson, K.D.R. Setchell and R. Ross, *Biomed. Chromatogr.*, 1991, **5**, 153.
124 R.J. Vreeken, M. Honing, B.L.M. Vanbaar, R.T. Ghijsen, G.J. DeJong and U.A.T. Brinkman, *Biol. Mass Spectrom.*, 1993, **22**, 621.
125 B. Yeung, P. Vouros, M.L. Siucaldera and G.S. Reddy, *Biochem. Pharmacol.*, 1995, **49**, 1099.

126 S.R. Wilson, Q.Y. Lu, M.L. Tulchinsky and Y.H. Wu, *J. Chem. Soc. Chem. Commun.*, 1993, 664.

127 S.R. Wilson, M.L. Tulchinsky and Y.H. Wu, *Bioinorg. Med. Chem. Lett.*, 1993, **3**, 1805.

128 I. Bjorkhem and I. Holmberg, *Clin. Chim. Acta*, 1976, **68**, 215.

129 I. Bjorkhem and A. Larsson, *Clin. Chim. Acta*, 1978, **88**, 559.

130 I. Bjorkhem, I. Holmberg, T. Kristiansen and J.I. Pedersen, *Clin. Chem.*, 1979, **25**, 584.

131 A.P. DeLeenheer and A.A. Cruyl, *Anal. Biochem.*, 1978, **91**, 293.

132 B. Zagalak, H.Ch. Curtius, R. Foschi, G. Wipf, U. Redweik and M.J. Zagalak, *Experientia*, 1978, **34**, 1537.

133 D.A. Seamark, D.J.H. Trafford and H.L.J. Makin, *Clin. Chim. Acta*, 1980, **106**, 51.

134 R.D. Coldwell, D.J.H. Trafford, M.J. Varley, D.N. Kirk and H.L.J. Makin, *Steroids*, 1990, **55**, 418.

135 R.D. Coldwell, D.J.H. Trafford and H.L.J. Makin, *J. Mass Spectrom.*, 1995, **30**, 348.

136 M. Careri, M.T. Lugari, A. Mangia, P. Manini and S. Spagnoli, *Fresenius' J. Anal. Chem.*, 1995, **351**, 768.

137 B. Yeung and P. Vouros, *Mass Spectrom. Rev.*, 1995, **14**, 179.

138 P.A. Morrissey, P.J.A. Sheehy and P. Gaynor, *Int. J. Vit. Nutr. Res.*, 1993, **63**, 260.

139 D.W. Thomas, R.M. Parkhurst, D.S. Negi, K.D. Lunan, A.C. Wen, A.E. Brandt and R.J. Stephens, *J. Chromatogr.*, 1981, **225**, 433.

140 D.C. Liebler, J.A. Burr, L. Philips and A.J.L. Ham, *Anal. Biochem.*, 1996, **236**, 27.

141 P.J. Dutton, L.A. Hughes, D.O. Foster, G.W. Burton and K.U. Ingold, *Free Rad. Res. Commun.*, 1990, **9**, 435.

142 G.W. Burton, K.U. Ingold, K.H. Cheeseman and T.F. Slater, *Free Rad. Res. Commun.*, 1990, **11**, 99.

143 R.V. Acuff, S.S. Thedford, N.N. Hidiroglou, A.M. Papas and T.A. Odom, *Am. J. Clin. Nutr.*, 1994, **60**, 397.

144 R.J. Caimi and J.T. Brenna, *J. Mass Spectrom.*, 1995, **30**, 466.

145 J. van der Greef, A.C. Tas and M.C. Ten Noever de Brauw, *J. Chromatogr.*, 1985, **323**, 81.

146 J.M. Snyder, S.L. Taylor and J.W. King, *J. Am. Oil Chem. Soc.*, 1993, **70**, 349.

147 T.J. Walton, C.J. Mullins, R.P. Newton, A.G. Brenton and J.H. Beynon, *Biomed. Environ. Mass Spectrom.*, 1988, **16**, 289.

148 W. Vetter, M. Vecchi, H. Gutmann, R. Rüegg, W. Walther and P. Meyer, *Helv. Chim Acta*, 1967, **194**, 1866.

149 G. Fauler, H.J. Leis, J. Schalamon, W. Munteau and H. Gleispach, *J. Mass Spectrom.*, 1996, **31**, 655.

150 A. Kuliopulos, B.R. Hubbard, Z. Lam, I.J. Koski, B. Furie, B.C. Furie and C.T. Walsh, *Biochem.*, 1992, **31**, 7722.

CHAPTER 10

Stable Isotope Studies of Organic Macronutrient Metabolism

1 Introduction

Stable isotope methods applied to macronutrients are, in common with similar studies of micronutrients, now seen as the most rigorous, ethical and safe techniques for conducting measurements of nutrient absorption and metabolism in all subjects and population groups.[1] Macronutrient metabolism is a large topic, deserving a book in its own right, and metabolic studies in normal populations overlap significantly with medical applications. Consequently, the subject will be covered only briefly here, as an introduction to the topic, and with a few selected examples outlining what can be achieved using modern mass spectrometric techniques. Mass spectrometric methods for measuring energy expenditure are also covered here.

2 Mass Spectrometric Techniques Used in Macronutrient Metabolism Studies

The most important mass spectrometric techniques used to measure stable isotopes in macronutrient metabolism studies are conventional GC/MS (with EI or CI), LC/MS, and high-precision isotope-ratio mass spectrometry (IRMS). IRMS is used to analyse isotopically breath CO_2 samples, combusted samples isolated from tissue or body fluids or, as GC/combustion/IRMS, sample mixtures. The enriched isotope labels employed in these studies are ^{13}C, ^2H or ^{15}N. Practical considerations regarding the use of stable isotopes in clinical and basic studies of nutrient metabolism have been discussed.[2] This useful review describes instrumentation, label types, dosing methods, common problems, and discusses ethical aspects of stable isotope studies.

3 Protein Metabolism

Methods for determining dynamic aspects of protein metabolism have been discussed.[3] A two-pool model (Figure 10.1) can be used to estimate the

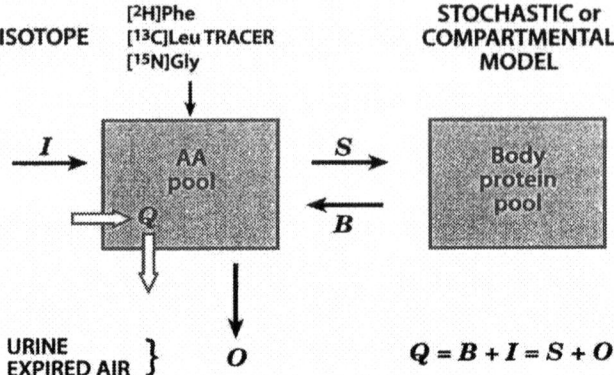

Figure 10.1 *General two-pool model for the determination of whole-body protein metabolism by stable isotope dilution: Q, flux; B, breakdown; I, dietary intake; S, synthesis, O, oxidation*
(Reproduced from D. Halliday, in *Stable Isotopes in Paediatric, Nutritional and Metabolic Research*, ed. T.E. Chapman, R. Berger, D.J. Reijngoud and A. Okken, Intercept, Andover, 1990, pp. 213–226, with permission of the author and publisher)

movement of free amino acids into and out of proteins, and to determine their involvement in oxidative metabolism.

It is beyond the scope of this book to describe the methodology in detail: the main measurements (following isotope administration by intravenous infusion or by bolus dose) are based either on measurements in plasma or in urinary products. The labelled amino acids that are available include [15N]glycine, [1-13C]leucine, [2H5]phenylalanine and deuterated lysine. Determination of protein synthesis rates in humans requires measurements of isotope tracers in precursor and product amino acids. Isotope ratios of plasma-free amino acids can be determined by conventional GC/MS, with a precision of 0.1%–1% RSD. However, protein-bound amino acids usually require higher precision measurements because isotopic enrichment is much lower (0.002–0.1 atom % excess), a range normally deemed suitable only for GIRMS. Nevertheless, conventional GC/MS measurements can be used for determining the incorporation of deuterated lysine and [2H5]phenylalanine into tissue proteins.[4-6] These studies exploit the use of multiply-, rather than singly-, labelled molecules in which the 'crosstalk', *i.e.* natural background of the tracer molecule, is several orders of magnitude lower. This is because the mass shift between the labelled and unlabelled molecules is several mass units. The measurements can be made using a simple, bench-top GC/MS system. In the case of lysine, the samples were measured as their heptafluorobutyryl derivatives.[4] Phenylalanine was converted to phenylethylamine by enzymatic decarboxlyation to reduce background interference in the mass channels of choice.[5,6]

The measurement of low isotopic enrichments of lysine and phenylethylamine in relation to protein turnover studies using a cheap, simple bench-top instrument shows the great potential of this methodology in food and nutrition research. However, both conventional GC/MS and GIRMS techniques may still be necessary (particularly the more sensitive continuous-flow IRMS methods) in appropriate cases. The final choice of method will depend on the study design and the requirements of the experiment. For example, some experiments may require the use of amino acids other than lysine or phenylalanine that may only be available in singly-labelled form (and would therefore require GIRMS measurement when incorporated into protein). An example of the state of the art in this field is the measurement of muscle protein fractional synthetic rate by capillary GC/GIRMS, using [1-^{13}C]leucine as the label.[7]

4   Carbohydrate Metabolism

Stable isotope techniques have been used for several years to study glucose metabolism and these types of application have recently been reviewed.[8] Several different types of mass spectrometric measurement can be made, depending on the aims and design of the study, and on the labels used. Appearance of ^{13}C in breath CO_2 is measured by GIRMS. Conventional SIM GC/MS of appropriate derivatives or breakdown products is suitable for measuring universally ^{13}C-labelled glucose or ^{2}H-labelled glucose.[9] Finally, GC/GIRMS of pentaacetate derivatives is the method of choice for single or highly diluted ^{13}C-labelled glucose. Conventional GC/MS can be used to quantify positional isomers of ^{2}H-labelled glucose[10] and to determine the labelling pattern of ^{13}C-labelled glucose.[11] Positional analysis of ^{13}C-labelled glucose by GC/MS has been used to determine gluconeogenesis in liver cells.[12] Figure 10.2 shows the EI mass spectrum of the bis(butylboronate) acetate derivative of glucose.

The [M − butyl]$^+$ isotopic cluster at *m/z* 297 and above is used to determine isotope ratios of labelled samples. The SIM GC/MS chromatogram of the isotopic peaks in glucose (from a blood sample taken from a volunteer given ^{13}C-labelled glucose) is shown in Figure 10.3. Isotopes are determined by integrating and determining ratios of mass ion intensities in this type of chromatogram. Combinations of several of the stable isotope techniques described above can be used to estimate gluconeogenesis *in vivo,* and to measure the oxidation of glucose to carbon dioxide.

Although these measurements are not directly relevant to food studies, they may be useful in selected cases, for example in investigating the effects of different foodstuffs on glucose metabolism in diabetics. However, the methodology can be adapted to study the metabolism of macronutrients presented in the form they are found in foodstuffs. This requires the production of ^{13}C-labelled plant foods by introducing $^{13}CO_2$ into enclosed chambers containing the growing plants (rice, in an early example[13]). Alternatively, naturally ^{13}C-enriched starch may be used.[14–16] C_4 plants that fractionate carbon isotopes metabolically generate this naturally enriched material. Although some pioneering studies concentrated on clinical problems (*e.g.* starch absorption in infants during acute gastro-enteritis[17]), metabolism of starch in normal humans has also been investigated.[16] In the latter study, volunteers were fed with extruded starch, pasta or polenta manufactured from maize starch. GIRMS and state of

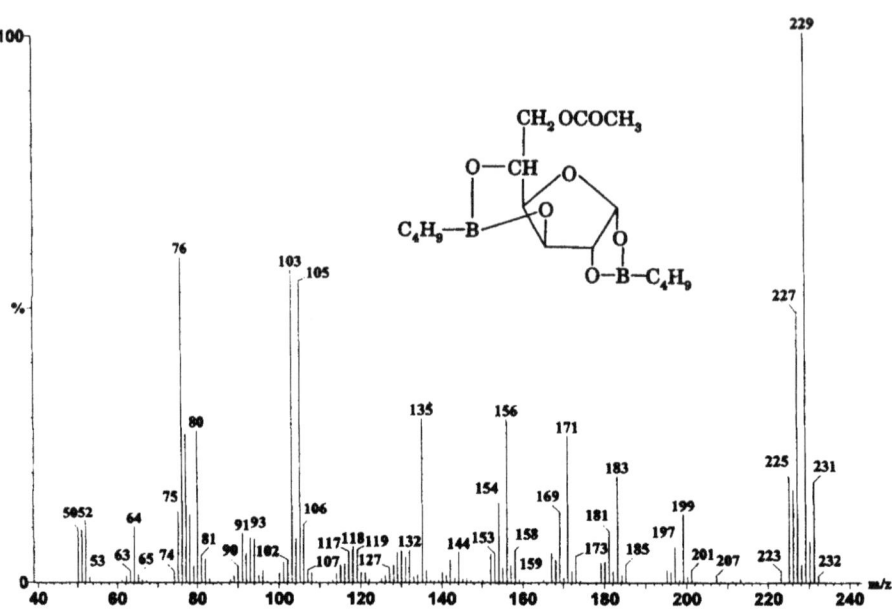

Figure 10.2   *Structure and EI mass spectrum of the bis(butylboronate acetate) derivative of glucose*

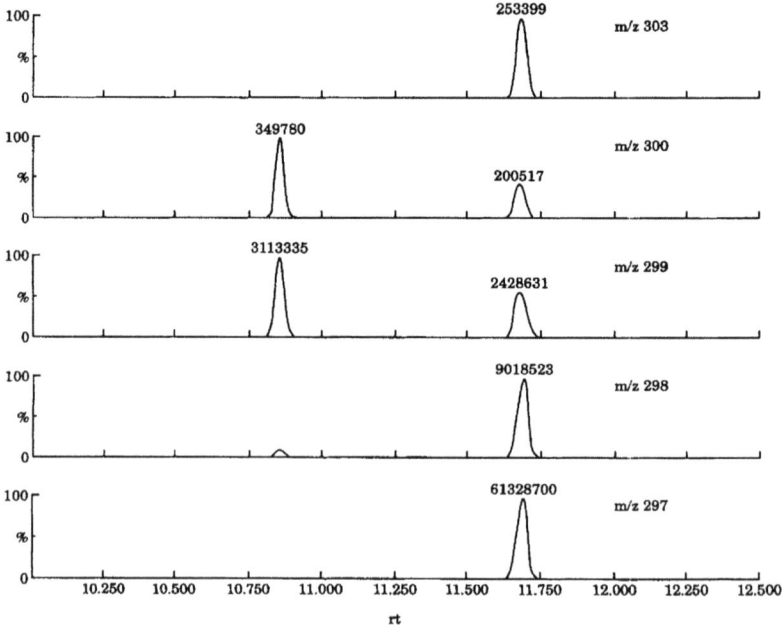

Figure 10.3   *SIM GC/MS chromatogram of the isotopic peaks in the bis(butylboronate) acetate derivative of glucose extracted from a blood sample taken from a volunteer who had ingested universally ^{13}C-labelled glucose*

the art GC/GIRMS (of glucose pentaacetate) measured the appearance of ^{13}C in breath CO_2 and plasma glucose respectively. This first measurement of the bioavailability of starch in differently prepared foods concluded that there were no significant differences between them during the period of the study. However, plasma glucose tended to be higher during feeding with polenta.

More recently, GIRMS, GC/GIRMS and conventional GC/MS techniques have been used to determine the bioavailability of different types of ^{13}C-labelled starch.[18] Glycaemia is regulated in the body by (i) the rate of hepatic glucose entry into the system and (ii) the rate of clearance from the circulation into insulin-sensitive tissues. Determining absorption from starchy foods solely from glycaemic response is therefore impossible. Because hepatic glucose recycling is minimal during glucose absorption, developing a method for determining glucose entering the circulation from ^{13}C-labelled starchy foods is possible. This is achieved by mass spectrometric measurement of both ^{13}C-enriched glucose and [6,6-$^2H_2$]-D-glucose. The ^2H-labelled glucose is used to measure the rate of glucose disposal. Peas were grown in an atmosphere containing $^{13}CO_2$: this yielded pea starch with enrichment 30% above natural ^{13}C levels. Volunteers had been primed with a continuous infusion of [6,6-$^2H_2$]-D-glucose before feeding with the enriched starch. Plasma samples were taken and GC/combustion/MS of glucose pentaacetate determined ^{13}C-enriched glucose: deuterium labelling was measured by conventional GC/MS of the butylboronate derivative. Breath $^{13}CO_2$ enrichment was measured by IRMS. These data were used to fit a non-steady-state one-compartment model of the kinetics. A representative absorption curve is shown in Figure 10.4.

The methodology has shown that glucose absorption from starch in peas can be very different in healthy adults.[19] This has led to speculation that glucose absorption rates may have a role in the aetiology of diseases such as diabetes

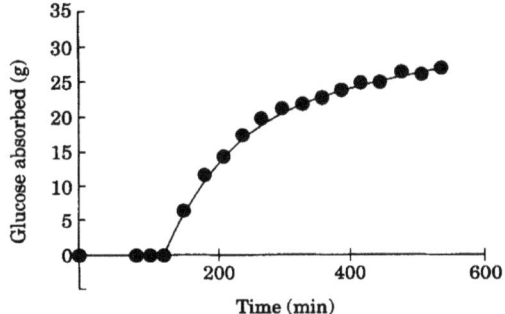

Figure 10.4 *Typical absorption curve for glucose derived from ^{13}C-labelled starch administered in pea soup to a single volunteer*
(Reproduced from G. Livesey, R. Faulks, P. Wilson, M. Roe, J. Brown, T. Newman, J. Eagles, F.A. Mellon, J. Dennis, I. Parker, R. Greenwood and D. Halliday, 'Development of a dual stable isotope method for determining glucose absorption from ^{13}C-enriched starchy foods', *Nutrition Society Summer Meeting*, University of Ulster, Coleraine, 24–28th June, 1996, with permission of the authors and publisher)

and glucose intolerance. The same techniques also show that insulin sensitivity is nearly doubled after prior consumption of mixed meals.[20] This suggests that carbohydrate consumption is tolerated more easily after mixed meals. These studies show that use of isotope labelling and 'multidisciplinary' mass spectrometric approaches can reveal hitherto unsuspected features of absorption and metabolism from foods. They also show the great potential of these methods in investigating diet–health relationships.

5   Lipid Metabolism

The use of stable isotopes and mass spectrometry in studying lipid metabolism, with particular emphasis on their role as energetic substrates, has been reviewed recently.[21] Dietary lipids are important as an energy source, as precursors of bioactive molecules (e.g. leukotrienes and prostaglandins) and as precursors of phospholipids incorporated into cellular membranes. Stable isotope mass spectrometry has been used to investigate aspects of all these roles. This is a rapidly expanding field and a few selected examples will be given here to show the potential of current methodology. Qualitative analysis of lipids by mass spectrometry is covered in Chapter 6 of this book.

The absorption of ^{13}C-labelled fatty acids may be measured by determining isotopic enrichment in breath CO_2 following administration of a bolus dose,[22,23] following correction for faecal losses. This type of methodology can be used in either clinical or normal dietary investigations. More sophisticated techniques of GC/combustion/IRMS allow investigations to be conducted on mixtures of polyunsaturated fatty acids.[24,25] Stable isotope methods, because of their combination of selectivity and sensitivity, are particularly suitable for conducting infant nutrition studies. A good example is a study of the conversion of U-^{13}C-labelled linoleic and α-linolenic acids into longer chain polyunsaturated derivatives in infants aged between 2 weeks and 11 months.[26] This showed a relative decrease in the efficiency of conversion with increasing age. IRMS equipment is not essential for all fatty acid metabolic studies. Conventional GC/MS may be used to determine $^{2}H_{2}$-, $^{2}H_{4}$- and $^{2}H_{6}$-labelled fatty acids by isobutane CI in measurement of desaturation and uptake of palmitic and stearic acids.[27]

When assessing the effect of different components in the diet (fats and carbohydrates for example) on health and disease in humans, being able to distinguish *de novo* lipogenesis from fats derived from other sources is important. The first reported direct measurement of *de novo* hepatic lipogenesis relied on an elegant combination of mass spectrometric techniques.[28] Because most lipogenesis is believed to occur in the liver, total hepatic lipogenesis can, in theory, be determined solely from blood samples (following administration of labelled substrates), *provided the isotopic labelling levels of hepatic cytosolic acetyl CoA are accessible*. This is because the true biosynthetic precursor of fatty acids is cytosolic acetyl CoA. Measurement of isotopically labelled acetyl CoA was achieved, after administration of ^{13}C-labelled acetate, by co-administration of sulfamethoxazole (SMX). This xenobiotic probe is acetylated by cytosolic acetyl CoA in liver cells, forming SMX-acetyl, which is then released into the bloodstream, allowing quantitative measurement of isotopic incorporation in the acetyl CoA. Isotope ratios of SMX-acetyl and C_{18} and C_{16} fatty acids (isolated from VLDL fractions) were measured by thermospray LC/MS and by GC/MS respectively. Use of an appropriate mathematical model then enabled the measurement of hepatic lipogenesis. The results of this study suggested that hepatic lipogenesis was a quantitatively minor pathway under the study conditions used.

Measurements of human serum cholesterol synthesis by determining deuterium incorporation following administration of D_2O appeared as long ago as 1966.[29] Many examples of the use of this methodology in clinical and dietary studies have appeared since. A recent example is the measurement of dietary fat saturation and cholesterol level on cholesterol synthesis.[30]

6 Whole Body Energy Expenditure

The 'doubly labelled water' method, first developed in animal models, is widely used to measure energy expenditure indirectly in free-living subjects.[31] Besides providing clinical data, it can be used to assess the effect of diet on energy expenditure. The principle of the technique is illustrated in Figures 10.5 and 10.6.

What follows is a simplified account of the method. A more comprehensive description is given in several publications, for example that of Heyman and Roberts.[32] Water labelled with ^2H and ^{18}O is administered orally to volunteers. This is rapidly distributed in body water and urinary collections are made over

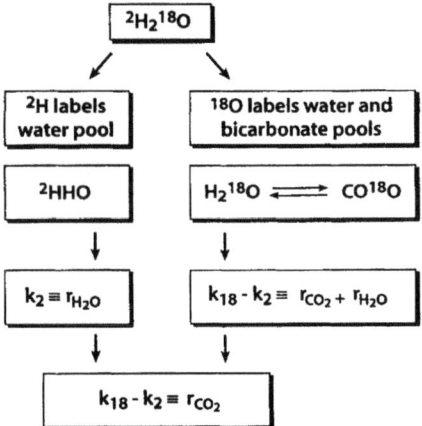

Figure 10.5 *Principle underlying the doubly-labelled water method. k = experimentally derived rate constant; r = production rate*
(Reproduced from K.A. Nagy, in *The Doubly-Labelled Water Method for Measuring Energy Expenditure: Technical Recommendations for Use in Humans*, ed. A.M. Prentice, International Atomic Energy Agency, Vienna, 1990, pp. 1–16, with permission of the author and publisher)

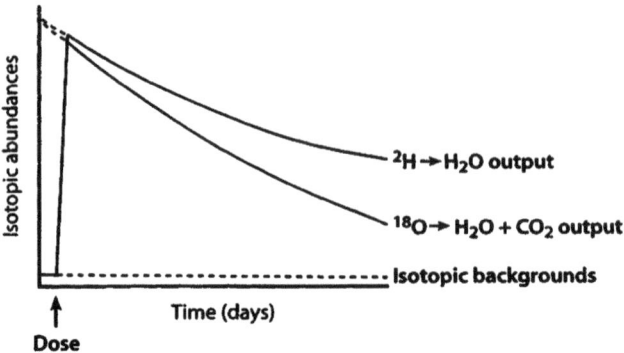

Figure 10.6 *Isotope disappearance curves for the doubly-labelled water method. ^2H and ^{18}O are administered at sufficient levels to raise isotope ratios above background. The difference between the disappearance rates of ^2H and ^{18}O approximates CO_2 output, which is then used to calculate energy expenditure* (Reproduced from M.B. Heyman and S.B. Roberts, in *Stable Isotopes in Paediatric, Nutritional and Metabolic Research*, ed. T.E. Chapman, R. Berger, D.J. Reijngoud and A. Okken, Intercept, Andover, 1990, pp. 51–66, with permission of the authors and publisher)

several days and the isotopic content is measured. The disappearance rate of deuterium is proportional to water output and ^{18}O loss is proportional to both water loss and CO_2 generation. The difference between the two values yields total CO_2 production. Energy expenditure can be calculated from this figure when the respiratory quotient of the subject is known or can be predicted. The method is reliable, provided sample processing and mass spectrometric measurement is conducted with care.[33] It has the major advantage over conventional methods that it allows the study of free-living subjects (*i.e.* they are not subject to the artificial constraints of living in a calorimetry room).

7   General Conclusions and Recommended Methods

Because body pools of macronutrients are so large, most studies of macronutrient metabolism rely on measuring small increases in isotope ratio above basal levels. GIRMS mass spectrometers are generally the best instruments for conducting these types of study, although GC/MS and LC/MS techniques are suitable in appropriate cases (see above). The recent development of sensitive, continuous flow GIRMS, GC/combustion/IRMS and robust, compact and dedicated GIRMS instruments are helping to expand the range and availability of isotope ratio techniques in nutrient metabolism studies. New techniques under development, particularly LC/combustion/IRMS for ^{13}C- and ^{15}N-labelled samples and GC/IRMS techniques for deuterium labelled samples, have the potential to extend the capabilities of mass spectrometric methods still further. It is confidently expected that the use of stable isotopes in macronutrient metabolism studies will expand rapidly in the next decade.

8 References

1 R. Self, F.A. Mellon, B.A. McGaw, A.G. Calder, G.E. Lobley and E. Milne in *Mass Spectrometry in Biomolecular Sciences*, ed. R.M. Caprioli, A. Malorni and G. Sindona, Kluwer Academic Publishers, Dordrecht, 1996, pp. 483–515.
2 G.N. Thompson, P.J. Pacy, G.C. Ford and D. Halliday, *Biomed. Environ. Mass Spectrom.*, 1989, **18**, 321.
3 D. Halliday, in *Stable Isotopes in Paediatric, Nutritional and Metabolic Research*, ed. T.E. Chapman, R. Berger, D.J. Reijngoud and A. Okken, Intercept, Andover, 1990, pp. 213–226.
4 B.W. Patterson, D.L. Hachey, G.L. Cook, J.M. Amann and P.D. Klein, *J. Lipid Res.*, 1991, **32**, 1063.
5 A.G. Calder, S.E. Anderson, I. Grant, M.A. McNurlan and P.J. Garlick, *Rapid Commun. Mass Spectrom.*, 1992, **6**, 421.
6 C. Slater, T. Preston, D.C. McMillan, J.S. Falconer and K.C.H. Fearon, *J. Mass Spectrom.*, 1995, **30**, 1325.
7 K.E. Yarasheski, K. Smith, M.J. Rennie and D.M. Bier, *Biol. Mass Spectrom.*, 1992, **21**, 486.
8 S.C. Kalhan, *J. Nutr.*, 1996, **126**, 362S.
9 A. Pickert, D. Overkamp, W. Renn, H. Liebich and M. Eggstein, *Biol. Mass Spectrom.*, 1991, **20**, 203.
10 Z. Guo, W.-N.P. Lee, J. Katz and A.E. Bergner, *Anal. Biochem.*, 1992, **204**, 273.
11 M. Beylot, S.F. Previs, F. David and H. Brunengraber, *Anal. Biochem.*, 1993, **212**, 526.
12 M. Desage, R. Guilluy, J.-L. Brazier, J.-P. Riou, M. Beylot, S. Normand and H. Vidal, *Biomed. Environ. Mass Spectrom.*, 1989, **18**, 1010.
13 T.W. Boutton, C.N. Bollich, B.D. Webb, S.L. Sekely, B.L. Nichols and P.D. Klein, *Am. J. Clin. Nutr.*, 1987, **45**, 844.
14 J.S. Garrow, P.F. Scott, S. Heela, K.S. Nair and D.A. Halliday, *Hum. Nutr. clin. Nutr.*, 1983, **37C**, 301.
15 M. Hiele, Y. Ghoos, P. Rutgeerts and G. Vantrappen, *Biomed. Environ. Mass Spectrom.*, 1988, **16**, 133.
16 S. Normand, C. Pachiaudi, Y. Khalfallah, R. Guilluy, R. Mornex and J.P. Riou, *Am. J. Clin. Nutr.*, 1992, **55**, 430.
17 C.H. Lifshitz, B. Torun, F. Chew, T.W. Boutton, C. Garza and P.D. Klein, *J. Pediatr.*, 1991, **118**, 526.
18 G. Livesey, R. Faulks, P. Wilson, M. Roe, J. Brown, T. Newman, J. Eagles, F.A. Mellon, J. Dennis, I. Parker, R. Greenwood and D. Hallliday, 'Development of a dual stable isotope method for determining glucose absorption from ^{13}C-enriched starchy foods', *Nutrition Society Summer Meeting*, University of Ulster, Coleraine, 24–28th June, 1996.
19 G. Livesey, R. Faulks, P. Wilson, M. Roe, J. Brown, T. Newman, J. Dennis, I. Parker, R. Greenwood and D. Halliday, 'Individual rates of glucose absorption from ^{13}C-starch in peas eaten by healthy adults can be very different: Is there a role in the aetiology of disease?', *Nutrition Society Summer Meeting*, University of Ulster, Coleraine, 24–28th June, 1996.
20 G. Livesey, R. Faulks, P. Wilson, J. Brown, M. Roe, T. Newman, K. Taylor, S. Hampton and R. Greenwood. 'Post-prandial insulin sensitivity of glucose disposal measured with 6,6-^{2}H-D-glucose is markedly dependent on prior meal consumption', *Nutrition Society Summer Meeting*, University of Ulster, Coleraine, 24–28th June, 1996.
21 M. Beylot, *Proc. Nutr. Soc.*, 1994, **53**, 355.
22 P.J.H. Jones, P.B. Pencharz and M.T. Clandinin, *J. Lab. Clin. Med.*, 1985, **105**, 647.
23 P.J.H. Jones, P.B. Pencharz and M.T. Clandinin, *Am. J. Clin. Nutr.*, 1985, **42**, 769.
24 S.T. Brookes, K.S. Craig and S.C. Cunnane, *Biochem. Soc. Trans.*, 1994, **22**, 164S.

25 N. Brossard, C. Pachiaudi, M. Croset, S. Normand, J. Lecerf, V. Chirouze, J.P. Riou, J.L. Tayot and M. Lagarde, *Anal. Biochem.*, 1994, **220**, 192.
26 H. Demmelmair, T. Sauerwald, B. Koletzko and T. Richter, *Eur. J. Pediatr.*, 1997, **156**, S70.
27 E.A. Emken, R.O. Adlof, W.K. Rohwedder and R.M. Gulley, *Biochim. Biophys. Acta*, 1993, **1170**, 173.
28 M.K. Hellerstein, M. Christiansen, S. Kaempfer, C. Kletke, K. Wu, J.S. Reid, K. Mulligan, N.S. Hellerstein and C.H.L. Shackleton, *J. Clin. Invest.*, 1991, **87**, 1841.
29 C. Bruce-Taylor, B. Mikkelson, J.A. Anderson and D.T. Forman, *Arch. Path.*, 1966, **81**, 213.
30 P.J.H. Jones, A.H. Lichtenstein and E.J. Schaefer, *J. Lipid Res.*, 1994, **35**, 1093.
31 N. Lifson and R. McLintock, *J. Theor. Biol.*, 1966, **12**, 46.
32 M.B. Heyman and S.B. Roberts, in *Stable Isotopes in Paediatric, Nutritional and Metabolic Research*, ed. T.E. Chapman, R. Berger, D.J. Reijngoud and A. Okken, Intercept, Andover, 1990, pp. 51–66.
33 D.A. Schoeller and J.M. Hnilicka, *J. Nutr.*, 1996, **126**, 348S.

CHAPTER 11

Pyrolysis Mass Spectrometry of Foods

1 Introduction

A comprehensive account of applied pyrolysis has been given by Wampler and is recommended to the interested reader.[1] Pyrolysis mass spectrometry (Py/MS) is a hyphenated technique that combines controlled thermal degradation of the analyte with the sensitivity and selectivity of mass spectrometry (see Chapter 1). It is most often applied to materials that are difficult to determine by conventional techniques because of their high molecular weight or involatility. Py/MS commonly requires minimal sample preparation and straightforward mass spectrometry: the most sophisticated stage of the analysis is usually the data processing. This frequently involves the use of advanced chemometric techniques to extract optimised information from the recorded spectra. The products of pyrolysis may be analysed directly in the mass spectrometer following in-source or near-source decomposition. Alternatively the mixtures formed, which are often very complex, may be separated and analysed by pyrolysis GC/MS. However, this technique has the disadvantage that higher molecular weight pyrolytic fragments might be lost because they are insufficiently volatile for successful gas chromatography.

A major review of Py/MS in food science was published in 1987.[2] When this review appeared, Py/MS had been applied to food microbiology, food authentication, food quality and food gums. The authors noted the lack of applications of Py/MS to whole foods and this situation persists. This topic will therefore be covered only briefly here; readers interested in a more comprehensive account are referred to the review mentioned above.[2] Despite the limited number of applications, Py/MS does have significant scope in food science. This is particularly the case in authentication studies where speed, sensitivity and minimal sample preparation could make it an attractive technique to food manufacturers.

2 Practical Aspects of Py/MS

Pyrolysis generates fragments of the starting material by thermal bond scission of the molecule or molecular mixtures that comprise the material of interest. These fragments are generally of low molecular weight and volatile. Pyrolytic fragments often yield useful qualitative or quantitative information about the structure and composition of the analyte. Modern pyrolysers allow controlled degradation of the analyte so pyrolysis products are highly reproducible in structure and relative concentration. These desirable characteristics are attained by three main operational and design characteristics: (i) increasing the pyrolyser temperature rapidly to an appropriate value in an inert atmosphere; (ii) using small sample sizes (micrograms or nanograms of material) and (iii) speedy removal of pyrolysis products from the pyrolyser to reduce secondary reactions.

3 Mass Spectrometric Techniques for Analytical Pyrolysis

As mentioned in Chapter 1, EI is still the most popular ionisation technique in Py/MS. Mass spectra are often acquired at low electron energy (25 eV or less) instead of the normal value of 70 eV, to reduce fragmentation and simplify the spectra. CI is also occasionally used for this reason. However, CI spectra are less reproducible between different models of mass spectrometer. This accounts for the unpopularity of this ionisation technique in Py/MS.

Specialised mass spectrometers dedicated to pyrolysis are available and have been described.[2,3] Although analytical pyrolysis is possible using conventional mass spectrometers, dedicated instruments have two major advantages: (a) the option of automated sample introduction and (b) an expansion chamber to delay the disappearance of pyrolysis products (and thus ensure that sufficient representative mass spectra are acquired) and reduce ion source contamination.

Conventional mass spectrometers also have a major role in Py/MS when they are used to analyse samples by pyrolysis gas chromatography/mass spectrometry (Py/GC/MS). Because the pyrolysis products are buffered and separated by the GC before mass spectrometric measurement, conventional (*i.e.* 70 eV) EI is the ionisation technique of choice. This yields the most reproducible spectra on the separated components and enables unknown peaks to be matched against any of the large databases of known mass spectra that are available, or against user-generated databases. Pyrolysers may be conveniently coupled to almost any type of conventional GC/MS instrument. However, there are obvious cost advantages in using cheap, bench-top quadrupole or ion trap mass spectrometers. Py/GC/MS has the disadvantage that some pyrolysis products may be too polar and/or involatile to chromatograph and will therefore be lost. Nevertheless, it does present the possibility of separating and identifying specific compounds that may be useful analytical markers. This is difficult to achieve if the analyte is pyrolysed directly into the mass spectrometer ion source on a dedicated instrument.

4 Analysis of Data

Spectra obtained by Py/MS are usually processed by chemometric techniques. This is because the spectra are generally very complex but, despite this, exhibit subtle variations between the different samples subjected to analysis. Because mass spectra are represented by the intensities of ions in integer mass channels, multivariate statistical approaches are particularly suitable for analysing Py/MS data.[3] Multivariate statistics can reduce the large quantity of complex mass spectrometric data and display it in a readily understandable form. Principal component analysis, canonical variate analysis, cluster analysis and factor analysis have all been used to interpret the chemical differences between Py/MS spectra.[2] Artificial neural networks are also useful in revealing the complex relationships in multivariate data.[3]

5 Selected Applications of Py/MS in Food Studies

Scotch Whisky Quality

Py/MS does, at first sight, appear to be an odd choice of technique for determining whisky quality. Many components that determine Scotch whisky quality are volatile, and therefore amenable to conventional mass spectrometric techniques such as GC/MS. However, it has been shown that the non-volatile fraction of whisky is an important factor in assessing whisky quality (or potential quality). This is because the maturation process is the result of chemical interactions between the whisky spirit and the maturation cask. Py/MS is a particularly suitable technique for assessing the quality of maturation cask oak because of the nature of the analyte.[4] A set of 14 malt whisky extracts from one distillery was first analysed by Py/MS to determine factors that affected quality. Principal component analysis (Figure 11.1) revealed excellent discrimination based on quality.

Further analysis of the data yielded the masses in the Py/MS spectra characteristic of quality (Figure 11.2). Most of these masses originated from the oak wood and included phenolic compounds (derived from lignin) and polysaccharide derived fragments. Among the masses identified as indicative of good quality were ions that corresponded to vanillin (m/z 151/152) and 5-hydroxymethyl-2-furaldehyde (m/z 126). Py/MS analysis of maturation cask woods revealed several pyrolysis fragments that were present in the whisky samples and were indicators of good quality. The data showed that cooperage wood containing increased proportions of earlywood growth was desirable for production of good quality Scotch whisky. Overall, the results of this study suggested that Py/MS has potential as a rapid analytical technique for measuring whisky quality and for assessing the oak woods used in whisky maturation casks. Py/MS also has potential as a rapid method for confirming Scotch whisky

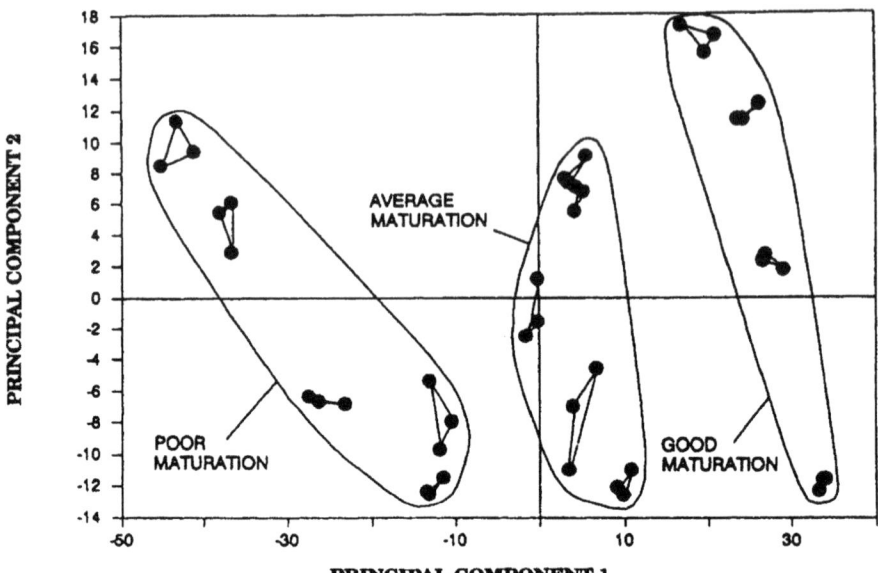

Figure 11.1   *Plot of the first two principal components obtained from the analysis of Distillery 1 whisky extracts*
(Reproduced from K.J.G. Reid, J.S. Swan and C.S. Gutteridge, *J. Anal. Appl. Pyrolysis*, 1993, **25**, 49, © 1993 with permission from Elsevier Science)

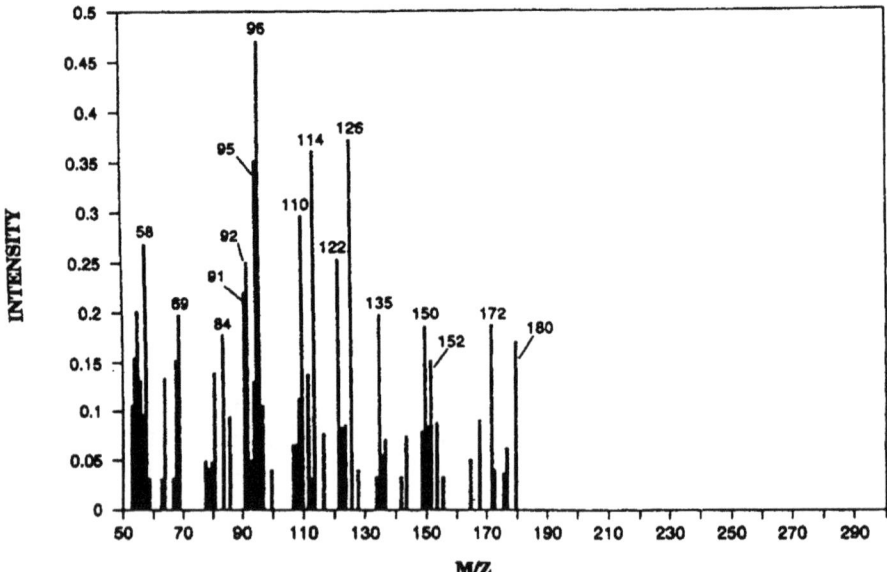

Figure 11.2   *Factor spectrum showing masses that characterised good quality Distillery 1 whisky extracts*
(Reproduced from K.J.G. Reid, J.S. Swan and C.S. Gutteridge, *J. Anal. Appl. Pyrolysis*, 1993, **25**, 49, © 1993 with permission from Elsevier Science)

authenticity.[5] Wine cask composition,[6] wine classification[7] wine bottle cork characterisation[8] and degradation[9] have also been studied by Py/MS, indicating that this technique may have wider potential in studying the quality and authenticity of high-value alcoholic drinks.

Analysis of Rice Lignin

Lignin is an important factor in determining the robustness of the rice plant under adverse environmental conditions. Despite the importance of rice as a staple food crop for a large proportion of the world population, little information is available relating lignin structure and rice yield. It is difficult to characterise lignin by conventional chemical and spectroscopic techniques, and degradative methods, including Py/GC/MS, have proved their value.[10] The Curie-point (500 °C) Py/GC/MS trace of a lignin fraction isolated from the leaf of the rice cultivar Koshikihari is shown in Figure 11.3 and the identity of the peaks in Table 11.1.

These data show that rice lignin is a mixed 4-hydroxyphenyl–guaiacyl–syringyl lignin. The main pyrolysis products were 4-vinylphenol and 4-vinylguaiacol. The experimental results suggested that the yield of 4-vinylphenol (compared with 4-vinylguaiacol) was related to the mechanical properties of rice plant. Pyrolysis has obvious potential as a rapid technique for determining the robustness of major food crops such as rice and is ripe for further exploitation.

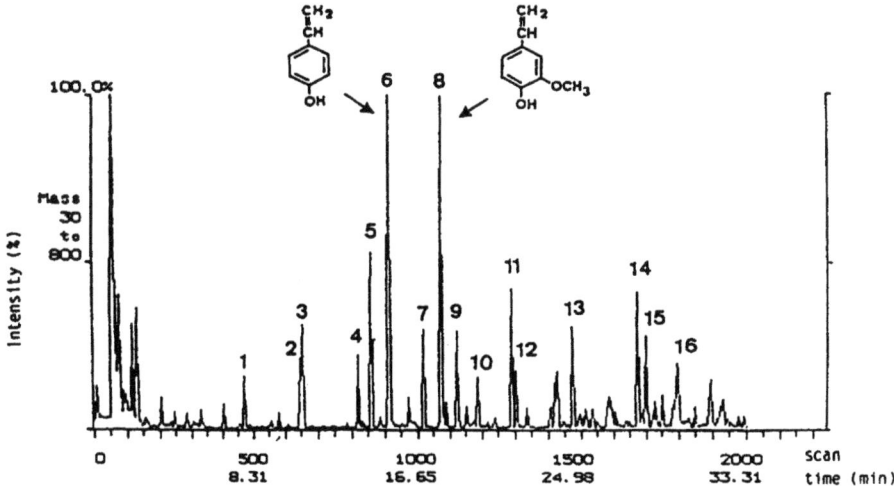

Figure 11.3   *Py/GC/MS trace of dioxane lignin of a Koshihikar leaf-sheath sampled at the full ripe stage. Peak numbers identified in Table 11.1*
(Reproduced from K. Kuroda, A. Suzuki, M. Kato and K. Imai, *J. Anal. Appl. Pyrolysis*, 1995, **34**, 1, © 1995 with permission from Elsevier Science)

Table 11.1 *Peak identities in Py/GC/MS of dioxane lignin of a Koshihikar leaf-*
sheath sampled at the full ripe stage
(Reproduced from K. Kuroda, A. Suzuki, M. Kato and K. Imai, *J.*
Anal. Appl. Pyrolysis, 1995, **34**, 1, © 1995 with permission from
Elsevier Science)

Peak no.	Assignment	Mw	m/z > 25% of base peak*	Relative intensity (%)
1	phenol	94	**94**, 66	13
2	4–cresol	108	108, **107**	20
3	guaiacol	124	124, **109**, 81	24
4	4-ethylphenol	122	122, **107**	16
5	4-methylguaiacol	138	138, **123**, 95, 67	38
6	4-vinylphenol	120	**120**, 91	100
7	4-ethylguaiacol	152	152, **137**	22
8	4-vinylguaiacol	150	**150**, 135, 107, 77	74
9	syringol	154	**154**, 139, 96, 93	22
10	vanillin	152	152, **151**, 81, 53	13
11	4-methylsyringol	168	**168**, 153, 125	31
12	4-propenylguaiacol	164	**164**, 149, 131, 103, 91, 77	16
13	4-vinylsyringol	180	**180**, 165, 137	22
14	4-propenylsyringol	194	**194**, 119, 91	29
15	acetosyringone	196	196, **181**	22
16	4-coumaric acid	164	**164**, 163, 147, 119, 118, 91, 73	15

* Base peak in bold.

Analysis of Caramel Colours

Caramels are widely used as one of the main brown food colourings and are
divided into four major classes. Gel filtration, the main analytical technique for
detecting caramel colours in foodstuffs, is unspecific and prone to interference
from other brown compounds of high molecular weight. Curie-point pyrolysis
GC/MS at 600 °C can differentiate the four classes of caramel colourings based
on their pyrolysis products.[11] The Py/GC/MS traces were very characteristic of
the class of caramel colouring present. Furthermore, the summed Py/GC/MS
spectra of class I–IV caramels, shown in Figure 11.4, suggest that Py/MS alone
may be suitable for their determination. As Py/MS is much faster than Py/GC/
MS, there is obvious potential for this method to be exploited as a rapid specific
analytical technique for determining these complex food additives.

6   Conclusions

Py/MS and Py/GC/MS still have great potential in the analysis of prepared
foods and food raw materials. The relatively low level of exploitation of the
methodology is perhaps surprising and probably reflects the caution and
conservatism of the food industry in adopting modern analytical techniques.
Mass spectrometry is still perceived as a high-cost technique and this may have

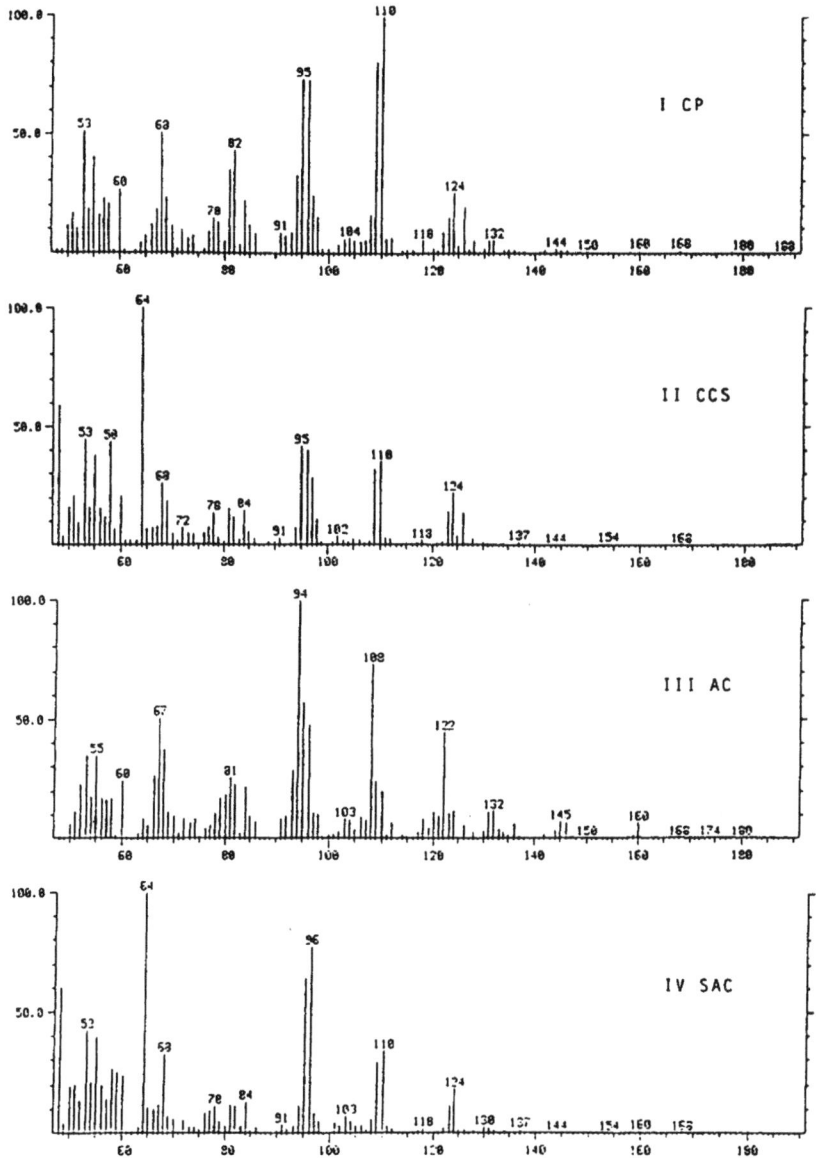

Figure 11.4 *Simulated pyrolysis mass spectra of caramel colours*
(Reproduced from R. Hardt and W. Baltes, *J. Anal. Appl. Pyrolysis*, 1989,
15, 159, © 1989 with permission from Elsevier Science)

been prejudicial to its use in monitoring food manufacturing. It may be
significant that one prominent area of application is in the authentication or
optimisation of higher-value products such as spirits and wines. However, the
wide range of cheap, powerful and portable bench-top instruments now
available may result in the more widespread adoption of Py/MS techniques in

the future. Recent examples indicating the potential of Py/MS in other food applications include classification of cocoa butters and other vegetable fats,[12] determination of the geographical origin of Italian extra virgin olive oils[13] and classification of backfat samples.[14]

7   References

1 T.P. Wampler (ed.), *Applied Pyrolysis Handbook*, Marcel Dekker, Inc., New York, 1995, 361 pp.
2 R.E. Aries and C.S. Gutteridge, in *Applications of Mass Spectrometry in Food Science*, ed. J. Gilbert, Elsevier Applied Science Publishers Ltd., Barking, 1987, pp. 377–430.
3 J . Maddock and T.W. Ottley, in *Applied Pyrolysis Handbook*, ed. T.P. Wampler, Marcel Dekker, Inc., New York, 1995, pp. 57–76.
4 K.J.G. Reid, J.S. Swan and C.S. Gutteridge, *J. Anal. Appl. Pyrolysis*, 1993, **25**, 49.
5 R.I. Aylott, A.H. Clyne, A.P. Fox and D.A. Walker, *Analyst*, 1994, **119**, 1741.
6 G.C. Galletti, A. Carnacini, P. Bocchini and A. Antonelli, *Rapid Commun. Mass Spectrom.*, 1995, **9**, 1331.
7 L. Montanarella, M.R. Bassani and O. Breas, *Rapid Commun. Mass Spectrom.*, 1995, **9**, 1589.
8 G.C. Galletti, P. Bocchini and A. Antonelli, *Ind. Bevande*, 1996, **25**, 229.
9 G.C. Galletti, P. Bocchini and A. Antonelli, *Rapid Commun. Mass Spectrom.*, 1996, **10**, 653.
10 K. Kuroda, A. Suzuki, M. Kato and K. Imai, *J. Anal. Appl. Pyrolysis*, 1995, **34**, 1.
11 R. Hardt and W. Baltes, *J. Anal. Appl. Pyrolysis*, 1989, **15**, 159.
12 E. Anklam, M.R. Bassani, T. Eiberger, S. Kriebel, M. Lipp and R. Matissek, *Fresenius' J. Anal. Chem.*, 1997, **357**, 981.
13 G.J. Salter, M. Lazzari, L. Giasante, R. Goodacre, A. Jones, G. Surrichio, D.B. Kell and G. Bianchi, *J. Anal. Appl. Pyrolysis*, 1997, **40–41**, 159.
14 J.-L. Berdague, C. Rabot and M. Bonneau, *Sci. Aliments*, 1996, **16**, 425.

Subject Index

www.ingramcontent.com/pod-product-compliance
Ingram Content Group UK Ltd.
Pitfield, Milton Keynes, MK11 3LW, UK
UKHW021836190226
468218UK00003B/5